BEAMFORMING
Sensor Signal Processing for Defence Applications

Communications and Signal Processing

Editors: Prof. A. Manikas & Prof. A. G. Constantinides
(Imperial College London, UK)

Communications and Signal Processing – Vol. 5

BEAMFORMING
Sensor Signal Processing for Defence Applications

editor

Athanassios Manikas
Imperial College London, UK

Imperial College Press

Published by

Imperial College Press
57 Shelton Street
Covent Garden
London WC2H 9HE

Distributed by

World Scientific Publishing Co. Pte. Ltd.

5 Toh Tuck Link, Singapore 596224

USA office: 27 Warren Street, Suite 401-402, Hackensack, NJ 07601

UK office: 57 Shelton Street, Covent Garden, London WC2H 9HE

Library of Congress Cataloging-in-Publication Data
Manikas, Athanassios.
    Beamforming : sensor signal processing for defence applications / Thanassis Manikas, Imperial College London, UK.
        pages cm. -- (Communications and signal processing ; volume 5)
    Includes bibliographical references and index.
    ISBN 978-1-78326-274-8 (hardcover : alk. paper)
    1. Radar transmitters. 2. Antenna radiation patterns. 3. Beam optics. 4. Radar--Military applications. I. Title.
    TK6587.M26 2015
    623.7'348--dc23

                                                2014044137

British Library Cataloguing-in-Publication Data
A catalogue record for this book is available from the British Library.

Typeset by Stallion Press
Email: enquiries@stallionpress.com

Printed in Singapore

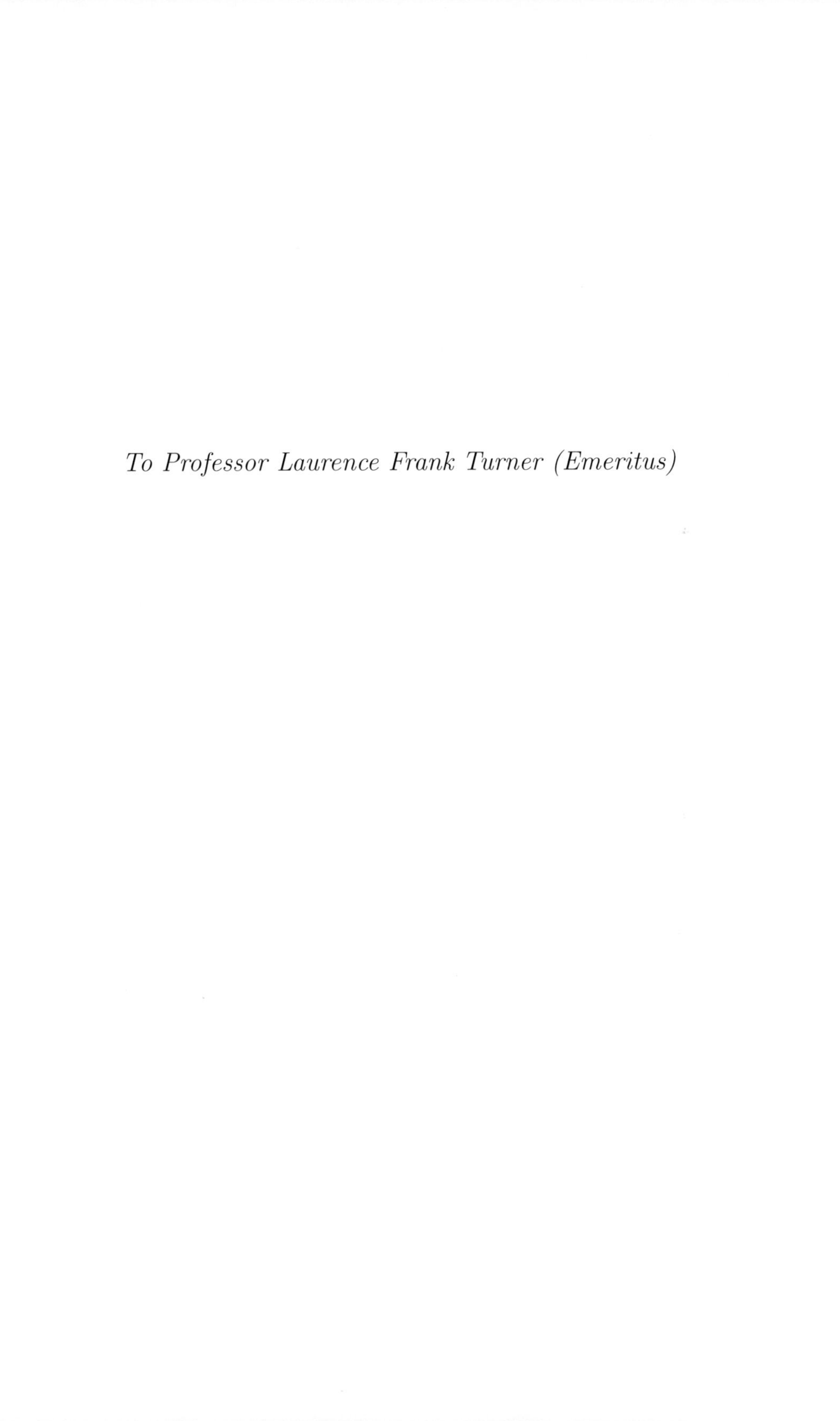

To Professor Laurence Frank Turner (Emeritus)

Preface

In recognition of the strategic importance of sensor signal processing for the UK Ministry of Defence (MOD), the University Defence Research Centre (UDRC) in Signal Processing was established in 2009 as a joint venture between the MOD and the Engineering and Physical Science Research Council (EPSRC).

The UDRC "Phase-1" ran until 2013 and incorporated 12 major UK universities led by Imperial College London. It has grown into a dynamic research centre and forum, enabling the cross-fertilisation of ideas and fostering a wider "community of practice" in signal processing.

This book presents a collection of research contributions from the UDRC "Phase-1" that address a number of topics broadly concerned with *beamforming* which is fundamental to many civilian applications but also to the capabilities of many defence systems. It is composed of eight chapters, the first five of which are concerned with various radar research problems and applications. In particular:

- Chapter 1 considers the recent work and advances in the area of space-time beamforming algorithms and their application to radar systems. Furthermore, it describes the most successful space-time adaptive processing (STAP) beamforming algorithms that exploit low-rank and sparsity properties as well as the use of prior knowledge to improve the performance of STAP algorithms in radar systems. Chapter 2 is concerned with "look-down" airborne radars and the employment of STAP beamforming. The focus of this chapter is on the non-homogeneity of STAP training data caused by the forward-looking radar platform as well as on robust beamforming. In Chapters 3 and 4 synthetic aperture radar (SAR) and multi-input multi-output (MIMO) radar are respectively investigated, while in Chapter 5 different types of ship wake waves and a number of common wake wave detection algorithms for two-dimensional SAR imagery are studied.

- Chapter 6 is related to ocean-towed arrays which find applications in a variety of areas such as defence, oil and gas exploration and geological and marine life studies. Here the major challenge is dealing with receiver positional uncertainties resulting from the array's flexible structure in combination with the ship's turning manoeuvers or water currents.

- Finally, the last two chapters are about handling array uncertainties with Chapter 7 dealing with geometrical and electrical uncertainties and Chapter 8 considering "pointing" error uncertaities and robustification issues.

Thanassis Manikas — London 2015

UDRC Technical Lead (2009–2013)

a.manikas@imperial.ac.uk

http://skynet.ee.imperial.ac.uk/manikas.html

Acknowledgments

I would like to thank all authors of the chapters for their contribution to this special book. I also wish to express my gratitude to Thibaud Gabillard, Zexi Fang and He Ren for reading various parts of the manuscript.

I am grateful to all UDRC colleagues from the Defence Science and Technology Laboratory (Dstl), especially Paul Thomas (Sensors & Countermeasures) and Bob Elsley (ISR Sensing & Processing) for their support and excellent collaboration in various aspects of the UDRC. Furthermore, I would also like to thank Nick Goddard (Naval Systems) for providing the real data set which was collected from a passive towed array during a trial in the Southwestern Approaches to the UK.

At Imperial College Press, I would like to thank my editor Thomas Stottor for his help and for showing a remarkable amount of patience with my slipping deadlines.

Contents

4. Arrayed MIMO Radar: Multi-target Parameter Estimation for Beamforming   119

Harry Commin, Kai Luo and Athanassios Manikas

5. Beamforming for Wake Wave Detection and Estimation — An Overview —

159

Karen Mak and Athanassios Manikas

6. **Towed Arrays: Channel Estimation, Tracking and Beamforming**   189

Vidhya Sridhar, Marc Willerton and Athanassios Manikas

7. **Array Uncertainties and Auto-calibration**   221

Marc Willerton, Evangelos Venieris and Athanassios Manikas

8. Robust Beamforming to Pointing Errors   263

Jie Zhuang and Athanassios Manikas

List of Notations

A, a	Scalar
$\underline{A}, \underline{a}$	Column vector
$\mathbb{A}$	Matrix
$(\cdot)^{T}$	Transpose
$(\cdot)^{H}$	Hermitian transpose
$(\cdot)^{*}$	Conjugate
$\|\mathbb{A}\|_{F}$	Frobenius norm of matrix $\mathbb{A}$
$\|\underline{A}\|$	Euclidean norm of vector $\underline{A}$
$|A|$	Absolute value
$\odot$	Hadamard product
$\oslash$	Hadamard division
$\otimes$	Kronecker product
$\mathcal{E}\{\cdot\}$	Expectation
$\mathbb{A}^{\#}$	Pseudoinverse of the matrix $\mathbb{A}$
$\underline{A}^{b}$	Element by element power
$\exp(\underline{A})$	Element by element exponential of vector $\underline{A}$
$\mathrm{trace}(\mathbb{A})$	Sum of the diagonal elements of matrix $\mathbb{A}$
$\underline{\mathrm{row}}_{i}\{\mathbb{A}\}$	column vector with elements the i-th row of $\mathbb{A}$
$\underline{\mathrm{diag}}(\mathbb{A})$	Column vector with elements the diagonal elements of matrix $\mathbb{A}$
$\mathrm{diag}(\underline{a})$	The diagonal matrix whose diagonal elements are the elements of $\underline{a}$
$\ln(a)$	Natural logarithm of a
$\log_{10}(a)$	Logarith of a relative to base 10
$\max\limits_{x}\left(\xi\left(x\right)\right)$	Maximum value of $\xi\left(x\right)$ over all x
$\min\limits_{x}\left(\xi\left(x\right)\right)$	Minimum value of $\xi\left(x\right)$ over all x

$\mathbb{P}_\mathbb{A}$	Projection operator on the subspace spanned by the columns of $\mathbb{A}$
$\mathbb{P}_\mathbb{A}^\perp$	Projection operator on the complement subspace of the subspace spanned by the columns of $\mathbb{A}$
$\underline{0}_N$	N-element column vector of all zeros
$\underline{1}_N$	N-element column vector of all ones
$\mathbb{I}_N$	Identity matrix of size $N \times N$
$\mathbb{O}_{M \times N}$	Matrix of zeros of size $M \times N$
$\mathrm{Re}\{x\}$	Real part of x
$\mathcal{O}(M)$	Order of M
$\mathcal{R}^{M \times N}$	Set of real matrices of M rows and N columns
$\mathcal{C}^{M \times N}$	Set of complex matrices of M rows and N columns
$\mathcal{B}$	Set of binary numbers
$\mathcal{N}$	Set of natural numbers
$\in$	Belongs to (or element of)
$\forall$	For every
$\perp$	Perpendicular
$\triangleq$	Is equal by definition to
$\angle$	Angle, phase

Chapter 1

Space-Time Adaptive Beamforming Algorithms for Airborne Radar Systems

Rodrigo de Lamare

Communications Group, Department of Electronics,
University of York

1.1 Introduction

Space-time adaptive processing (STAP) techniques [1, 2] have been thoroughly investigated over the last few decades as a key technology which enables advanced airborne radar applications, following the seminal work by Brennan and Reed [3]. A great deal of attention has been given to STAP algorithms and to different strategies for the design of space-time beamformers to mitigate the effect of clutter and jamming signals [4,5]. It is well understood that STAP techniques can improve slow-moving target detection through better mainlobe clutter suppression, can provide better detection in combined clutter and jamming environments, and offer a significant increase in output signal-to-interference-plus-noise ratio (SINR). Moreover, it is also well understood that clutter and jamming signals often reside in a signal subspace whose dimension is typically much lower than the dimension M of the "observation" space which is related to the number of degrees of freedom of the array and to the associated space-time beamformer. Due to the large computational complexity of the matrix inversion operation, the optimum STAP processor for SINR maximization is prohibitive for practical implementation. Another very challenging issue that is encountered by the optimal STAP technique is the case when the number of elements M of a spatiotemporal snapshot in the spatiotemporal beamformer is large. It is well known that $K \geq 2M$ independent and identically distributed

1

(i.i.d.) training samples (spatiotemporal snapshots) are required for the beamformer to achieve steady-state performance [6]. Thus, in dynamic scenarios the optimal STAP with large M usually fails or provides poor performance in tracking target signals contaminated by interference and noise.

In recent years, a number of innovative space-time beamforming algorithms have been added to the literature for clutter and interference mitigation in radar systems. These algorithms include low-rank and reduced-dimension techniques [7, 8], which employ a two-stage processing framework to exploit the low-rank property of the clutter and jamming signals. The first stage performs dimension reduction and is followed by a second stage that employs a beamforming algorithm with a reduced dimensional filter. Another class of important space-time beamforming algorithms adopts the strategy of compressive sensing and sparsity-awareness, which exploit the fact that space-time beamformers do not need all their degrees of freedom to mitigate clutter and jamming signals. These algorithms implement sparse space-time beamformers which can converge faster and are effective for STAP in radar systems. By exploiting the low-rank properties of the interference and devising sparse STAP algorithms, designers make use of prior knowledge about the clutter and the jamming signals. It has been shown recently that it is also beneficial in terms of performance to exploit prior knowledge about the environment and the data in the form of a known covariance data matrix. Space-time beamforming algorithms that exploit different forms of prior knowledge are called knowledge-aided STAP (KA-STAP) algorithms.

The goal of this chapter is to review the recent work and advances in the area of space-time beamforming algorithms and their application to radar systems. These systems include phased-array [2] and multi-input multi-output (MIMO) radar systems [9], monostatic and bistatic radar systems and other configurations [2]. Furthermore, this chapter also describes in detail some of the most successful space-time beamforming algorithms that exploit low-rank and sparsity properties as well as prior knowledge to improve the performance of STAP algorithms in radar systems.

This chapter is structured as follows. In Section 1.2 the radar system under consideration is mathematically described by means of a signal model. In Section 1.3 the problem of designing space-time beamformers is formulated and conventional space-time beamforming algorithms are reviewed. In Section 1.4 low-rank space-time beamforming algorithms are examined, whereas in Section 1.5 the concept of sparsity-aware space-time beamforming algorithms is explored. In Section 1.6 knowledge-aided beamforming algorithms, and how these techniques can be adopted in existing radar systems, are discussed. In Section 1.7 a number of existing algorithms are compared and simulation results are presented. The chapter ends with concluding remarks in Section 1.8.

1.2   Pulsed Doppler radar: System and signal models

The system under consideration is a pulsed Doppler radar residing on an airborne platform. The radar antenna is a uniformly spaced linear antenna array of half wavelength spacing, consisting of N elements. The radar returns are collected in a coherent processing interval (CPI), which is referred to as the 3D radar datacube shown in Fig. 1.1(a), where L denotes the number of samples (snapshots) collected to cover the range interval. The data is then processed at one range of interest, which corresponds to a slice of the CPI datacube. This slice is a $N \times N_p$ matrix which consists of $N \times 1$ spatial snapshots for N_p pulses at the range of interest. It is convenient to stack the matrix column-wise (i.e. vec$(\cdot)$ operator) to form the $M \times 1$ vector $\underline{x}[i]$, termed as the i-th range gate space-time snapshot (for $1 < i \leq L$) [1], where

$$M = NN_p. \tag{1.1}$$

The objective of a radar is to ascertain whether targets are present in the data. Thus, given a space-time snapshot $\underline{x}[i]$, radar detection is a binary hypothesis problem, where hypothesis H_0 corresponds to the absence of a

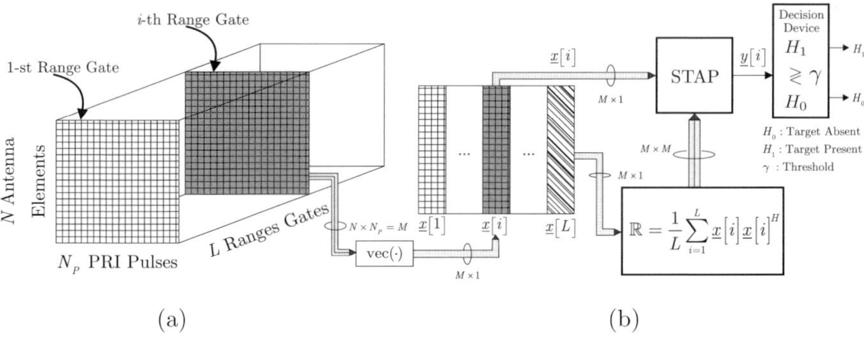

Fig. 1.1   (a) The Radar CPI datacube. (b) The STAP schematic.

target and hypothesis H_1 corresponds to the presence of a target. The radar $(M \times 1)$ space-time snapshot vector is then expressed for each of the two hypotheses in the following form

$$
\begin{aligned}
H_0 &: \underline{x}[i] = \underline{v}[i] \\
H_1 &: \underline{x}[i] = a\underline{h} + \underline{v}[i]
\end{aligned}
\tag{1.2}
$$

where a is a zero-mean complex Gaussian random variable with variance σ_s^2, $\underline{h}$ denotes the statiotemporal manifold vector (this will be defined later), $\underline{v}[i]$ is a vector which consists of the clutter $\underline{x}_c[i]$, the jamming signal vector $\underline{x}_j[i]$ and the complex white Gaussian noise vector $\underline{n}[i]$ — all of dimensions $M \times 1$. That is,

$$
\underline{v}[i] = \underline{x}_c[i] + \underline{x}_j[i] + \underline{n}[i].
\tag{1.3}
$$

These three components are assumed to be mutually uncorrelated. Thus, the $M \times M$ covariance matrix $\mathbb{R}_v$ of the undesired clutter-plus-jammer-plus-noise component can be modeled as

$$
\begin{aligned}
\mathbb{R}_v &= \mathcal{E}\left\{\underline{v}[i]\underline{v}[i]^H\right\} \\
&= \mathbb{R}_c + \mathbb{R}_j + \mathbb{R}_n
\end{aligned}
\tag{1.4}
$$

where $(\cdot)^H$ represents the Hermitian transpose and $\mathcal{E}\{\cdot\}$ denotes the expectation operator. The noise covariance matrix is given by

$$
\mathbb{R}_n = \mathcal{E}\left\{\underline{n}[i]\underline{n}[i]^H\right\} = \sigma_n^2 \mathbb{I}_M
\tag{1.5}
$$

where σ_n^2 is the variance of the noise and $\mathbb{I}_M$ is an $(M \times M)$ identity matrix. The clutter signal can be modeled as the superposition of a large number of independent clutter patches which are evenly distributed in azimuth about the receiver. Thus, the clutter covariance matrix can be expressed as

$$
\mathbb{R}_c = \mathcal{E}\left\{\underline{x}_c[i]\underline{x}_c[i]^H\right\}
\tag{1.6}
$$

$$
= \sum_{k=1}^{N_\rho} \sum_{l=1}^{N_{cp}} P_{k,l}^c \left[\underline{S}\left(\vartheta_{k,l}^c\right) \underline{S}\left(\vartheta_{k,l}^c\right)^H\right] \otimes \left[\underline{\mathcal{T}}\left(\mathcal{F}_{k,l}^c\right) \underline{\mathcal{T}}\left(\mathcal{F}_{k,l}^c\right)^H\right],
\tag{1.7}
$$

where N_ρ denotes the number of *range ambiguities* and N_{cp} denotes the number of *clutter patches*. The factor $P_{k,l}^c$ is the power of the signal reflected by the (k,l)-th clutter patch. The symbol $\otimes$ denotes Kronecker product and the vectors $\underline{S}(\vartheta_{k,l}^c)$ and $\underline{\mathcal{T}}(\mathcal{F}_{k,l}^c)$ represent the $N \times 1$ spatial steering vector with the spatial frequency $\vartheta_{k,l}^c$ and the $N_p \times 1$ temporal steering vector with the normalized Doppler frequency $\mathcal{F}_{k,l}^c$ for the (k,l)-th clutter patch respectively, and they can be expressed as follows

$$\underline{S}(\vartheta) = \begin{bmatrix} 1 \\ \exp\left(-j2\pi\vartheta\right) \\ \exp\left(-j2\pi 2\vartheta\right) \\ \vdots \\ \exp\left(-j2\pi\left(N-1\right)\vartheta\right) \end{bmatrix}, \quad \underline{\mathcal{I}}(\mathcal{F}) = \begin{bmatrix} 1 \\ \exp\left(-j2\pi\mathcal{F}\right) \\ \exp\left(-j2\pi 2\mathcal{F}\right) \\ \vdots \\ \exp\left(-j2\pi\left(N_p-1\right)\mathcal{F}\right) \end{bmatrix},$$

$$(1.8)$$

where $\vartheta = \frac{d}{\lambda}\cos(\phi)\cos(\theta)$ and $\mathcal{F} = f_d/f_r$, λ is the wavelength, d is the interelement spacing, which is normally set to half wavelength and ϕ and θ are the elevation and the azimuth angles respectively, with azimuth measured with respect to the x-axis. The quantities f_d and f_r are the Doppler frequency and the pulse repetition frequency respectively. The $M \times M$ jamming covariance matrix $\mathbb{R}_j = \mathcal{E}\{\underline{x}_j[i]\underline{x}_j^H[i]\}$ can be written as

$$\mathbb{R}_j = \sum_{q=1}^{N_j} P_q^j \left[\underline{S}\left(\vartheta_q^j\right)\underline{S}\left(\vartheta_q^j\right)^H\right] \otimes \mathbb{I}_{N_p} \tag{1.9}$$

where P_q^j is the power of the q-th jammer. The vector $\underline{S}(\vartheta_q^j)$ is the $N \times 1$ spatial steering vector with the spatial frequency ϑ_q^j of the q-th jammer and N_j is the number of jamming signals.

The vector $\underline{h}$ is the $M \times 1$ space-time steering vector in the space-time look-direction, which can be defined as

$$\underline{h} = \underline{S}(\vartheta_T) \otimes \underline{\mathcal{I}}(\mathcal{F}_T) \tag{1.10}$$

where $\underline{S}(\vartheta_T)$ is the $N \times 1$ spatial steering vector in the direction provided by the target spatial frequency ϑ_T and $\underline{\mathcal{I}}(\mathcal{F}_T)$ is the $N_p \times 1$ temporal steering vector at the target Doppler frequency $\mathcal{F}_T$. Finally, P_T denotes the power of the target.

1.3   Conventional beamforming

In order to detect the presence of targets, each range bin (see Fig. 1.1(b)) is processed by an adaptive space-time beamformer, which is typically designed to achieve maximum output SINR, followed by a hypothesis test to determine the target presence or absence. The secondary data $\underline{x}[i]$ are taken from training samples, which should be ideally i.i.d. training samples but are often non-heterogeneous [1]. The optimum full-rank STAP that maximizes the SINR can be obtained by solving the following minimum variance distortionless response (MVDR) constrained optimization problem given by:

$$\underline{w}_{\mathrm{opt}} = \arg\min_{\underline{w}} \left(\underline{w}^{H}\mathbb{R}\underline{w}\right) \qquad (1.11)$$

$$\text{subject to } \underline{w}^{H}\underline{h} = 1 \qquad (1.12)$$

where $\mathbb{R} = \mathcal{E}\left\{\underline{x}[i]\,\underline{x}[i]^{H}\right\}$ is the $M \times M$ data covariance matrix and the $M \times 1$ optimal space-time MVDR beamformer $\underline{w}_{\mathrm{opt}}$ is designed to maximize the SINR and to maintain a normalized response in the target spatial-Doppler look-direction. The solution to the optimization problem above is described by:

$$\underline{w}_{\mathrm{opt}} = \frac{\mathbb{R}^{-1}\underline{h}}{\underline{h}^{H}\mathbb{R}^{-1}\underline{h}}. \qquad (1.13)$$

Thus, the space-time beamformer $\underline{w}_{\mathrm{opt}}$ can be computed by using Eq. (1.13). Alternatively, the space-time beamformer can be estimated using adaptive algorithms [6]. These algorithms include the least mean-square (LMS), the conjugate gradient (CG) and the recursive least-squares (RLS) techniques. The computational complexity of these algorithms ranges from a linear function of M for the LMS to a quadratic function of M for the CG and RLS algorithms. A common problem with conventional adaptive algorithms is that the laws that govern their convergence and tracking behaviors imply that they depend on M and on the eigenvalue spread of $\mathbb{R}$. This means that their performance may degrade significantly when the space-time beamformer has many parameters for adaptation, which makes the computation of the parameters of the beamformer slow and costly. This problem can be addressed by some recent techniques reported in the literature, namely

- low-rank,
- sparsity-aware and
- knowledge-aided

beamforming algorithms.

1.4   Low-rank beamforming algorithms

Low-rank adaptive signal processing is considered a key technique for dealing with large systems. The basic idea of low-rank algorithms is to reduce the number of adaptive coefficients by projecting the received data vectors onto a subspace of smaller dimension, which consists of a set of basis vectors. The adaptation of the low-order filter within the reduced-dimension subspace results in significant computational savings, faster convergence speed and better tracking performance. The first statistical low-rank method was based on a principal components (PC) decomposition of the target-free covariance matrix [7]. Another class of eigen-decomposition methods was based on the cross-spectral metric (CSM) [10, 11].

Furthermore, the family of Krylov subspace methods has been investigated thoroughly in recent years. This class of low-rank algorithms includes the multi-stage Wiener filter (MSWF) [5, 12, 13], which projects the observation data onto a reduced-dimension Krylov subspace, and the auxiliary-vector filters (AVF) [14–16]. These methods are relatively complex to implement in practice and may suffer from numerical problems despite their improved convergence and tracking performance.

The joint domain localized (JDL) approach, which is a beamspace reduced-dimension algorithm, was proposed by Wang and Cai [17] and investigated in both homogeneous and non-homogeneous environments in [18] and [19] respectively. Recently, reduced-rank adaptive processing algorithms were proposed based on

- joint iterative optimization of adaptive filters [20–24] (for homogeneous) and
- an adaptive diversity-combined decimation and interpolation scheme [8, 25–27] (for non-homogeneous).

As stated, the basic idea behind low-rank algorithms is to reduce the number of adaptive coefficients by projecting the received vectors onto a reduced-dimension subspace. Let $\mathbb{B}_D$ denote the $M \times D$ rank-reduction matrix with column vectors which form an $M \times 1$ basis for a D-dimensional subspace, where $D < M$. Thus, the received signal $\underline{x}[i]$ is transformed into its reduced-rank (low-rank) version $\underline{x}_D[i]$ given by

$$\underline{x}_D[i] = \mathbb{B}_D^H \underline{x}[i]. \tag{1.14}$$

The low-rank signal $\underline{x}_D[i]$ is processed by an adaptive low-rank space-time beamformer $\underline{w}_D$ with D coefficients — i.e. a $D \times 1$ vector. This is illustrated in Fig. 1.2. Subsequently, the decision is made based on the output of the beamformer

$$y[i] = \underline{w}_D^H \mathbb{B}_D^H \underline{x}[i]. \tag{1.15}$$

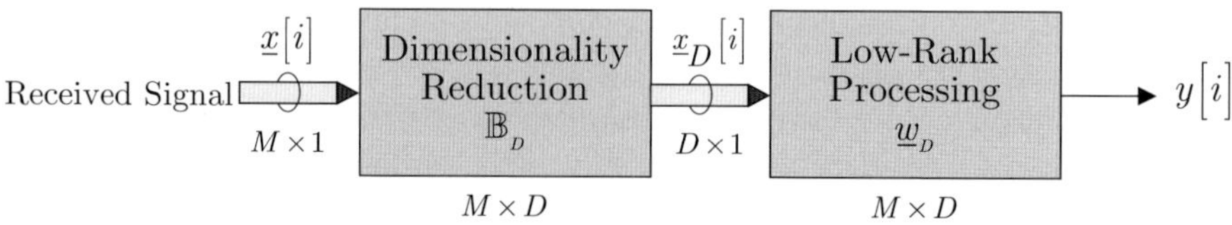

Fig. 1.2   Low-rank signal processing scheme.

A designer can compute the parameters of the beamformer by solving the following constrained optimization problem:

$$\underline{w}_{D,\text{opt}} = \arg \min_{\underline{w}_D} \left(\underline{w}_D^H \mathbb{B}_D^H \mathbb{R} \mathbb{B}_D \underline{w}_D \right) \tag{1.16}$$

$$\text{subject to } \underline{w}_D^H \mathbb{B}_D^H \underline{h} = 1. \tag{1.17}$$

The optimal low-rank MVDR solution for the above problem is given by

$$\underline{w}_{D,\text{opt}} = \frac{(\mathbb{B}_D^H \mathbb{R} \mathbb{B}_D)^{-1} \mathbb{B}_D^H \underline{h}}{\underline{h}^H \mathbb{B}_D (\mathbb{B}_D^H \mathbb{R} \mathbb{B}_D)^{-1} \mathbb{B}_D^H \underline{h}} = \frac{\mathbb{R}_D^{-1} \underline{h}_D}{\underline{h}_D^H \mathbb{R}_D^{-1} \underline{h}_D} \tag{1.18}$$

where

$$\mathbb{R}_D = \mathbb{B}_D^H \mathbb{R} \mathbb{B}_D \tag{1.19}$$

denotes the low-rank covariance matrix and

$$\underline{h}_D = \mathbb{B}_D^H \underline{h} \tag{1.20}$$

denotes the low-rank space-time steering vector. The key challenge in the design of low-rank STAP algorithms is to find a cost-effective method to compute the rank-reduction matrix $\mathbb{B}_D$.

1.4.1   *Eigenvalue-decomposition-based algorithms*

The eigenvalue-decomposition (EVD)-based beamforming algorithms are also known as PC-based algorithms and have been originally reported as the eigencanceler method. These PC-based algorithms refer to the beamformers constructed with a subset of the eigenvectors of the interference-only covariance matrix associated with the eigenvalues of largest magnitude. The first application of this method to radar systems was reported in [7].

The basic idea of the EVD-based beamformer is to approximate the $M \times M$ covariance matrix $\mathbb{R}$ of the received data vector $\underline{x}[i]$ as follows:

$$\mathbb{R} = \sum_{d=1}^{D} \lambda_d \underline{E}_d \underline{E}_d^H \tag{1.21}$$

where the $M \times 1$ vector $\underline{E}_d$ denotes the d-th eigenvector of $\mathbb{R}$ and λ_d is the d-th eigenvalue of $\mathbb{R}$. That is, by assuming that the eigenvalues are obtained in decreasing order of magnitude, the EVD-based method approximates $\mathbb{R}$ using its D dominant eigenvectors. The rank-reduction matrix is constructed by using the D dominant eigenvectors $(D < M)$ as described by

$$\mathbb{B}_D = [\underline{E}_1,\ \underline{E}_2,\ \ldots,\ \underline{E}_D]. \tag{1.22}$$

The low-rank MVDR solution for the above problem is given by

$$\underline{w}_D = \frac{\mathbb{B}_D(\mathbb{B}_D^H \mathbb{R} \mathbb{B}_D)^{-1}\mathbb{B}_D^H \underline{h}}{\underline{h}^H \mathbb{B}_D(\mathbb{B}_D^H \mathbb{R} \mathbb{B}_D)^{-1}\mathbb{B}_D^H \underline{h}} = \frac{\left(\sum\limits_{d=1}^{D} \lambda_d^{-1} \underline{E}_d \underline{E}_d^H \right) \underline{h}}{\underline{h}^H \left(\sum\limits_{d=1}^{D} \lambda_d^{-1} \underline{E}_d \underline{E}_d^H \right) \underline{h}}. \tag{1.23}$$

The EVD-based low-rank MVDR space-time beamformer described above does not take into account the target space-time steering vector $\underline{h}$ when selecting a suitable subspace representation of the interference. Clearly, this low-rank space-time beamformer requires the computation of an EVD, which has a computational cost that is cubic with M [28]. In order to reduce this computational complexity, a designer can resort to subspace tracking algorithms which bring the cost down to $\mathcal{O}(M^2)$ [29, 30]. Another technique associated with EVD-based beamforming that can improve the performance of low-rank MVDR space-time beamformers is the cross-spectral metric (CSM) [10]. The CSM approach chooses the set of D eigenvectors for the rank-reduction matrix that optimizes the desired criterion, namely the maximization of the SINR, as opposed to the PC method, which always chooses the dominant eigenvectors.

1.4.2 *Krylov subspace-based algorithms*

The first Krylov methods, namely the conjugate gradient (CG) method [31] and the Lanczos algorithm [32], were originally proposed for solving large systems of linear equations. These algorithms, used in numerical linear algebra, are mathematically identical to each other and have been derived for Hermitian and positive definite system matrices. Other techniques have been reported for solving these problems. The Arnoldi algorithm [33] is a computationally efficient procedure for arbitrarily invertible system matrices. The MSWF [10] and the AVF [14] algorithms are based on a multistage decomposition of the linear MMSE estimator. A key feature of these methods is that they do not require an EVD and have a very good level of performance. It turns out that Krylov subspace algorithms that are used for solving very large and sparse systems of linear equations are highly suitable alternatives for designing low-rank space-time beamforming algorithms in radar systems. The basic idea behind Krylov subspace algorithms is to construct the rank-reduction matrix $\mathbb{B}_D$ with the following structure:

$$\mathbb{B}_D = \left[\underline{q},\ \mathbb{R}\underline{q},\ \ldots,\ \mathbb{R}^{D-1}\underline{q}\right], \tag{1.24}$$

where $\underline{q} = \frac{\underline{h}}{||\underline{h}||}$ and $||\cdot||$ denotes the Euclidean norm (or the 2-norm) of a vector. In order to compute the basis vectors of the Krylov subspace (the column vectors of $\mathbb{B}_D$), a designer can either directly employ the expression in (1.24) or resort to more sophisticated approaches such as the Arnoldi iteration [33]. The low-rank MVDR solution for the space-time beamformer using the Krylov subspace is also given by Eq. (1.18). That is,

$$\underline{w}_D = \frac{(\mathbb{B}_D^H \mathbb{R} \mathbb{B}_D)^{-1} \mathbb{B}_D^H \underline{h}}{\underline{h}^H \mathbb{B}_D (\mathbb{B}_D^H \mathbb{R} \mathbb{B}_D)^{-1} \mathbb{B}_D^H \underline{h}}. \tag{1.25}$$

An appealing feature of the Krylov subspace algorithms is that the required model order D does not scale with the system size. Indeed, when M goes to infinity, the required D remains a finite and relatively small value. This result was established in [34]. Among the disadvantages of Krylov subspace methods are the relatively high computational cost of constructing $\mathbb{B}_D$ ($\mathcal{O}(DM^2)$), the numerical instability of some implementations and the lack of flexibility for imposing constraints on the design of the basis vectors.

1.4.3 *Joint iterative optimization (JIO)-based algorithms*

The aim of this section is to introduce low-rank beamforming algorithms based on JIO techniques. The idea behind these methods is to design the main components of a low-rank space-time beamforming scheme via a general optimization approach. The basic ideas of JIO techniques have been reported in [20–24]. Amongst the advantages of JIO techniques is the flexibility to choose the optimization algorithm and to impose constraints, which provides a significant advantage over eigen-based and Krylov subspace methods. One disadvantage that is shared amongst the JIO techniques, eigen-based and Krylov subspace methods is the complexity associated with the design of the matrix $\mathbb{B}_D$. For instance, for the design of a beamforming algorithm with a very large M, the problem of having to design an $M \times D$ rank-reduction matrix $\mathbb{B}_D$ remains.

In the framework of JIO techniques, the design of the matrix $\mathbb{B}_D$ and the beamforming vector $\underline{w}_D$ for a fixed model order D will be dictated by the optimization problem and by the algorithm chosen to compute the solution.

To this end, we will focus on a generic $\mathbb{B}_D = [\underline{b}_1, \underline{b}_2, \ldots, \underline{b}_D]$, in which the basis vectors $\underline{b}_d$, $d = 1, 2, \ldots, D$ will be obtained via an optimization algorithm and iterations between the $\mathbb{B}_D$ and $\underline{w}_D$ will be performed. The JIO method consists of solving the following optimization problem:

$$\left[\mathbb{B}_{D,\text{opt}}, \underline{w}_{D,\text{opt}}\right] = \arg \min_{\mathbb{B}_D, \underline{w}_D} \left(\underline{w}_D^H \mathbb{B}_D \mathbb{R} \mathbb{B}_D \underline{w}_D\right) \tag{1.26a}$$

$$\text{subject to } \underline{w}_D^H \mathbb{B}_D^H \underline{h} = 1 \tag{1.26b}$$

where it should be remarked that, although the optimization problem in Eq. (1.26a) is non-convex, the algorithms do not present convergence problems. Numerical studies with JIO methods indicate that the minima are identical and global. Proofs of global convergence have been established with different versions of JIO schemes [20–24], which demonstrate that a least-squares (LS) algorithm converges to the reduced-rank Wiener filter.

In order to solve the above problem, we resort to the method of Lagrange multipliers [6] and transform the constrained optimization into an unconstrained one expressed by the Lagrangian

$$\xi(\mathbb{B}_D, \underline{w}_D) = \underline{w}_D^H \mathbb{B}_D \mathbb{R} \mathbb{B}_D \underline{w}_D + \lambda(\underline{w}_D^H \mathbb{B}_D^H \underline{h} - 1) \tag{1.27}$$

where λ is a scalar Lagrange multiplier. By fixing $\underline{w}_D$, minimizing (1.27) with respect to $\mathbb{B}_D$ and solving for λ, we obtain

$$\mathbb{B}_D = \frac{\mathbb{R}^{-1} \underline{h}\, \underline{w}_D^H \mathbb{R}_w^{-1}}{\underline{w}_D^H \mathbb{R}_w^{-1} \underline{w}_D \underline{h}^H \mathbb{R}^{-1} \underline{h}} \tag{1.28}$$

where $\mathbb{R} = \mathcal{E}\left\{\underline{x}[i]\underline{x}^H[i]\right\}$ and $\mathbb{R}_w = \mathcal{E}\left\{\underline{w}_D \underline{w}_D^H\right\}$. By fixing $\mathbb{B}_D$, minimizing (1.27) with respect to $\underline{w}_D$ and solving for λ, we arrive at the expression

$$\underline{w}_D = \frac{\mathbb{R}_D^{-1} \underline{h}}{\underline{h}^H \mathbb{R}_D^{-1} \underline{h}} \tag{1.29}$$

where $\mathbb{R}_D = \mathcal{E}\left\{\underline{x}_D[i]\underline{x}_D^H[i]\right\} = \mathcal{E}\left\{\mathbb{B}_D^H \underline{x}[i]\underline{x}^H[i]\mathbb{B}_D\right\}$, $\underline{x}_D[i] = \mathbb{B}_D^H \underline{x}[i]$. Note that the expressions in (1.28) and (1.29) are not closed-form solutions for $\underline{w}_D$ and $\mathbb{B}_D$ since (1.28) is a function of $\underline{w}_D$ and (1.29) depends on $\mathbb{B}_D$. Thus, it is necessary to iterate (1.28) and (1.29) with initial values to obtain a solution. Unlike the Krylov subspace-based methods [11] and the AVF [15] methods, the JIO scheme provides an iterative exchange of information between the low-rank beamformer and the rank-reduction matrix and leads to a simpler adaptive implementation. The key strategy lies in the joint optimization of the filters. The rank D must be set by the designer to ensure appropriate performance or can be estimated via another algorithm. In terms of complexity, the JIO techniques have a computational cost that is related to the optimization algorithm. With recursive LS algorithms the complexity is quadratic with M ($\mathcal{O}(M^2)$), whereas the complexity can be as low as linear with M when stochastic gradient algorithms are adopted [8].

1.4.4 *Joint interpolation, decimation and filtering (JIDF)-based algorithms*

A low-rank space-time beamforming technique, based on the joint interpolation, decimation and filtering (JIDF) concept [8,25,26], allows a designer to compute the parameters of the rank-reduction matrix and of the low-rank space-time beamformer with a low complexity. The motivation for designing a rank-reduction matrix based on interpolation and decimation comes from two observations. The first is that rank reduction can be performed by constructing new samples with interpolators and eliminating (decimating) samples that are not useful in the STAP design. The second is the structure of the rank-reduction matrix, whose columns are a set of vectors formed by the interpolators and the decimators.

In the JIDF scheme, the number of elements for adaptive processing is substantially reduced, resulting in considerable computational savings and very fast convergence performance for the radar applications. The $M \times 1$ received vector $\underline{x}[i]$ is processed by a multiple processing branch (MPB) scheme with B branches, where each spatiotemporal processing branch contains an interpolator, a decimation unit and a low-rank space-time beamformer. In the b-th branch, the received vector $\underline{x}[i]$ is filtered by the interpolator vector $\underline{V}_b = [V_{b,1},\ V_{b,2},\ \ldots,\ V_{b,I}]^T$ of I coefficients, resulting in an interpolated received vector $\underline{x}_b[i]$ with M elements (samples), which is expressed by

$$\underline{x}_b[i] = \mathbb{V}_b^H \underline{x}[i] \tag{1.30}$$

where the $M \times M$ Toeplitz convolution matrix $\mathbb{V}_b$ is given by

$$\mathbb{V}_b = \begin{bmatrix} V_{b,1}, & 0, & \ldots, & 0 \\ V_{b,2}, & V_{b,1}, & \ldots, & 0 \\ \vdots & \vdots & \ddots & \vdots \\ V_{b,I}, & V_{b,I-1}, & \ldots, & 0 \\ 0, & V_{b,I}, & \ldots, & 0 \\ 0, & 0, & \ldots, & 0 \\ \vdots & \vdots & \ddots & \vdots \\ 0, & 0, & \ldots, & 0 \\ 0, & 0, & \ldots, & V_{b,1} \end{bmatrix}. \tag{1.31}$$

The vector $\underline{x}_b[i]$ can be expressed in an alternative way that is useful for the design of the JIDF scheme and is described by

$$\underline{x}_b[i] = \mathbb{V}_b^H \underline{x}[i] = \mathbb{X}_0[i]\underline{V}_b \tag{1.32}$$

where the $M \times I$ matrix $\mathbb{X}_0[i]$ with the samples of $\underline{x}(i)$ has a Hankel structure and is described by

$$\mathbb{X}_0[i] = \begin{bmatrix} x_1[i], & x_2[i], & \cdots, & x_I[i] \\ x_2[i], & x_3[i], & \cdots, & x_{I+1}[i] \\ \vdots & \vdots & \ddots & \vdots \\ x_{M-1}[i], & x_M[i], & \cdots, & 0 \\ x_M[i], & 0, & \cdots, & 0 \end{bmatrix}. \tag{1.33}$$

The dimensionality reduction is performed by a decimation unit with $D \times M$ decimation matrices $\mathbb{D}_{D,b}$ that transforms $\underline{x}_b[i]$ into $D \times 1$ vectors $\underline{x}_{D,b}[i]$ with $b = 1, \ldots, B$, where B is a parameter to be set by the designer. Furthermore, $D = M/Q$ is the rank of the resulting system of equations that will be generated and Q is the decimation factor. The $D \times 1$ vector $\underline{x}_{D,b}[i]$ for branch b is expressed by

$$\begin{aligned} \underline{x}_{D,b}[i] &= \mathbb{B}_{D,b}^H \underline{x}[i] = \mathbb{D}_{D,b}\mathbb{V}_b^H \underline{x}[i] \\ &= \mathbb{D}_{D,b}\mathbb{X}_o[i]\underline{V}_b \end{aligned} \tag{1.34}$$

where $\mathbb{B}_{D,b}$ is the rank-reduction matrix and the vector $\underline{x}_{D,b}[i]$ for branch b is used in the minimization of the output power for branch b. The output at the end of the JIDF scheme $y[i]$ is selected according to

$$y[i] = y_{b_s}[i] \quad \text{when } b_s = \arg\min_b |y_b|^2. \tag{1.35}$$

For the computation of the parameters of the JIDF scheme, it is fundamental to express the output $y_b[i]$ in terms of the interpolator $\underline{V}_b$, of the decimation matrix $\mathbb{D}_{D,b}$ and of the low-rank space-time beamformer $\underline{w}_{D,b}$ as follows:

$$\begin{aligned} y_b[i] &= \underline{w}_{D,b}^H \mathbb{B}_{D,b}^H \underline{x}[i] \\ &= \underline{w}_{D,b}^H \mathbb{D}_{D,b}\mathbb{X}_o[i]\underline{V}_b \end{aligned} \tag{1.36}$$

where the expression (1.36) indicates that the dimensionality reduction carried out by the JIDF scheme depends on finding appropriate $\underline{V}_b$, $\mathbb{D}_{D,b}$ and $\underline{w}_{D,b}$. Unlike the previously discussed low-rank beamforming techniques, the JIDF is able to substantially reduce the cost of the rank-reduction matrix.

The parameters of the JIDF scheme that perform low-rank space-time MVDR beamforming can be computed by solving the following optimization problem

$$[w_{D,\text{opt}}, V_{\text{opt}}, \mathbb{D}_{D,b_s}] = \arg \min_{\substack{\underline{w}_{D,b}, \underline{V}_b, \\ \mathbb{D}_{D,b}}} \underline{w}_{D,b}^H \mathcal{E}\left\{ \mathbb{D}_{D,b} \mathbb{X}_o[i] \underline{V}_b \underline{V}_b^H \mathbb{X}_o^H[i] \mathbb{D}_{D,b}^H \right\} \underline{w}_{D,b}$$

$$\text{subject to } \underline{w}_{D,b}^H \mathbb{D}_{D,b} \mathbb{H}_o[i] \underline{V}_b = 1$$

$$(1.37)$$

where $\mathbb{H}_o[i]$ is a $M \times I$ space-time steering matrix with a Hankel structure consisting of the elements of the space-time steering vector $\underline{h}[i]$ and given by

$$\mathbb{H}_o[i] = \begin{bmatrix} h_1[i], & h_2[i], & \ldots, & h_I[i] \\ h_2[i], & h_3[i], & \ldots, & h_{I+1}[i] \\ \vdots & \vdots & \ddots & \vdots \\ h_{M-1}[i], & h_M[i], & \ldots, & 0 \\ h_M[i], & 0, & \ldots, & 0 \end{bmatrix}. \tag{1.38}$$

The constrained optimization in (1.37) can be transformed into an unconstrained optimization problem by using the method of Lagrange multipliers, which results in

$$\xi(\underline{w}_{D,b}, \underline{V}_b, \mathbb{D}_{D,b}) = \underline{w}_{D,b}^H \mathcal{E}\left\{ \mathbb{D}_{D,b} \mathbb{X}_o[i] \underline{V}_b \underline{V}_b^H \mathbb{X}_o^H[i] \mathbb{D}_{D,b}^H \right\} \underline{w}_{D,b} \quad (1.39)$$

$$+ \lambda(\underline{w}_{D,b}^H \mathbb{D}_{D,b} \mathbb{H}_o[i] \underline{V}_b - 1)$$

where λ is a Lagrange multiplier.

The strategy of computing the parameters of the low-rank space-time beamformer based on the JIDF scheme is to minimize the cost function given by Eq. (1.39) with respect to a subset of parameters and fix the remaining parameters. By minimizing (1.39) with respect to $\underline{V}_b$, we obtain

$$\underline{V}_b = \frac{\mathbb{R}_{v,b}^{-1} \underline{h}_{v,b}}{\underline{h}_{v,b}^H \mathbb{R}_{v,b}^{-1} \underline{h}_{v,b}} \tag{1.40}$$

where $\mathbb{R}_{v,b} = \mathcal{E}\left\{ \underline{x}_{v,b} \underline{x}_{v,b}^H \right\}$ is the $I \times I$ auto-correlation matrix of $\underline{x}_{v,b} = \mathbb{D}_{D,b}^H \mathbb{R}_o^H \underline{w}_{D,b}$, and $\underline{h}_{v,b} = \mathbb{D}_{D,b}^H \mathbb{H}_o^H \underline{w}_{D,b}$ is the $I \times 1$ low-rank steering vector. By minimizing (1.39) with respect to $\underline{w}_{D,b}$, we have

$$\underline{w}_{D,b} = \frac{\mathbb{R}_{w,b}^{-1} \underline{h}_{w,b}}{\underline{h}_{w,b}^H \mathbb{R}_{w,b}^{-1} \underline{h}_{w,b}} \tag{1.41}$$

where $\mathbb{R}_{w,b} = \mathcal{E}\{ \underline{x}_{w,b} \underline{x}_{w,b}^H \}$ is the $D \times D$ covariance matrix of $\underline{x}_{w,b} = \mathbb{D}_{D,b} \mathbb{X}_o \underline{V}_b$, and $\underline{h}_{w,b} = \mathbb{D}_{D,b} \mathbb{H}_o \underline{V}_b$ is the $D \times 1$ low-rank steering vector. In order to compute $\underline{V}_b$ and $\underline{w}_{D,b}$, a designer needs to iterate them for each processing branch b.

The decimation matrix $\mathbb{D}_{D,b}$ is selected to minimize the square of the output of the beamformer $y_b[i]$ obtained for all the B branches

$$\mathbb{D}_{D,b} = \mathbb{D}_{D,b_s}[i] \quad \text{when} \quad b_s = \arg \min_{1 \leq b \leq B} |y_b[i]|^2. \tag{1.42}$$

The design of the decimation matrix $\mathbb{D}_{D,b}$ imposes the constraint on the values of the elements of the matrix such that they only take the value 0 or 1. Since the optimal approach for the design of $\mathbb{D}_{D,b}$ corresponds to an exhaustive search, we consider a suboptimal technique that employs pre-stored patterns. The decimation scheme employs a structure formed in the following way

$$\mathbb{S}_{D,b} = [\underline{\delta}_{b,1}, \ \underline{\delta}_{b,2}, \dots, \ \underline{\delta}_{b,D}] \tag{1.43}$$

where $\underline{\delta}_{b,d}$ is an $M \times 1$ vector composed of a single 1 and 0s as described by

$$\underline{\delta}_{b,i} = [\underbrace{0, \ \dots, \ 0}_{i-1}, \ 1, \ \underbrace{0, \ \dots, \ 0}_{M-i}]^T, \tag{1.44}$$

where the i-th element is the only element equal to one. We set the value of i in a deterministic way which can be expressed as

$$i = \frac{M}{D} \times (i - 1) + b \tag{1.45}$$

with $i \geqslant 2$. In order to obtain the solution, it is necessary to iterate (1.40), (1.41) and (1.42) successively (one followed by the other) with an initial value. The expectations can be estimated either via time averages or by instantaneous estimates and with the help of adaptive algorithms.

1.5 Sparsity-aware beamforming algorithms

Recently, motivated by compressive sensing (CS) techniques used in radar, several authors have considered CS ideas for moving target indication (MTI) and STAP problems [35, 36]. The core notion in CS is to regularize a linear inverse problem by including prior knowledge that the signal of interest is sparse [37]. These works on space-time beamforming techniques based on CS rely on the recovery of the clutter power in the angle-Doppler plane, which is usually carried out via convex optimization tools. However, these methods are based on linear programming and have quite a high computational complexity ($\mathcal{O}(K^3)$), where K is the dimension of the angle-Doppler plane. In this section, we describe the concept of a sparsity-aware STAP (SA-STAP) algorithm that can improve the detection capability using a small number of snapshots. To overcome the high complexity of the CS-STAP type algorithm, we design the STAP algorithm with another strategy, by imposing the sparse regularization to the minimum variance (MV) cost function. Since the interference variance often has a low-rank property, we assume that a number of samples of the datacube are not meaningful for processing and the optimal STAP beamformer is sparse, or nearly sparse. Then, we exploit this feature by using a l_1-norm regularization. With this motivation, the STAP algorithm design becomes a mixed l_1-norm and l_2-norm optimization problem.

The conventional space-time beamforming algorithms do not exploit the sparsity of the received signals. In this exposition, it is assumed that a number of samples of the datacube are not meaningful for processing and a reduced number of active weights of the space-time beamformer can effectively suppress the clutter and the jamming signals. Specifically, a sparse regularization is imposed on the space-time MVDR beamforming design. Thus, the space-time beamformer design can be described as the following optimization problem

$$\underline{w}_{\mathrm{opt}} = \arg\min_{\underline{w}} \left(\underline{w}^H \mathbb{R} \underline{w} \right) \tag{1.46}$$

$$\text{subject to } \underline{w}^H \underline{h} = 1 \quad \text{and} \quad \|\underline{w}\|_1 = 0 \tag{1.47}$$

where the objective of the l_1-norm regularization is to force the components of the space-time beamformer $\underline{w}$ to zero [38]. This problem can be solved using the method of Lagrange multipliers, which results in the following unconstrained cost function

$$\xi(\underline{w}, \lambda_1, \lambda_2) = \underline{w}^H \mathbb{R} \underline{w} + \lambda_1 (\underline{w}^H \underline{h} - 1) + \lambda_2 (\|\underline{w}\|_1), \tag{1.48}$$

where λ_1, λ_2 are the Lagrange multipliers. The unconstrained cost function above is convex; however it is non-differentiable, which makes it difficult for one to use the method of Lagrange multipliers directly and obtain an expression for the space-time beamformer. To this end, the following approximation to the regularization term is employed

$$\|\underline{w}\|_1 \approx \underline{w}^H \mathbf{\Lambda} \underline{w}, \tag{1.49}$$

where

$$\mathbf{\Lambda} = \mathrm{diag} \left(\left[\frac{1}{|w_1| + \epsilon}, \frac{1}{|w_2| + \epsilon}, \cdots, \frac{1}{|w_M| + \epsilon} \right]^T \right), \tag{1.50}$$

where ϵ is a small positive constant. Furthermore, the partial derivative of $\underline{w}^H \mathbf{\Lambda} \underline{w}$ with respect to $\underline{w}$ is given by

$$\frac{\partial \underline{w}^H \mathbf{\Lambda} \underline{w}}{\partial \underline{w}} = \mathbf{\Lambda} \underline{w}. \tag{1.51}$$

With the above development, an approximation of the unconstrained cost function given by Eq. (1.48) can be employed as follows

$$\xi(\underline{w}, \lambda_1, \lambda_2) \approx \underline{w}^H \mathbb{R} \underline{w} + \lambda_1 \left(\underline{w}^H \underline{h} - 1 \right) + \lambda_2 \underline{w}^H \mathbf{\Lambda} \underline{w}. \tag{1.52}$$

By computing the gradient terms with respect to $\underline{w}^*$ and λ_1 and equating them to zero, we obtain the following expression for the space-time beamformer

$$\underline{w} = \frac{\left(\mathbb{R} + \lambda_2 \mathbf{\Lambda} \right)^{-1} \underline{h}}{\underline{h}^H \left(\mathbb{R} + \lambda_2 \mathbf{\Lambda} \right)^{-1} \underline{h}}. \tag{1.53}$$

Comparing (1.53) with the conventional optimal space-time beamformer in (1.13), we find that there is an additional term $\lambda_2 \mathbf{\Lambda}$ in the inverse of the interference covariance matrix $\mathbb{R}$, which is due to the l_1-norm regularization. The term λ_2 is a positive scalar which provides a trade-off between the sparsity and the output interference power. The larger the chosen λ_2, the more components are shrunk to zero [39]. It should also be remarked that the expression for the beamformer in (1.53) is not a closed-form solution since $\mathbf{\Lambda}$ is a function of $\underline{w}$. Thus it is necessary to develop an iterative procedure to compute the parameters of the space-time beamformer.

1.6   Knowledge-aided beamforming algorithms

Although STAP techniques are considered efficient tools for the detection of slow targets by airborne radar systems in strong clutter environments [1], due to the very large number of degrees of freedom (DoFs), conventional space-time beamformers have a slow convergence and require about twice the DoFs of the independent and identically distributed (i.i.d.) training snapshots to yield an average performance loss of roughly 3 dB [40]. In real scenarios, it is hard to obtain so many i.i.d. training snapshots, especially in heterogeneous environments. Low-rank [5–8] and sparsity-aware [35–41] methods have been considered to counteract the slow convergence of the conventional space-time beamformers. Nevertheless, there are other alternatives to improve the training and performance of STAP algorithms. These other methods can also be combined with the techniques previously discussed. Recently developed knowledge-aided (KA) STAP algorithms have sparked a growing interest and become a key concept for the next generation of adaptive radar systems [42, 43]. The core idea of KA-STAP is to incorporate prior knowledge, provided by digital elevation maps, land cover databases, road maps, the Global Positioning System (GPS), previous scanning data and other known features, to compute estimates of the clutter covariance matrix with high accuracy [44]. Previous work on KA-STAP algorithms includes the exploitation of prior knowledge of the clutter ridge to form the STAP filter weights [45], the use of prior knowledge about the terrain [46] and of prior knowledge about the covariance matrix of the clutter and the jamming signals [43, 47].

In this section, we discuss a strategy to mitigate the deleterious effects of the heterogeneity in the secondary data, which makes use of *a priori* knowledge of the clutter covariance matrix and has recently gained significant attention in the literature [42, 43]. In KA-STAP techniques, there are two basic tasks that need to be addressed. The first one is how to obtain prior knowledge from the terrain knowledge of the clutter and how to estimate the real interference covariance matrix with this prior knowledge [42–46] and the second is how to apply the covariance matrix estimates in the design of the space-time beamforming algorithm [43, 47]. We first review how a designer can obtain prior knowledge of the clutter and employ this knowledge to build a known covariance matrix $\mathbb{R}_o$. Then, we present a method to combine this prior knowledge with commonly used estimation techniques to compute the covariance matrix of the received vector $\underline{x}[i]$, resulting in a combined covariance matrix estimate $\hat{\mathbb{R}}_c$ for use in the space-time beamformer that is more accurate and has an enhanced performance.

The optimal space-time beamformer, given by (1.13), in practice employs the following weight vector

$$\underline{w} = \frac{\hat{\mathbb{R}}^{-1}\underline{h}}{\underline{h}^H\hat{\mathbb{R}}^{-1}\underline{h}} \tag{1.54}$$

where an estimate of the covariance matrix is typically obtained by

$$\hat{\mathbb{R}} = \frac{1}{L}\sum_{k=1}^{L}\underline{x}[k]\underline{x}^H[k] \tag{1.55}$$

with the $M \times 1$ vector $\underline{x}[k]$ taken from secondary data. The estimate $\hat{\mathbb{R}}$ can be sufficiently accurate when the number of snapshots L is at least twice as great as M [3] and the training samples are assumed i.i.d. However, it is by now well understood that the clutter environments are often heterogeneous and this leads to performance degradation in space-time beamforming. KA-STAP techniques can significantly help to combat the heterogeneity [48].

With KA techniques the clutter covariance matrix $\mathbb{R}_c$ is estimated by combining an initial guess of the covariance matrix $\mathbb{R}_o$ derived from the digital terrain database or the data probed by radar in previous scans and the sample average covariance matrix estimate in the present scan $\hat{\mathbb{R}}$ so that

$$\mathbb{R}_c = \alpha \mathbb{R}_o + (1 - \alpha)\hat{\mathbb{R}} \tag{1.56}$$

where $0 \leq \alpha \leq 1$. Alternatively, this principle can be applied to the inverse of the covariance matrix estimate

$$\mathbb{R}_c^{-1} = \eta \mathbb{R}_o^{-1} + (1 - \eta)\hat{\mathbb{R}}^{-1} \tag{1.57}$$

where $0 \leq \eta \leq 1$.

In order to compute the parameter η, we need to consider the optimization problem

$$\eta_{\text{opt}} = \arg \min_{\eta} \left(\underline{w}^H \mathbb{R} \underline{w} \right), \tag{1.58}$$

where we use the relation

$$\underline{w} = \eta \underline{w}_o + (1 - \eta)\hat{\underline{w}} \tag{1.59}$$

where $\underline{w} = \mathbb{R}^{-1}\underline{h}$, $\underline{w}_o = \mathbb{R}_o^{-1}\underline{h}$ and $\hat{\underline{w}} = \hat{\mathbb{R}}^{-1}\underline{h}$. We can obtain the optimal value for η by equating the gradient of the cost function in (1.58) to zero, which results in [48]

$$\eta_{\text{opt}} = \frac{\text{Re}\left\{ \underline{h}^H (\hat{\mathbb{R}}^{-1} - \mathbb{R}_o^{-1})\mathbb{R}\hat{\mathbb{R}}^{-1}\underline{h} \right\}}{\underline{h}^H (\mathbb{R}_o^{-1} - \hat{\mathbb{R}}^{-1})\mathbb{R}(\mathbb{R}_o^{-1} - \hat{\mathbb{R}}^{-1})\underline{h}}. \tag{1.60}$$

Since $\mathbb{R}$ above is unknown, we have to estimate it in real time using either time averages or adaptive algorithms.

1.7   Simulations

In this section, the performance of space-time beamforming algorithms discussed in this chapter is assessed using simulated radar data. Specifically, we consider the optimal MVDR space-time beamforming algorithm that assumes perfect knowledge of the covariance matrix of the received data and the MVDR space-time beamformer using the sample matrix inversion (SMI-MVDR). The low-rank space-time beamformers using the following algorithms:

- EVD (LR-EVD),
- the Krylov subspace approach (LR-Krylov),
- the JIO (LR-JIO) and
- the JIDF (LR-JIDF)

are also considered as having a rank equal to D. We also consider the space-time beamforming algorithms:

- the sparsity-aware (SA-MVDR) and
- the knowledge-aided (KA-MVDR).

All the analyzed algorithms estimate the statistical quantities via time-averages in a similar way to a LS method. The parameters of the simulated radar platform, using a sideway-looking array (SLA), are shown in Table 1.1. For all simulations, the presence of a mixture of two broadband jammers at $-45°$ and $60°$ is assumed with jammer-to-noise ratio (JNR) equal to 40 dB. The clutter-to-noise-ratio (CNR) is fixed at 40 dB. All the results presented are averages over 1000 independent Monte-Carlo runs.

In the first experiment, we assess the output SINR performance of the different space-time beamforming algorithms, which are simulated over 800 snapshots and the input signal-to-noise ratio (SNR) is set to 10 dB. The results are shown in Fig. 1.3 indicating that the LR-JIDF algorithm

Table 1.1   Airborne radar system parameters.

Parameter	Symbol	Value
Antenna array	SLA	-
Number of antennas	N	8
Carrier frequency	F_c	450 MHz
Transmit pattern	AP	Uniform
PRF	f_r	300 Hz
Platform velocity	v	75 m/s
Platform height	h	9000 m
Clutter-to-noise ratio	CNR	40 dB
Number of pulses	N_p	8

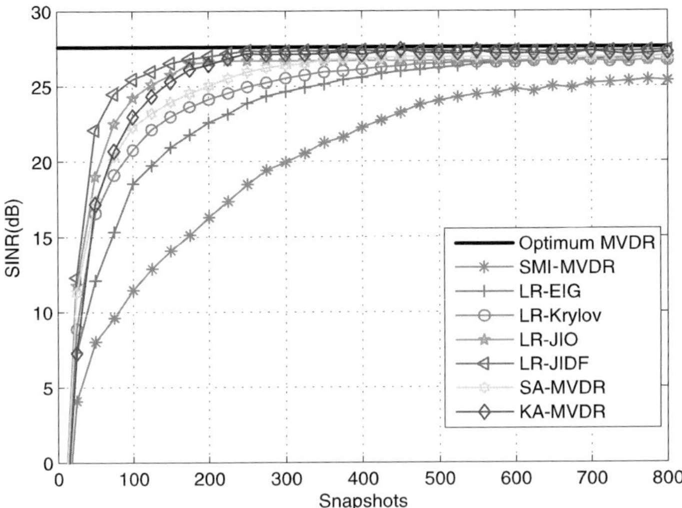

Fig. 1.3   SINR output performance against the number of snapshots. Parameters: $B = 8$, $I = 8$ and $D = 6$, $M = 64$, $\mathbb{R}_o = 0.01\mathbb{I}_M$, $\epsilon = 0.1$.

achieves the best results, followed by the LR-JIO, the KA-MVDR, the SA-MVDR, the LR-Krylov, the LR-EIG and the SMI-MVDR algorithms. The curves in Fig. 1.3 indicate that the use of low-rank algorithms is highly beneficial to the performance of space-time beamforming algorithms in radar systems. In particular, the LR-JIDF and LR-JIO algorithms have a very fast convergence. It should also be remarked that the SA-MVDR and KA-MVDR algorithms achieve a performance that is significantly better than that of the conventional SMI MVDR algorithm. Since the SA-MVDR and KA-MVDR techniques are modifications of the SMI-MVDR techniques exploiting sparsity and prior knowledge about the covariance matrix respectively, it is interesting to note that by exploiting these properties it is possible to significantly improve the performance of beamforming algorithms.

In addition, the output SINR performance is evaluated against the target Doppler frequency at the main beam look-angle for the various algorithms and the results are illustrated in Fig. 1.4. The potential Doppler frequency space from -100 Hz to 100 Hz is examined and 100 snapshots are used to train the beamformers. The plots show that the analyzed algorithms converge and approach the optimum in a short time (see Fig. 1.3), and form a deep null to cancel the main beam clutter (see Fig. 1.4). Again, the LR-JIDF algorithm outperforms the other analyzed algorithms.

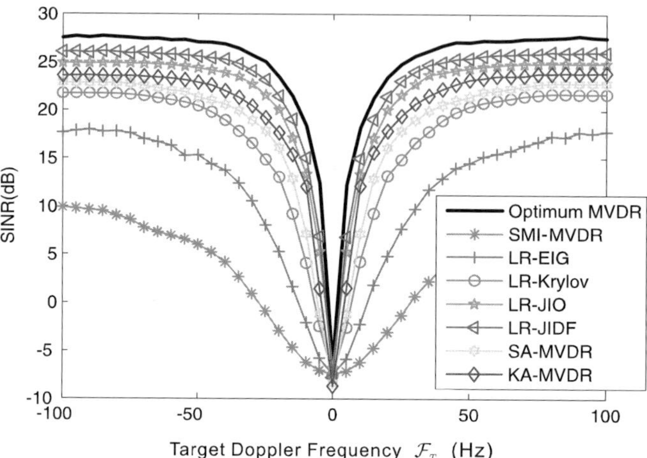

Fig. 1.4　SINR output performance against the target Doppler frequency. Parameters: $B = 8$, $I = 8$ and $D = 6$, $M = 64$, $\mathbb{R}_o = 0.01\mathbb{I}_M$, $\epsilon = 0.1$.

In the third example, the probability of detection P_D versus input SNR performance is examined for all schemes using 200 snapshots as the training data and the results are shown in Fig. 1.5. The probability of false alarm (PFA) is set to 10^{-6} and we consider the target injected in the boresight ($0°$) with Doppler frequency 100Hz. Figure 1.5 illustrates that the analyzed algorithms provide suboptimal detection performance using short support data. Note that for $P_D = 0.9$ (90% percent), the LR-JIDF and LR-JIO schemes are within less than 1 dB from the performance of the optimal MVDR algorithm. The remaining techniques exhibit increasing performance losses as compared to the optimal MVDR algorithm. In addition, it should be noted that the conventional SMI-MVDR method has a performance degradation of up to 5 dB for the same performance measured in terms of P_D. This suggests that the application of more sophisticated space-time beamforming algorithms is of paramount importance to achieving an improved performance.

Finally, the computational complexity is examined in terms of the multiplications of the analyzed schemes and the results are shown in Fig. 1.6. The curves show that the computational complexity of the LR-JIDF and LR-Krylov algorithms is significantly lower than that of the remaining algorithms. Indeed, there is a significant computational advantage obtained by using the LR-JIDF and LR-Krylov algorithms and this advantage becomes

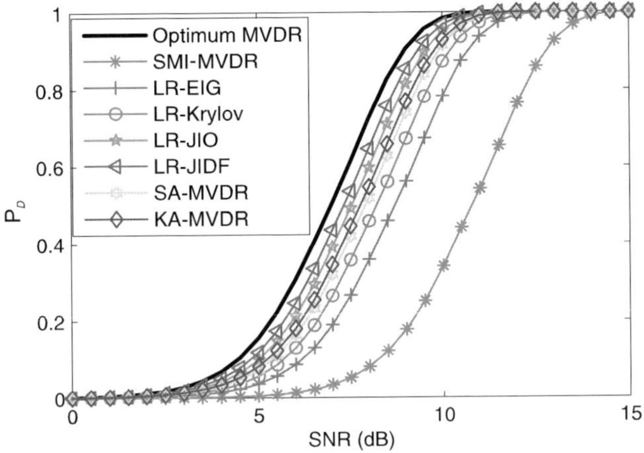

Fig. 1.5   Probability of detection against input SNR. Parameters: $B = 8$, $I = 8$ and $D = 6$, $M = 64$, $\mathbb{R}_o = 0.01\mathbb{I}_M$, $\epsilon = 0.1$.

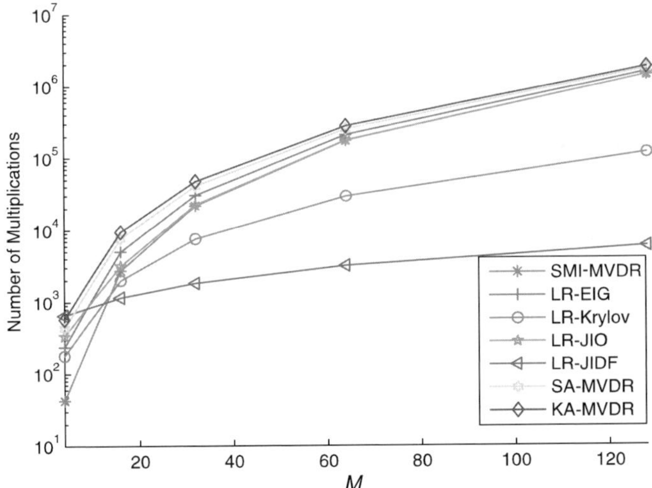

Fig. 1.6   Computational complexity in terms of multiplications of the analyzed space-time beamforming algorithms. Parameters: $B = 8$, $I = 8$ and $D = 6$.

more pronounced as M is increased. The other analyzed algorithms have a complexity that scales as a cubic function of M. This high complexity can be mitigated by the use of adaptive algorithms, which can reduce the computational cost by at least one order of magnitude.

1.8   Concluding remarks

In this chapter, the recent advances in the field of space-time beamforming algorithms for phased-array radar systems have been presented and the performance of these algorithms has been investigated via computer simulation studies. Specifically, some of the most successful space-time beamforming algorithms that exploit low-rank and sparsity properties as well as the use of prior knowledge to improve the performance of STAP algorithms have been presented in detail. The results of our studies suggest that low-rank algorithms have a substantial performance advantage over conventional MVDR space-time beamforming algorithms. Furthermore, the use of sparsity-aware and knowledge-aided strategies is also able to improve the performance of space-beamforming algorithms and can be combined with low-rank schemes. These beamforming algorithms can be also applied to MIMO radar systems, monostatic and bistatic radar systems, and other sensing applications such as sonar systems.

References

[1] R. Klemm, *Principles of Space-Time Adaptive Processing.*   Bodmin, UK: IET Press, 2002.

[2] W. Melvin, "A STAP overview," *IEEE Aerospace and Electronic Systems Magazine*, vol. 19, no. 1, pp. 19–35, Jan. 2004.

[3] L. Brennan and I. Reed, "Theory of adaptive radar," *IEEE Transactions on Aerospace and Electronic Systems*, vol. 9, no. 2, pp. 237–252, Mar. 1973.

[4] I. Reed, J. Mallett, and L. Brennan, "Rapid convergence rate in adaptive arrays," *IEEE Transactions on Aerospace and Electronic Systems*, vol. 10, no. 6, pp. 853–863, Nov. 1974.

[5] J. Guerci, J. Goldstein, and I. Reed, "Optimal and adaptive reduced-rank STAP," *IEEE Transactions on Aerospace and Electronic Systems*, vol. 36, no. 2, pp. 647–663, Apr. 2000.

[6] S. Haykin, *Adaptive Filter Theory*, 4th ed.   Upper Saddle River, NJ, USA: Prentice-Hall, 2002.

[7] A. Haimovich and Y. Bar-Ness, "An eigenanalysis interference canceler," *IEEE Transactions on Signal Processing*, vol. 39, no. 1, pp. 76–84, Jan. 1991.

[8] R. de Lamare and R. Sampaio-Neto, "Adaptive reduced-rank processing based on joint and iterative interpolation, decimation, and filtering," *IEEE Transactions on Signal Processing*, vol. 57, no. 7, pp. 2503–2514, Jul. 2009.

[9] A. M. Haimovich, R. S. Blum, and L. J. Cimini, "MIMO radar with widely separated antennas," *IEEE Signal Processing Magazine*, vol. 25, no. 1, pp. 116–129, Jan. 2008.

[10] J. Goldstein and I. Reed, "Subspace selection for partially adaptive sensor array processing," *IEEE Transactions on Aerospace and Electronic Systems*, vol. 33, no. 2, pp. 539–544, Apr. 1997.

[11] ——, "Reduced-rank adaptive filtering," *IEEE Transactions on Signal Processing*, vol. 45, no. 2, pp. 492–496, Feb. 1997.

[12] J. Goldstein, I. Reed, and L. Scharf, "A multistage representation of the Wiener filter based on orthogonal projections," *IEEE Transactions on Information Theory*, vol. 44, no. 7, pp. 2943–2959, Nov. 1998.

[13] Y.-L. Gau and I. Reed, "An improved reduced-rank CFAR space-time adaptive radar detection algorithm," *IEEE Transactions on Signal Processing*, vol. 46, no. 8, pp. 2139–2146, Aug. 1998.

[14] D. Pados and S. Batalama, "Joint space-time auxiliary-vector filtering for DS/CDMA systems with antenna arrays," *IEEE Transactions on Communications*, vol. 47, no. 9, pp. 1406–1415, Sep. 1999.

[15] D. Pados and G. Karystinos, "An iterative algorithm for the computation of the MVDR filter," *IEEE Transactions on Signal Processing*, vol. 49, no. 2, pp. 290–300, Feb. 2001.

[16] D. Pados, G. Karystinos, S. Batalama, and J. Matyjas, "Short-data-record adaptive detection," in *Proceedings of IEEE Radar Conference*, Apr. 2007, pp. 357–361.

[17] H. Wang and L. Cai, "On adaptive spatial-temporal processing for airborne surveillance radar systems," *IEEE Transactions on Aerospace and Electronic Systems*, vol. 30, no. 3, pp. 660–670, Jul. 1994.

[18] R. Adve, T. Hale, and M. Wicks, "Practical joint domain localised adaptive processing in homogeneous and nonhomogeneous environments. Part 1: Homogeneous environments," *IEE Proceedings - Radar, Sonar and Navigation*, vol. 147, no. 2, pp. 57–65, Apr. 2000.

[19] ——, "Practical joint domain localised adaptive processing in homogeneous and nonhomogeneous environments. Part 2. nonhomogeneous environments," *IEE Proceedings - Radar, Sonar and Navigation*, vol. 147, no. 2, pp. 66–74, Apr. 2000.

[20] R. de Lamare and R. Sampaio-Neto, "Reduced-rank adaptive filtering based on joint iterative optimization of adaptive filters," *IEEE Signal Processing Letters*, vol. 14, no. 12, pp. 980–983, Dec. 2007.

[21] R. Fa, R. de Lamare, and D. Zanatta-Filho, "Reduced-rank STAP algorithm for adaptive radar based on joint iterative optimization of adaptive filters," in *Asilomar Conference on Signals, Systems and Computers*, Oct. 2008, pp. 533–537.

[22] R. de Lamare, "Adaptive reduced-rank LCMV beamforming algorithms based on joint iterative optimisation of filters," *Electronics Letters*, vol. 44, no. 9, pp. 565–566, Apr. 2008.

[23] R. C. de Lamare, L. Wang, and R. Fa, "Adaptive reduced-rank beamforming algorithms based on joint iterative optimization of filters: design and analysis," *Signal Processing*, vol. 90, no. 2, pp. 640–652, Feb. 2010.

[24] R. Fa and R. de Lamare, "Reduced-rank STAP algorithms using joint itera-tive optimization of filters," *IEEE Transactions on Aerospace and Electronic Systems*, vol. 47, no. 3, pp. 1668–1684, Jul. 2011.

[25] R. de Lamare and R. Sampaio-Neto, "Adaptive reduced-rank MMSE param-eter estimation based on an adaptive diversity-combined decimation and interpolation scheme," in *IEEE International Conference on Acoustics, Speech and Signal Processing*, vol. 3, Apr. 2007, pp. 1317–1320.

[26] R. Fa, R. de Lamare, and L. Wang, "Reduced-rank STAP schemes for air-borne radar based on switched joint interpolation, decimation and filtering algorithm," *IEEE Transactions on Signal Processing*, vol. 58, no. 8, pp. 4182–4194, Aug. 2010.

[27] R. de Lamare, R. Sampaio-Neto, and M. Haardt, "Blind adaptive con-strained constant-modulus reduced-rank interference suppression algorithms based on interpolation and switched decimation," *IEEE Transactions on Signal Processing*, vol. 59, no. 2, pp. 681–695, Feb. 2011.

[28] G. H. Golub and C. F. Van Loan, *Matrix Computations*.   Baltimore, US: The Johns Hopkins University Press, Oct. 1996.

[29] B. Yang, "Projection approximation subspace tracking," *IEEE Transactions on Signal Processing*, vol. 43, no. 1, pp. 95–107, Jan. 1995.

[30] R. Badeau, B. David, and G. Richard, "Fast approximated power iteration subspace tracking," *IEEE Transactions on Signal Processing*, vol. 53, no. 8, pp. 2931–2941, Aug. 2005.

[31] M. R. Hestenes and E. Stiefel, "Methods of conjugate gradients for solving linear systems," *Journal of Research of the National Bureau of Standards*, vol. 49, no. 6, pp. 409–436, Dec. 1952.

[32] C. Lanczos, "Solution of systems of linear equations byminimized iterations," *Journal of Research of the National Bureau of Standards*, vol. 49, no. 1, pp. 33–53, Jul. 1952.

[33] W. E. Arnoldi, "The principle of minimized iterations in the solution of the matrix eigenvalue problem," *Quarterly of Applied Mathematics*, vol. 9, pp. 17–29, 1951.

[34] W. Xiao and M. Honig, "Large system transient analysis of adaptive least squares filtering," *IEEE Transactions on Information Theory*, vol. 51, no. 7, pp. 2447–2474, Jul. 2005.

[35] S. Maria and J.-J. Fuchs, "Application of the global matched filter to STAP data an efficient algorithmic approach," in *IEEE International Conference on Acoustics, Speech and Signal Processing*, vol. 4, May 2006.

[36] I. Selesnick, S. Pillai, K. Y. Li, and B. Himed, "Angle-Doppler processing using sparse regularization," in *IEEE International Conference on Acoustics Speech and Signal Processing*, Mar. 2010, pp. 2750–2753.

[37] J. Parker and L. Potter, "A Bayesian perspective on sparse regularization for STAP post-processing," in *Proc. IEEE Radar Conference*, May 2010, pp. 1471–1475.

[38] D. Angelosante, J. Bazerque, and G. Giannakis, "Online adaptive estimation of sparse signals: where RLS meets the l_1-norm," *IEEE Transactions on Signal Processing*, vol. 58, no. 7, pp. 3436–3447, Jul. 2010.

[39] M. Zibulevsky and M. Elad, "L1-L2 optimization in signal and image processing," *IEEE Signal Processing Magazine*, vol. 27, no. 3, pp. 76–88, May 2010.

[40] J. Ward, "Space-time adaptive processing for airborne radar," MIT Lincoln Laboratory, Lexington, MA, Tech. Rep., Dec. 1994.

[41] Z. Yang, R. de Lamare, and X. Li, "l_1-regularized STAP algorithms with a generalized sidelobe canceler architecture for airborne radar," *IEEE Transactions on Signal Processing*, vol. 60, no. 2, pp. 674–686, Feb. 2012.

[42] M. Wicks, M. Rangaswamy, R. Adve, and T. Hale, "Space-time adaptive processing: a knowledge-based perspective for airborne radar," *IEEE Signal Processing Magazine*, vol. 23, no. 1, pp. 51–65, Jan. 2006.

[43] R. Fa, R. de Lamare, and P. Clarke, "Reduced-rank STAP for MIMO radar based on joint iterative optimization of knowledge-aided adaptive filters," in *Asilomar Conference on Signals, Systems and Computers*, Nov. 2009, pp. 496–500.

[44] W. Melvin and J. Guerci, "Knowledge-aided signal processing: a new paradigm for radar and other advanced sensors," *IEEE Transactions on Aerospace and Electronic Systems*, vol. 42, no. 3, pp. 983–996, Jul. 2006.

[45] W. Melvin and G. Showman, "An approach to knowledge-aided covariance estimation," *IEEE Transactions on Aerospace and Electronic Systems*, vol. 42, no. 3, pp. 1021–1042, Jul. 2006.

[46] C. Capraro, G. Capraro, I. Bradaric, D. Weiner, M. Wicks, and W. Baldygo, "Implementing digital terrain data in knowledge-aided space-time adaptive processing," *IEEE Transactions on Aerospace and Electronic Systems*, vol. 42, no. 3, pp. 1080–1099, Jul. 2006.

[47] S. Blunt, K. Gerlach, and M. Rangaswamy, "STAP using knowledge-aided covariance estimation and the fracta algorithm," *IEEE Transactions on Aerospace and Electronic Systems*, vol. 42, no. 3, pp. 1043–1057, Jul. 2006.

[48] P. Stoica, J. Li, X. Zhu, and J. Guerci, "On using a priori knowledge in space-time adaptive processing," *IEEE Transactions on Signal Processing*, vol. 56, no. 6, pp. 2598–2602, Jun. 2008.

Chapter 2

Transmit Beamforming for Forward-Looking Space-Time Radars

Mathini Sellathurai and David Wilcox

Heriot-Watt University

Look-down airborne radars illuminate significant amounts of ground clutter, which due to the radar platform motion has a geometrically defined angle-Doppler characteristic. Space-time adaptive processing (STAP) is often employed at the receiver array to capitalise on the clutter angle-Doppler relationship and to detect targets, but there are significant problems with this approach in real-world scenarios, such as clutter non-homogeneity, which severely degrade STAP performance. In this chapter, the use of transmit beamforming is explored to alleviate some of these issues. The focus of the chapter will be the non-homogeneity of STAP training data caused by the forward-looking radar platform. In particular, the signal model for this scenario is developed and it is shown that there exist circumstances where transmit beamforming can significantly increase performance, particularly in range-ambiguous scenarios. Robust beamforming under uncertainties about the geometry will also be explored.

2.1 Introduction

Radar platforms moving at high velocities need to distinguish between reflections from stationary ground clutter and those from slow moving targets. In order to make this distinction the difference in the Doppler frequency of the reflected signal is used. The small difference in the clutter and target velocities relative to the platform makes the differentiation of

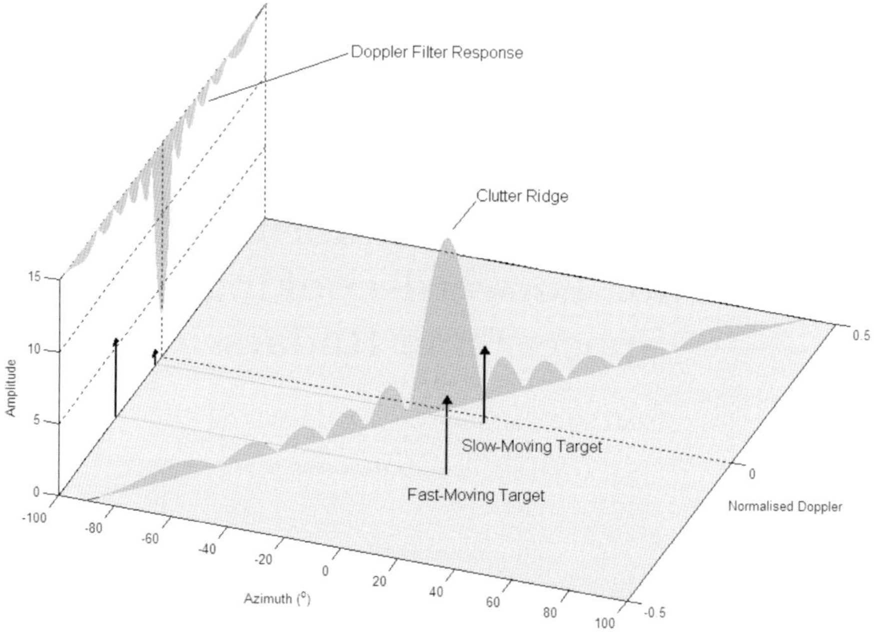

Fig. 2.1   Angle-Doppler map showing stationary clutter ridge and two targets, one fast-moving and one slow-moving.

these Doppler signals a challenging signal processing problem. Furthermore, clutter illuminated in different directions will have different Doppler shifts, some of which will coincide with the target Doppler. Therefore it is necessary to make distinctions in the angular domain as well as in the Doppler domain.

Figure 2.1 shows a representative angle-Doppler map of a side-looking airborne radar scenario, with a clutter ridge along the diagonal whose power is modulated by the transmit beampattern, two targets moving with different velocities and a Doppler filter designed to null the interference spectrum produced by the clutter. The 2D form of the signal components can be seen plainly. A Doppler clutter filter designed to null the clutter ridge may be able to detect the fast-moving target, but will null the slow moving target and therefore this will not be detected. However, the target is clearly distinct from the clutter ridge and this is the basis for using joint spatial and temporal processing to discern the presence of targets.

Space-time adaptive processing (STAP) is a moving target indication (MTI) processing technique which combines temporal and spatial (Doppler and angular) processing in a single 2D filter to capitalise on the angle-Doppler characteristics of the clutter reflections. This improves the detection capability of the pulse-Doppler radar far beyond that of sequential spatial and Doppler processing techniques. STAP has been described extensively in the two excellent sources by Klemm [1] and Ward [2].

Full dimensional space-time adaptive processing, i.e. that which operates across all antennas and radar pulses, can be computationally intensive, so suboptimum techniques with lower complexity are of great importance to practical applications. Fortuitously, in some idealised scenarios, the clutter takes on a low-dimensional form, and therefore significant reductions in computation can be made with little reduction in performance. However, in practice several factors can degrade the performance of STAP. One of the major assumptions that are often violated is that of homogeneous clutter statistics. This is often because the physical characteristics of the terrain that the radar illuminates change over the illuminated area. In Section 2.2, where an introduction to STAP concepts and the signal modelling is given, it will be shown that the radar itself has a large effect on the clutter statistics via the array orientation. In particular, under conditions other than uniform linear arrays which are side-looking with respect to the platform's motion vector, the illuminated clutter statistics become heterogeneous with range. This chapter focuses on the variation in clutter angle-Doppler statistics caused by the radar's orientation and the application of transmit diversity to help overcome these effects.

Much of the STAP processing literature has concentrated on the receiver signal processing, which is generally sufficient for side-looking radar scenarios. However, this ignores half of the degrees of freedom (DOFs) available to the radar which are present in the transmitter. These additional DOFs could be particularly useful in the case of forward-looking radar scenarios, where clutter rejection is more challenging. Utilising transmitter DOFs offers a number of opportunities, which have as yet remained largely unexplored. There have been some preliminary investigations into the use of transmit diversity for STAP radar in [3–5] for example. In Section 2.3 some of the possibilities available to the STAP radar are investigated by exploiting these transmit DOFs.

Finally, in Section 2.4, a technique to significantly improve STAP performance in a range-ambiguous forward-looking scenario is presented. This technique capitalises on the known relationship between range and elevation angle in the platform geometry such that elevation angles which are range-ambiguous are nulled in the transmit beampattern [6].

2.2   Principles of STAP

The STAP radar transmits N_P consecutive pulses of a signal waveform, which radiate into the environment, reflecting off ground clutter and possibly off targets of interest, and the returned radiation is measured through the N elements of a phased array antenna. The received signal also contains thermal noise and may also include jamming signals; however in this work the effects of jamming will not be studied.

2.2.1 *Array response vectors*

Without loss of generality, let the motion vector of the array be aligned with the x-axis of a 3D Cartesian coordinate system. A direction from the array is defined by an azimuth angle θ and elevation angle ϕ as shown in Fig. 2.2, while the angle θ_{CRAB} is the crab angle due to misalignment between the array and the platform motion vector.

The additional distance travelled by a photon from direction $(\theta,\ \phi)$ to the i-th element of the array with position vector $(x_i,\ y_i,\ z_i)$ relative to the

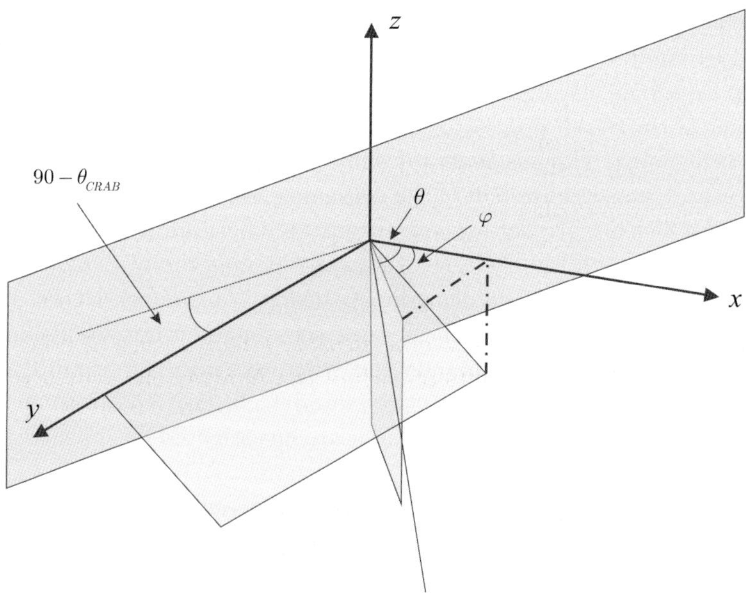

Fig. 2.2   Airborne radar array geometry.

origin of the Cartesian coordinate system is

$$d_i(\theta, \phi) = (x_i \cos(\theta) + y_i \sin(\theta)) \cos(\varphi) + z_i \sin(\varphi). \tag{2.1}$$

Thus the spatial steering vector of the array may be written

$$\underline{S}(\theta, \phi) = \left[\exp\left(j\frac{2\pi}{\lambda_0} d_1(\theta, \phi) \right), \ldots, \exp\left(j\frac{2\pi}{\lambda_0} d_N(\theta, \phi) \right) \right]^T \tag{2.2}$$

where λ_0 denotes the wavelength of the signal carrier.

Assume the radar and an illuminated scatterer move towards each other with a velocity v. Between pulses, the round-trip distance between the two changes by

$$\Delta R = 2vT_r = \frac{2v}{f_r} \tag{2.3}$$

where T_r is the pulse repetition time and f_r is the pulse repetition frequency. Therefore between successive pulses the scatterer changes phase by $2\pi\frac{\Delta R}{\lambda_0}$, and the baseband representation of the Doppler is

$$\exp\left(j\frac{2\pi}{\lambda_0} \Delta R \right) = \exp\left(j\frac{2\pi 2v}{\lambda_0} T_r \right) = \exp\left(j2\pi f_d T_r \right) \tag{2.4}$$

where $f_d = \frac{2v}{\lambda_0}$ is the Doppler frequency. We define the normalised Doppler frequency as

$$\mathcal{F} = \frac{f_d}{f_r} \tag{2.5}$$

where $f_r = \frac{1}{T_r}$ is the pulse repetition frequency and $\mathcal{F}$ gives the number of full rotations of the complex phasor associated with the Doppler between consecutive pulses of the radar. Therefore we can define a temporal steering vector of phase changes of the scatterer through the N_P pulses as

$$\underline{T}(\mathcal{F}) = [1, \; e^{j2\pi\mathcal{F}}, \; \cdots, \; e^{j2\pi(N_P-1)\mathcal{F}}]^T. \tag{2.6}$$

We may then define a *space-time steering vector*, $\underline{h}(\theta, \phi, \mathcal{F}) \in \mathcal{C}^{NN_P \times 1}$ which contains the phase shifts of each of the N array elements across the N_P pulses, as follows

$$\underline{h}(\theta, \phi, \mathcal{F}) = \underline{S}(\theta, \phi) \otimes \underline{T}(\mathcal{F}) \tag{2.7}$$

where use of the Kronecker product $\otimes$ has been made. For this chapter, we will be mainly concerned with a planar rectangular array with N_H horizontal elements and N_V vertical elements, i.e. $N = N_V N_H$, arranged in a grid aligned with the z and y co-ordinate axes, as this affords some significant simplifications due to the structure of the planar array steering vector.

However, many of the results are general and can be applied to any array geometry. The spatial steering vector for planar array can be decomposed into vertical and horizontal parts, $\underline{S}_V(\theta)$ and $\underline{S}_H(\theta, \phi)$ respectively, as

$$\underline{S}(\theta, \phi) = \underline{S}_H(\theta) \otimes \underline{S}_V(\theta, \phi). \tag{2.8}$$

2.2.2   *Scatterer response*

The per element signal power received from a scatterer illuminated by the radar is given by the *radar equation*:

$$\zeta = \frac{P_T \mathrm{AF} |g_T(\theta, \phi)|^2 |g_R(\theta, \phi)|^2 \lambda_0^2 \sigma}{(4\pi)^3 L_S R^4} \tag{2.9}$$

where P_T is the element transmit power in Watts, AF is the array factor, $g_T(\theta, \phi)$ and $g_R(\theta, \phi)$ are respectively the transmit and receive element radiation patterns in the direction (θ, ϕ), σ is the radar cross section (RCS) of the scatterer, L_S denotes system losses and R is the range of the scatterer.

The velocity of the general stationary point C of direction (θ_c, ϕ_c) illuminated by the radar due to the platform motion is

$$v_c = v_r \cos(\theta_c) \cos(\phi_c) \tag{2.10}$$

where v_r is the velocity of the radar platform. If the scatterer is moving, the radial velocity in the direction of the radar can be subtracted from the above value. The Doppler frequency of the scatterer at a point C is obtained by substituting (2.10) into (2.4).

2.2.3   *Clutter*

In the airborne radar environment, to model the clutter response, the ground may be divided into clutter patches which can each be considered as an individual scatterer with a complex reflectivity α_i for the i-th clutter patch. The properties of the clutter patches can also be modeled statistically. Common assumptions on the scatterer complex amplitudes for ground clutter, which are used in this work, are that they are zero mean

$$\mathcal{E}\{\alpha_i\} = 0 \tag{2.11}$$

and clutter patches are statistically independent

$$\mathcal{E}\{\alpha_i \alpha_j^*\} = 0 \quad i \neq j. \tag{2.12}$$

The signal power received from each patch is

$$\mathcal{E}\{\alpha_i \alpha_i^*\} = \zeta_i \tag{2.13}$$

where ζ_i is calculated via the *radar equation* (2.9) with the appropriate parameters (θ_i, ϕ_i), σ_i and R_i relating to the i-th clutter patch being substituted.

Assume the ground plane is divided into N_R range rings, and each range ring is divided into N_{AB} azimuthal patches such that each patch has a unique index (l, i), where $l \in [1, \cdots, N_R]$ and $i \in [1, \cdots, N_{AB}]$. The signal from the (l, i)-th clutter patch (i.e. from the i-th patch of the l-th range ring) can then be written down as

$$\underline{x}_c(l, i) = \alpha_{l,i}\underline{h}(\theta_{l,i}, \phi_{l,i}, \mathcal{F}_{l,i}). \tag{2.14}$$

2.2.4 *Optimum STAP receiver processing*

After demodulation, pulse compression filtering,[1] amplification and analogue to digital conversion, the received signal space-time l-th snapshot is written as an $NN_P \times 1$ vector $\underline{x}[l]$ which contains the contributions from the l-th range cell under test (RCUT), namely the clutter $\underline{x}_c[l]$, jamming $\underline{x}_j[l]$ and noise $\underline{x}_n[l]$ components mentioned above, along with the possible target signal $\underline{x}_T[l]$. Thus, when no target is present (hypothesis H_0),

$$\underline{x}[l] = \underline{x}_{H_0}[l] = \underline{x}_c[l] + \underline{x}_j[l] + \underline{x}_n[l] \tag{2.15}$$

and when there is a target signal (hypothesis H_1),

$$\underline{x}[l] = \underline{x}_{H_1}[l] = \underline{x}_T[l] + \underline{x}_c[l] + \underline{x}_j[l] + \underline{x}_n[l]. \tag{2.16}$$

The contributions from the range bins of the radar are concatenated in the so-called STAP datacube, depicted in Fig. 2.3. The range bins are sequentially tested for target contributions by removing the interference contributions, and performing a binary hypothesis test on the resulting detection statistic.

The optimal weight vector for rejecting the interference is given by the minimum variance solution

$$\underline{w}_{opt} = \mathbb{R}_u^{-1}\underline{h} \tag{2.17}$$

---

[1]Care should be taken when referring to a pulse compression filter as a matched filter, as while the terms can be used interchangeably in general radar literature, in STAP literature the term "matched filter" has a different connotation. The same term is often used for both with no distinction which confuses confusion. As such we make the distinction here of using "pulse compression" filter to refer to the conventional filter, and "matched filter" to refer to the STAP matched filter to remain in line with the STAP literature.

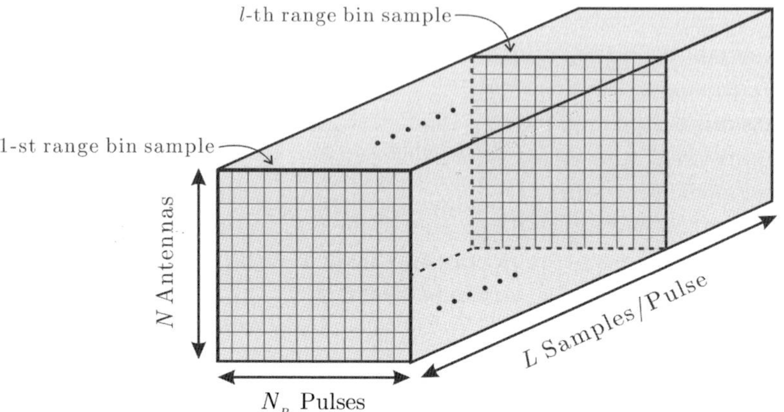

Fig. 2.3   The STAP datacube.

where $\underline{h}$ is a space-time steering vector in the direction and Doppler of interest and $\mathbb{R}_u$ is the "unwanted" covariance matrix, that is

$$\mathbb{R}_u = \mathcal{E}\{\underline{x}_{H_0}[l]\underline{x}_{H_0}^H[l]\}. \tag{2.18}$$

Equation (2.17) is also termed as the matched filter processor. This space-time filter places a clutter null (or "notch") in the angle-Doppler plane at positions containing clutter energy to remove this energy from the signal. The performance of STAP is largely dependent on how well this clutter notch matches the true clutter characteristic. A significant portion of the STAP literature addresses techniques for reducing the computation involved in Eq. (2.17) and for dealing with real-world non-ideal data, without degrading performance significantly, that is, making the clutter notch of the filter match the true clutter energy characteristic. In this chapter a brief overview of STAP concepts is given for our purposes. An interested reader may refer to [1] for more details. Assuming independence between clutter, jamming and noise signals, which is a commonly used and not unreasonable assumption, the following holds

$$\mathbb{R}_u = \underbrace{\mathcal{E}\{\underline{x}_c[l]\underline{x}_c^H[l]\}}_{\mathbb{R}_c} + \underbrace{\mathcal{E}\{\underline{x}_j[l]\underline{x}_j^H[l]\}}_{\mathbb{R}_j} + \underbrace{\mathcal{E}\{\underline{n}[l]\underline{n}^H[l]\}}_{\mathbb{R}_n} \tag{2.19}$$

$$= \mathbb{R}_c + \mathbb{R}_j + \underbrace{\sigma_n^2 \mathbb{I}_{NN_p}}_{\triangleq \mathbb{R}_n} \tag{2.20}$$

where σ_n^2 is the noise power and $\mathbb{I}_{NN_p}$ is the $NN_p \times NN_p$ identity matrix. The last equation introduces the assumption that noise samples over the

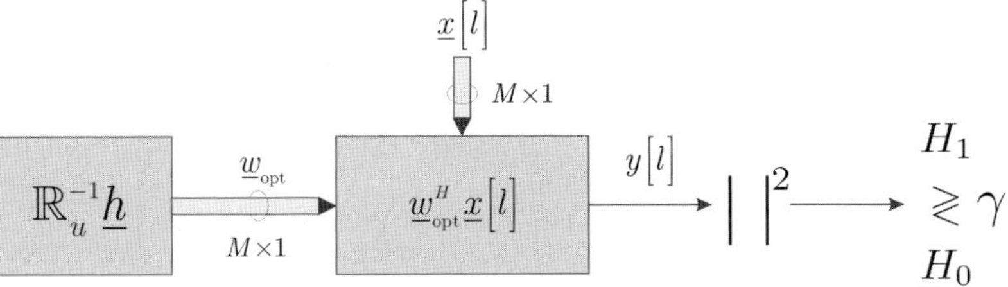

Fig. 2.4 Optimal target detection processing with $M = NN_p$.

elements of the array and across pulses are independent and identically distributed (i.i.d), and it is further stated here that they are assumed to be complex Gaussian distributed.

The detection process for the RCUT is based on the inner product of $\underline{w}_{\mathrm{opt}}$ and the received data vector for the RCUT is

$$y[l] = \underline{w}_{\mathrm{opt}}^{H}\underline{x}[l]. \tag{2.21}$$

The squared magnitude of the test decision variable $y[l]$ is compared to a threshold γ to make the decision between "target present" and "target absent", i.e.

$$|y[l]|^2 < \gamma \quad H_0 : \text{ Target Absent} \tag{2.22}$$

$$|y[l]|^2 > \gamma \quad H_1 : \text{ Target Present.} \tag{2.23}$$

This process is depicted in Fig. 2.4.

The interference covariance matrix is not known *a priori* and must therefore be estimated in some manner. A common method is by assuming that range gates adjacent to the RCUT have similar interference characteristics to the RCUT, but do not contain target reflections. A seminal paper by Reed, Mallet and Brennan [7] presents the now well-known sample matrix inversion (SMI) technique for the weight vector calculation based on the estimated covariance matrix $\hat{\mathbb{R}}_u$. The maximum likelihood estimate of $\hat{\mathbb{R}}_u$ given L range samples with identical statistics to the RCUT is:

$$\hat{\mathbb{R}}_u = \frac{1}{L}\sum_{l=1}^{L} \underline{x}_{H_0}[l]\underline{x}_{H_0}^{H}[l]. \tag{2.24}$$

Using this estimate, an approximation to $\underline{w}_{opt}$ may be obtained

$$\underline{w}_{\mathrm{SMI}} = \hat{\mathbb{R}}_u^{-1}\underline{h}. \tag{2.25}$$

The SMI receiver processing is depicted in Fig. 2.5.

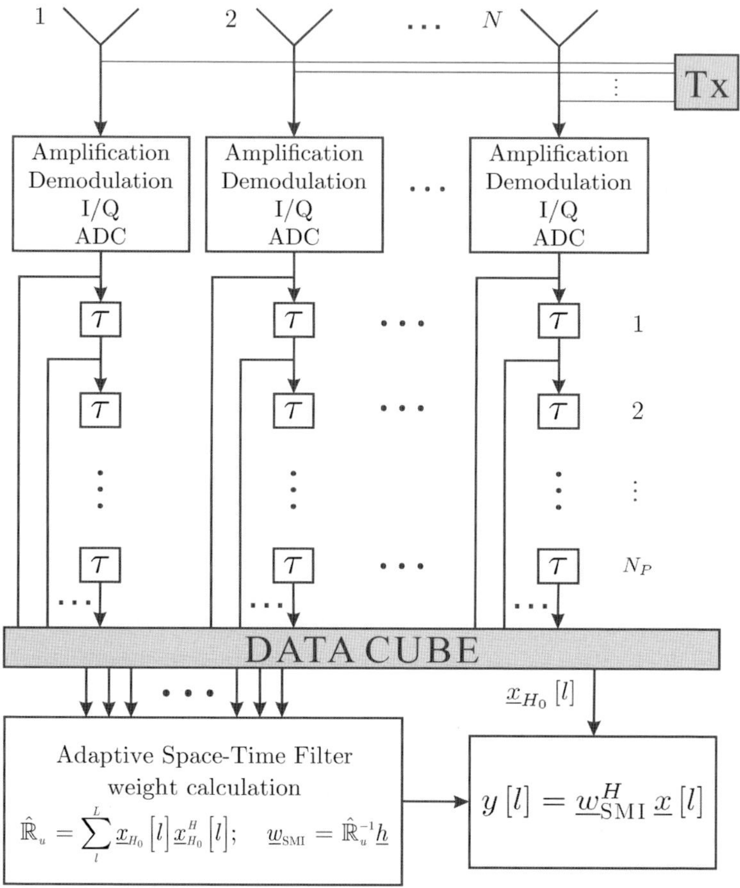

Fig. 2.5   Sample matrix inversion receiver processing.

The number of i.i.d. range samples required to estimate $\hat{\mathbb{R}}_u$ so that the performance of the STAP using SMI comes within -3 dB of the optimum processor is $2NN_P$, a fact termed as the RMB rule using the initials of the authors in [7]. The fact that these range samples come from different regions can degrade STAP performance. As an example, suppose a radar system has:

- a bandwidth B of 1 MHz,
- 20 receiving antenna elements and
- five Doppler taps.

The range resolution of the radar is given by

$$\Delta R = \frac{c}{2B} \tag{2.26}$$

where c is the speed of light, so the range resolution ΔR is 150 m. Therefore the range extent of the ground from which reflections are collected to estimate the clutter statistics of the RCUT is $2 \times 20 \times 5 \times 150 = 30\,\text{km}$. The assumption that the clutter statistics will remain the same over such a range is an unlikely one since over large ranges the terrain is unlikely to be homogeneous, though this can be partly mitigated by using non-homogeneity detection techniques. In Section 2.2.6 a second reason for performance degradation, namely the variation in clutter angle-Doppler characteristics, is discussed.

2.2.5 *Side-looking radar*

When the radar phased array normal is perpendicular to the direction of platform motion, the Doppler frequency of the reflection from a clutter patch has a simple relationship with the direction of arrival of that clutter signal. The clutter occupies a diagonal ridge in the normalised angle-Doppler plane. Figure 2.6 shows the minimum variance (MV) spectrum of simulated clutter illuminated by a side-looking moving platform from a single range gate. With knowledge of the clutter statistics, STAP processing is able to effectively null the diagonal clutter ridge and to detect moving targets that are located away from the clutter ridge. The major advantage of this configuration, however, is that the clutter is range independent. Clutter at long ranges (or alternatively, higher elevation angles) and clutter at shorter ranges (or, more negative elevation angles), both have angle-Doppler characteristics that occupy the diagonal ridge shown in Fig. 2.6. The benefit is that range samples either side of a RCUT, which are used to estimate the interference covariance of the RCUT, will have the same spatial-temporal characteristics as that RCUT, which means the same filter can be used for multiple ranges and energy from range-ambiguous clutter returns does not significantly affect performance.

2.2.6 *Forward-looking radar*

When the array is at an angle to the platform's motion vector (termed as the crab angle, θ_{CRAB}) the clutter no longer occupies the diagonal ridge of the side-looking case, but forms an ellipse in the angle-Doppler

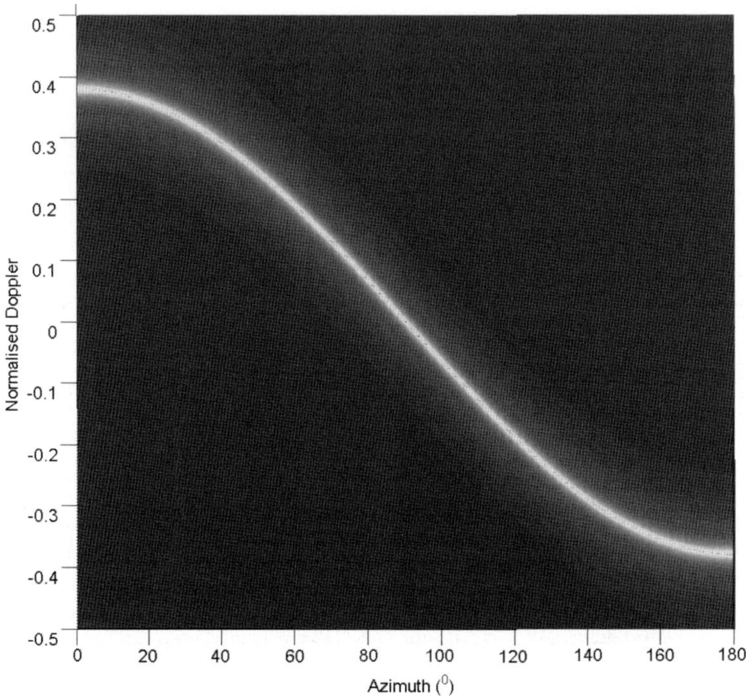

Fig. 2.6   Side-looking radar: spectrum showing angle-Doppler characteristics of station-ary clutter.

plane. The eccentricity of the ellipse is inversely related to the crab angle, such that when the array is perpendicular to the platform's motion vector (the forward-looking case) the clutter ridge forms a circle. The resulting shape of the clutter in the angle-Doppler domain is shown in Fig. 2.7. The angle-Doppler characteristic of the clutter is now *range dependent*, so that clutter from different range cells forms concentric ellipses. Therefore, as the spatio-temporal statistics change between ranges, each RCUT must have an individual matched filter designed to greatly increase the processing burden.

To make matters worse, the clutter range dependency destroys the i.i.d. sample requirement on the auxiliary range cells that are used to estimate the sample covariance matrix. The effect of using heterogeneous data is that the received clutter signal vectors span diverse subspaces, which increases the rank of the sample clutter covariance matrix in Eq. (2.24). This can

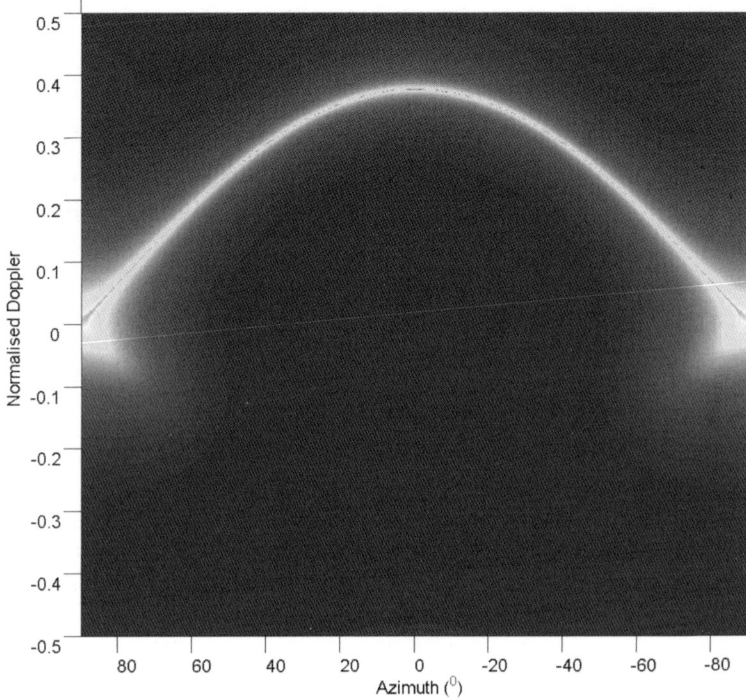

Fig. 2.7 Forward-looking radar: spectrum showing angle-Doppler characteristics of stationary clutter.

significantly degrade target detection performance by widening the clutter filter notch and thereby also nulling target energy. In forward-looking radars, radar returns from range-ambiguous clutter compounds the heterogeneity of the range samples further, thus reducing STAP performance, which will be addressed in Section 2.4. The following sections investigate the application of transmit diversity, with a view to simplifying the range dependency problem in forward-looking radars.

2.3 Adaptive transmit diversity STAP

In this section the concept of adaptive transmit diversity for the STAP radar is introduced. It has so far been assumed that the steering vector applied to the transmitter is a conventional beamformer (i.e. space-only beamformer). The beamformer can be designed to direct energy from the transmitter to a particular direction, and also by pulse coding, to normalise a particular Doppler frequency. However, these techniques use very few of the DOFs available at the transmitter.

The paper by De Maio *et al.* [3] also uses transmit pulse coding and does not vary the transmit phase in a linear manner as does Doppler normalisation. Instead, it constrains the pulse coding to be similar to a desired phase coded sequence, while maximising signal-to-interference-plus-noise ratio (SINR). This utilises up to N_P DOFs of the transmitter. There are, however, NN_P DOFs available. These have been used in the work by Corbell *et al.* [4, 5, 8]. In this work, STAP clutter nulling is performed by applying a 'filter' to the transmitters (i.e. a space-time beamforming vector) which performs nulling of clutter on transmit. The space-time beamformer is referred to as a space-time illumination pattern (STIP). Next, the signal model which forms the basis for applying transmit diversity is introduced.

2.3.1 *Signal model*

In the adaptive transmit STAP model, a different transmit steering vector is applied to the transmit array at each pulse, meaning that the phase and magnitude of the beampattern towards the location (θ, ϕ) varies with each pulse. Let the transmit beamformer applied to the n-th pulse be $\underline{\bar{w}}_n \in \mathcal{C}^{N \times 1}$, then let the transmit beamforming matrix $\bar{\mathbb{W}} \in \mathcal{C}^{N \times N_P}$ be defined as

$$\bar{\mathbb{W}} = [\underline{\bar{w}}_1, \underline{\bar{w}}_2 \cdots, \ \underline{\bar{w}}_{N_P}]. \tag{2.27}$$

Then the complex beampattern response toward the direction (θ, ϕ) at each pulse may be written down as:

$$\mathrm{AP}_i(\theta, \ \phi) = \underline{\bar{w}}_i^H \underline{S}(\theta, \ \phi), \forall i \tag{2.28}$$

$$\Rightarrow \underline{\mathrm{AP}}(\theta, \phi) = \bar{\mathbb{W}}^H \underline{S}(\theta, \ \phi). \tag{2.29}$$

It is convenient to incorporate these phase and magnitude variations into the temporal steering vector used to determine the response of a scatterer to illumination by the radar[2] (see for example Eq. (2.14)). Let

$$\underline{\tilde{\mathcal{I}}}(\theta, \ \phi, \ \mathcal{F}) \triangleq \underline{\mathrm{AP}}(\theta, \ \phi) \odot \underline{\mathcal{I}}(\mathcal{F}) \tag{2.30}$$

then each term (pulse) of this augmented temporal steering vector now also varies with the transmitter's beampattern. The augmented steering vector

---

[2] Here we assume simple line of sight propagation on transmit and receive. This signal model may not be applicable to more complex multi-path reflection scenarios.

in the direction (θ, ϕ) is now:

$$\underline{\tilde{h}}(\theta,\,\phi,\,\mathcal{F}) = \underline{S}(\theta,\,\phi) \otimes \underline{\tilde{\mathcal{I}}}(\theta,\,\phi,\,\mathcal{F}). \tag{2.31}$$

Note that for the simplified case of precisely known geometry, the dependence on $\mathcal{F}$ may be dropped as this can be calculated from the angular location of the scatterer through Eqs. (2.10) and (2.4).

2.3.2 *Space-time illumination patterns*

Space-time illumination patterns have the ability to null clutter through space-time transmit beamforming. In order to achieve this, we take the optimum STAP filter vector given by Eq. (2.17) and divide it into N_P equal parts relating to the pulses of the radar. These vectors are time-reversed and applied as transmit beamforming vectors in the matrix $\mathbb{W}$ (2.27).

Figure 2.8 shows the azimuth-Doppler beampattern for the transmit and receive filters for the conventional STAP with forward-looking geometry. The transmit pattern is a conventional beampattern centred at $0°$ azimuth and 0 Hz Doppler, whereas the receive STAP filter forms a null along the curved clutter ridge, with a mainbeam directed at the target Doppler and azimuth.

Figure 2.9 shows the same graphs for the STAP radar employing the STIP instead. This clearly shows how the clutter is nulled by the transmitter, as indicated by the curved null in the transmit beampattern (the inversion of the Doppler axis is due to the time-reversal) and the proof that this is successful is that the optimum receive filter has a conventional angle-Doppler response. This implies that there is no clutter energy left to null, so a conventional beamformer produces the optimum performance. That is, the receive filter only needs to steer a conventional beam in the target locations and Doppler to obtain the optimum response.

While this is promising in terms of the ability to null clutter on transmit, there are a number of practical issues with this methodology. Firstly, the processing burden is simply transferred to the transmitter from the receiver and therefore does not become any simpler computationally nor is any performance improvement offered. In fact, because the optimal receiver filter can have weights greater than unity and the transmitter weights are limited by the elemental power constraints, there would actually be a reduction in performance. The interference covariance matrix used to design the transmit filter must be known prior to transmission and is obviously not available. However, this is overcome by first transmitting conventionally to

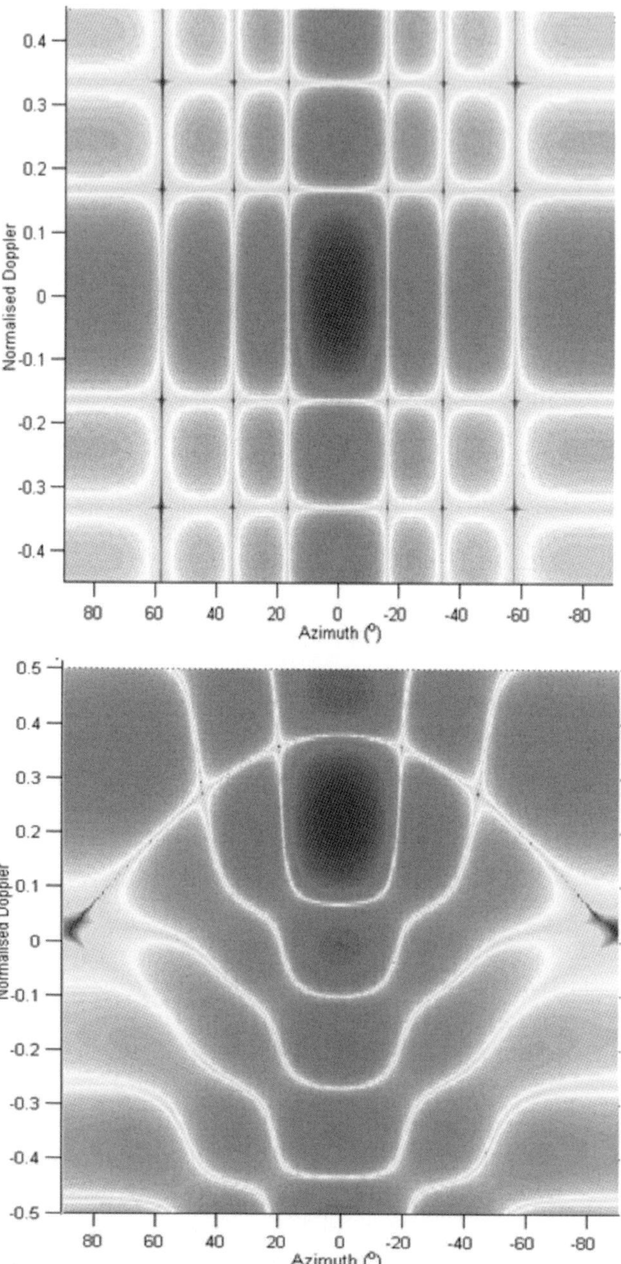

Fig. 2.8   Conventional STAP: angle-Doppler response of filter applied to the transmitter (top) and receiver (bottom).

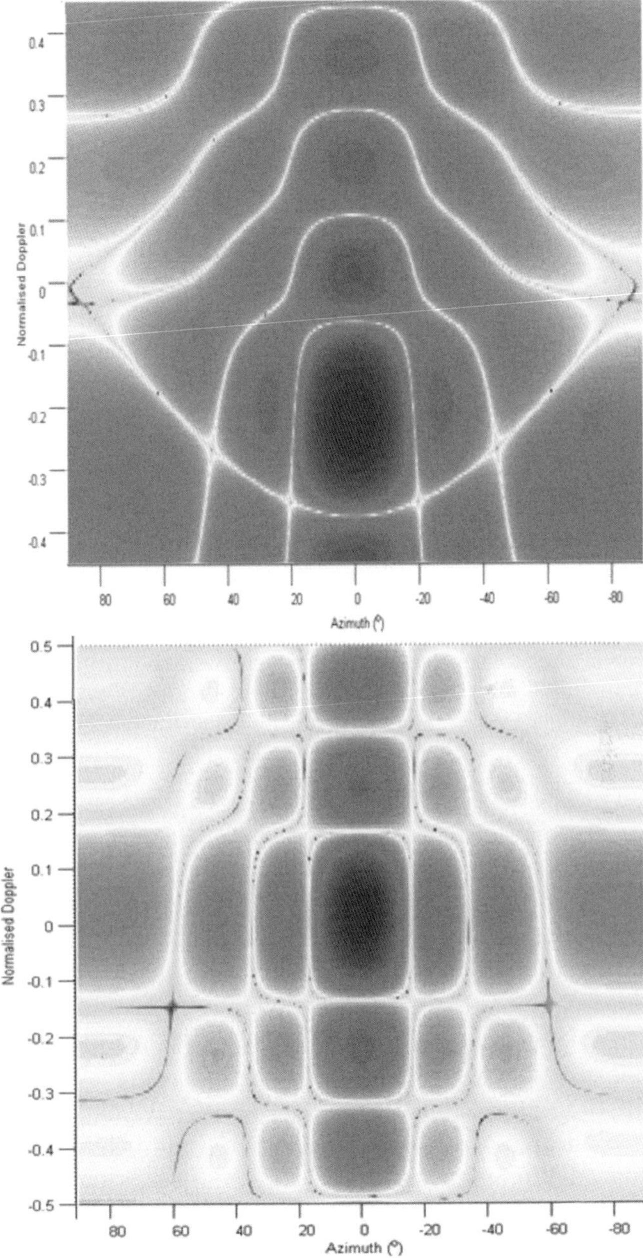

Fig. 2.9   Conventional STIP: angle-Doppler response of filter applied to the transmitter (top) and receiver (bottom).

obtain an estimate to be used for the next CPI. Furthermore, the clutter nulling is only achieved for a single range bin when done at transmit, whereas when in the receiver, multiple filters can be designed and applied to the data independently. This means that each CPI can only concentrate on a single range bin, which greatly limits the capability of a STAP radar.

The goal of the STAP radar is to null the clutter contribution to the signal such that targets can be detected. As we have seen in Section 2.2, the clutter is in fact no more hamful to the forward-looking radar than to the sideways-looking case, provided there are no range ambiguities. The main facilitator to successfully nulling clutter is knowledge of the clutter statistics at the range bin of interest. It is this knowledge which is in fact missing in the forward-looking radar (FLR) scenario and is the main reason for reduced performance. The goal of this section is to investigate the application of transmit diversity so that the characteristics of the clutter are improved and the receiver improved the knowledge on which it bases its processing.

As discussed in Section 2.2, the angle-angle-Doppler clutter characteristics vary with the elevation angle for a FLR. Considering a FLR with a small negative elevation (depression) angle such that the transmit beam points towards the horizon range, the range bins around the centre of the mainlobe will be grouped very closely in elevation angles due to the tangential nature of the line of sight to the earth's surface. When the radar depression angle is large so that the radar looks down, the range bins are more spread out in the elevation domain. The relationship between depression angle and range is demonstrated in Fig. 2.10. Therefore we see that at long ranges where the radar looks toward the horizon, there will be little benefit from elevation diversity because the clutter statistics will not change significantly across the range of the training data. However, in lookdown scenarios, the clutter statistics will change greatly across the range of the training data and therefore elevation diversity could show significant benefits.

The goal of utilising transmit diversity is the homogenisation of the clutter statistics. This manifests itself in a reduction in the rank of the sample-based interference covariance matrix estimate, as fewer eigenmodes of the clutter are illuminated. While the optimal performance from using a clairvoyant covariance matrix for STAP is degraded by using transmit diversity, the practical performance using the SMI technique would be improved by clutter homogenisation. Therefore, this is the metric that will be used to assess the efficacy of the techniques.

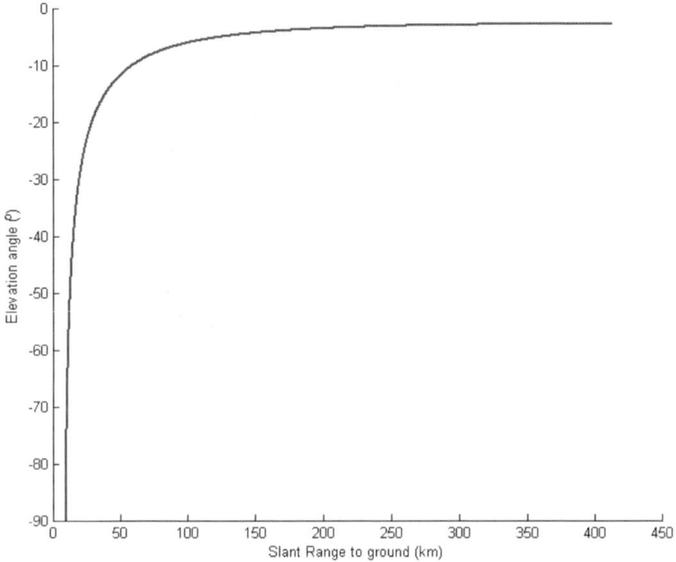

Fig. 2.10 Range–elevation relationship for radar platform at $h_r = 10000$ using $\frac{4}{3}$ earth radius.

2.3.3 3D Doppler compensation

In this section a non-data-adaptive method of homogenising clutter statistics, which amounts to performing Doppler compensation across the azimuth-elevation-time domain will be investigated.

The received signal vector from the l-th range ring is given by summing over the clutter patches making up that range ring. Therefore from Eq. (2.14) the vector signal $\underline{x}_c[l]$ can be written as follows

$$\underline{x}_c[l] = \sum_{(\theta_{l,i},\,\phi_{l,i})\in\mathcal{A}_l} \alpha_{l,i}\underline{\tilde{h}}(\theta_{l,i},\,\phi_{l,i},\,\mathcal{F}_{l,i}) \tag{2.32}$$

where $\mathcal{A}_l$ is the set of ordered angle pairs in the directions of the clutter patches on the l-th range ring.

Furthermore, the equation for $\underline{x}_c[l]$ may be rearranged as

$$\underline{x}_c[l] = \sum_{(\theta_{l,i},\,\phi_{l,i})\in\mathcal{A}_l} \alpha_{l,i}\underline{S}(\theta_{l,i},\,\phi_{l,i}) \otimes$$

$$\left[\underline{\mathcal{I}}(\mathcal{F}_{l,i}) \odot \left(\bar{\mathbb{W}}^H\underline{S}(\theta_{l,i},\,\phi_{l,i})\right)\right] \tag{2.33}$$

$$= \sum_{\mathcal{A}_l} \alpha_{l,i}\underline{S}_{l,i} \otimes \left[\underline{\mathcal{I}}_{l,i} \odot \left(\bar{\mathbb{W}}^H\underline{S}_{l,i}\right)\right] \tag{2.34}$$

where the parameter dependence in the last line has been dropped to make the relationship clearer, with the subscripts (l, i) referring to the i-th clutter patch in the l-th range ring.

If the transmit beamformer were designed such that the phase of the transmit beampattern cancels the Doppler shift perfectly at each of the clutter patches across the pulses of the CPI, such that

$$\bar{\mathbb{W}}^H \underline{S}_{l,i} = \underline{\mathcal{I}}^*_{l,i}; \quad \forall(l, i) \tag{2.35}$$

then the second factor in Eq. (2.34) becomes

$$\underline{\mathcal{I}}_{l,i} \odot \underline{\mathcal{I}}^*_{l,i} = \underline{1}_{N_P}. \tag{2.36}$$

This effectively means that all the clutter appears stationary to the radar, or alternatively, that the platform is no longer moving.

Considering the effect on the rank of the interference covariance matrix, the covariance matrix $\mathbb{R}_c$ may be written as

$$\mathbb{R}_c = \mathcal{E}\left\{\underline{x}_c[l]\underline{x}_c^H[l]\right\} \tag{2.37}$$

$$= \mathcal{E}\left\{\left(\sum_{\mathcal{A}_l} \alpha_{l,i}\underline{S}_{l,i} \otimes \underline{1}_{N_P}\right)\left(\sum_{\mathcal{A}_l} \alpha_{l,i}\underline{S}_{l,i} \otimes \underline{1}_{N_P}\right)^H\right\}$$

$$= \mathcal{E}\left\{\left(\sum_{\mathcal{A}_l} \alpha_{l,i}\underline{S}_{l,i}\right)\left(\sum_{\mathcal{A}_l} \alpha_{l,i}\underline{S}_{l,i}\right)^H \otimes \underline{1}_{N_P}\underline{1}_{N_P}^T\right\}$$

$$= \mathcal{E}\left\{\left(\sum_{\mathcal{A}_l} \alpha_{l,i}\underline{S}_{l,i}\right)\left(\sum_{\mathcal{A}_l} \alpha_{l,i}\underline{S}_{l,i}\right)^H\right\} \otimes \underline{1}_{N_P}\underline{1}_{N_P}^T$$

where in the penultimate step the following identity was used

$$(\mathbb{A} \otimes \mathbb{C})(\mathbb{B} \otimes \mathbb{D}) = \mathbb{A}\mathbb{B} \otimes \mathbb{C}\mathbb{D}. \tag{2.38}$$

Employing the inequality

$$\text{rank}\{\mathbb{A} \otimes \mathbb{B}\} \leq \text{rank}\{\mathbb{A}\}.\text{rank}\{\mathbb{B}\} \tag{2.39}$$

it can be seen that

$$\text{rank}\left\{\mathbb{R}_c\right\} \leq \text{rank}\left\{\mathcal{E}\left\{\left[\sum_{\mathcal{A}_l} \alpha_{l,i}\underline{S}_{l,i}\right]\left[\sum_{\mathcal{A}_l} \alpha_{l,i}\underline{S}_{l,i}\right]^H\right\}\right\}$$

$$\times \underbrace{\text{rank}\left\{\underline{1}_{N_P}\underline{1}_{N_P}^T\right\}}_{=1} \tag{2.40}$$

$$\Rightarrow \quad \text{rank}\left\{\mathbb{R}_c\right\} \leq \text{rank}\left\{\mathcal{E}\left\{\left[\sum_{\mathcal{A}_l} \alpha_{l,i}\underline{S}_{l,i}\right]\left[\sum_{\mathcal{A}_l} \alpha_{l,i}\underline{S}_{l,i}\right]^H\right\}\right\}. \tag{2.41}$$

The spatial steering vectors of the clutter patches are the same, irrespective of whether transmit diversity is employed or not. Therefore, perfect Doppler compensation as achieved through Eq. (2.35) is the rank minimising transmit beamformer of the sample interference covariance matrix, giving

$$\text{rank}\{\mathbb{R}_c\} \leq N. \tag{2.42}$$

Essentially, this would occur when the ground appears stationary to the radar.

The problem then is to design a transmit beamformer which produces such a beampattern, which would be challenging in itself as phase-control of the beampattern is required. However, it has been found that the synthesis of a Doppler compensation transmit beampattern also requires a significant level of diversity. That is, achieving the spatial-angular phase control necessary reduces the main beam amplitude response to almost zero. The size of the array required to achieve a significant mainbeam response would be impractically large and, furthermore, the optimisation of the beampattern phase would be too complex for real-time implementation.

However, there are several open problems, illuminated by the above result, which could have an impact on the sample support problems of forward-looking STAP radars. Firstly, while a conventional array cannot provide the necessary diversity to compensate for the angular variation in Doppler shift, recent advances in multiple-input multiple-output (MIMO) technology and also frequency diverse arrays (FDAs) may have applications to this problem, as they have significantly more beampattern diversity available to them. Secondly, Doppler compensation is easy to perform on the receiver across different range gates, i.e. in the elevation dimension, a topic studied in [9]. Therefore transmit beamforming could be applied only to the azimuthal domain, splitting the burden between transmitter and receiver. Furthermore, it is unnecessary to completely normalise the Doppler and it may be more appropriate to concentrate on only compensating for differences between range samples to reinstate the i.i.d. assumption.

2.4   Ambiguous range transmit nulling

As it was previously discussed, range-ambiguous radar returns (i.e. energy from previous transmitted pulses that arrives concurrently with the current pulse from a greater range) cause clutter energy with significantly different angle-Doppler characteristics to contribute to the signal. In this section, the performance reduction this causes is considered and, furthermore, a transmit beamforming solution to the range ambiguity problem is taken into account, where angular regions whose range is ambiguous to the range of interest are nulled in the transmit beampattern. Thus, energy from these areas will not contribute significantly to the received signal.

The ranges that are ambiguous with respect to a region of interest can be calculated from the scenario geometry and known radar parameters, such as pulse repetition frequency (PRF). However, when there are errors or uncertainties in the geometry, for example due to unknown ground elevations, calculations of the range-ambiguous regions may be inaccurate. To overcome this issue, a data-adaptive procedure to detect ambiguous ranges using data from a previous CPI is developed. The knowledge of the range-ambiguous regions is then used to efficiently design beampatterns that null contributions from the elevation angles, causing interference to the area of interest. In particular, for rectangular planar arrays the optimisation of the beampattern has a low complexity form due to the separability of the vertical phase progression from the horizontal and Doppler components. The benefits of this approach are threefold:

- there is a reduction in the clutter energy which is returned to the receiver, thereby reducing clutter-to-noise ratio;
- the training data used to design the STAP filter has a low rank, resulting in faster convergence of the filter and lower sample support requirements;
- the clutter is more homogeneous within the target range cell under consideration and, therefore, this approach represents the clutter to be rejected more accurately.

It is demonstrated that the combined effect of these improvements results in better target detection performance. Similar range-nulling approaches have been applied to the receiver elevation beampattern [10, 11], though these require vertical beamforming of the received data. In this case, elevation information is lost, so it is no longer possible to perform 3D STAP techniques, which are well known to provide superior performance to 2D processing for forward-looking radars [8]. Furthermore, if the beamforming is carried out in the digital part of the receiver, it needs to be done at each range gate, whereas a transmit beamforming approach only needs to be applied once at the transmitter.

Figure 2.11 shows the eigenvalues of the "ideal" and sample clutter covariance matrices formed for an example airborne STAP scenario using

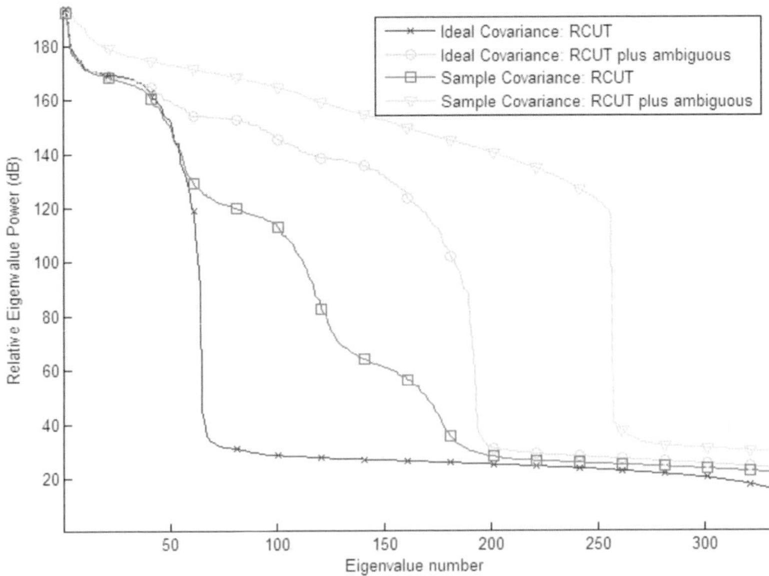

Fig. 2.11   Eigenvalue magnitudes of ideal and sample clutter covariance matrices, with and without ambiguous regions.

data with and without the N_r ambiguous ranges included. The effect of the ambiguous clutter is to significantly increase both the clutter power incident at the receiver and the rank of the covariance matrix, in both the ideal and sample versions. Higher clutter ranks show that the clutter snapshots making up the sample covariance estimate are inhomogeneous. When full 3D (angle-angle-Doppler) processing is used with the ideal covariance matrix, this does not affect STAP performance, because the ambiguous clutter regions are far from the target in the angle-angle-Doppler space and so there is little effect on the clutter notch around the target. However, in practice, the higher clutter rank will increase the sample support requirements and the processing burden for calculating the STAP filter weights. Furthermore, under some transformations of the data, the ambiguous clutter can fall close to the target in the reduced-dimension space (see Section 2.4.3 for an example of this). In addition, note that the sample covariance matrix with ambiguous ranges cuts off at 256 on the abscissa because the rank of this matrix is limited by the number of range samples L used to form the estimate. This means that the clutter in the received signal will not be fully characterised by the covariance matrix, which will reduce performance.

The motivation for transmit beamforming to reduce the effect of ambiguous ranges on STAP performance is therefore threefold:

- the rank of the clutter covariance will be reduced, which will aid practical STAP approaches by allowing lower computational burden for the STAP filter;
- this will mean that there will be lower sample support requirements to describe the clutter characteristics as the illuminated clutter will be more homogeneous;
- the clutter signal power will be reduced, thereby reducing the clutter-to-noise ratio (CNR).

All of these stem from the fact that the clutter is not illuminated by the transmitter and this both improves STAP performance and simplifies processing at the receiver.

2.4.1 *Angular location of ambiguous ranges*

In this section a non-adaptive geometry-based elevation estimation technique to determine the angular location of ambiguous ranges is presented, which can be used in the beampattern design procedure developed in the next section.

Under the conditions of known geometry, the elevation angles of ambiguous ranges can be calculated. In this section the formulation for a spherical earth model is presented, though more complicated solutions could be developed to include land altitude maps.

The unambiguous range of the radar is given by

$$R_U = \frac{c}{2f_r} \tag{2.43}$$

where c is the speed of light and f_r is the pulse repetition frequency. With reference to Fig. 2.12, therefore, ranges R_a which are ambiguous with the target range R_t satisfy $R_a = R_t + nR_U$ where n is an integer such that $h_r < R_a < R_h$, with R_h denoting the range to the horizon, which is

$$R_h = \sqrt{h_r^2 + 2h_r r_e}. \tag{2.44}$$

Here, h_r is the height of the radar and r_e is the effective earth radius, taken as four-thirds of the true earth radius to account for atmospheric refraction. The number of ambiguous range rings is denoted as N_{rings}.

The elevation angle $\varphi_a(i)$ of the i-th ambiguous range $R_a(i)$ can be found simply from the solution to the intersection of two circles of radii

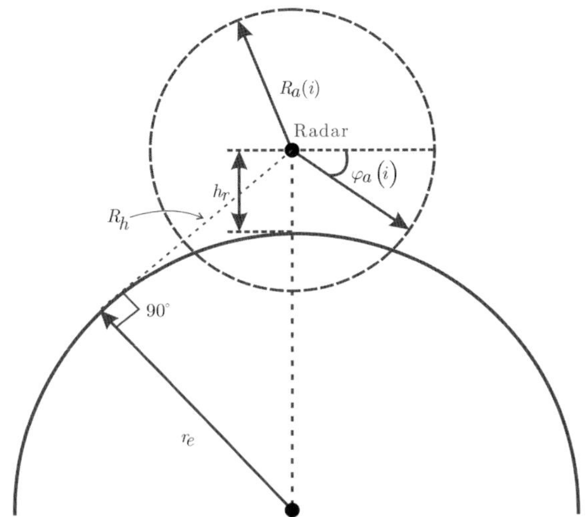

Fig. 2.12 Geometry of the range-ambiguous STAP radar.

$R_a(i)$ and r_e whose centres are displaced by a distance $r_e + h_r$. This is given by

$$\varphi_a(i) = \sin^{-1}\left(\frac{r_e^2 - R_a(i)^2 - (h_r + r_e)^2}{2(h_r + r_e)R_a(i)}\right) \tag{2.45}$$

and is shown in Fig. 2.12, together with other relevant geometrical parameters.

2.4.2 *Beampattern design*

In this section the elevation beampattern design principle for mitigation of range-ambiguous clutter is developed. Our goal is to design a space-time beamformer which places nulls in directions of ambiguous range rings on transmit. Similar to a conventional STAP receiver filter, a space-time transmit beamformer $\underline{w} \in \mathcal{C}^{N_H N_V N_P \times 1}$ can be considered [4,5], applied to the transmit array elements of an $N_H \times N_V$ grid array geometry across the N_P pulses. This elicits a transmit space-time array pattern, given by

$$\mathrm{AP}(\theta, \phi, \mathcal{F}) = \underline{w}^H \underline{h}(\theta, \phi, \mathcal{F}). \tag{2.46}$$

From Eq. (2.7) it is apparent that this transmit weight vector $\underline{w}$ can be decomposed so that the design can be done solely in the spatial domain. Moreover, if the desired pattern nulls can be localised to elevation angles only, from Eq. (2.8) the structure of the planar array manifold vector allows to concentrate only on the vertical steering vector to further reduce the dimension of the problem. This simplification does not necessarily hold for other array configurations. Then a full $N_H N_V N_P \times 1$ space-time transmit weight vector is created to illuminate the desired space as:

$$\underline{\bar{w}} = \underbrace{\underline{\bar{w}}_H \otimes \underline{\bar{w}}_V}_{=\underline{\bar{w}}_s} \otimes \underline{\bar{w}}_D \tag{2.47}$$

$$= \underline{\bar{w}}_s \otimes \underline{\bar{w}}_D, \tag{2.48}$$

where $\underline{\bar{w}}_D$ is the transmit temporal weight vector that is equal to the temporal (Doppler) steering vector given by Eq. (2.6).

Furthermore, the condition for forming a beampattern null at the location (θ_a, ϕ_a) is

$$\underline{\bar{w}}_s^H \underline{S}(\theta_a, \phi_a) = 0. \tag{2.49}$$

Let $\mathbb{S}(\underline{\theta}_a, \underline{\phi}_a)$ be a matrix containing as columns the steering vectors at which nulls are to be placed. That is,

$$\mathbb{S}_s(\underline{\theta}_a, \underline{\phi}_a) = [\underline{S}(\theta_{a1}, \phi_{a1}), \underline{S}(\theta_{a2}, \phi_{a2}), ..., \underline{S}(\theta_{aQ}, \phi_{aQ})] \tag{2.50}$$

where Q is the number of null locations, which are placed in a suitably spaced grid.

To avoid a trivial solution, the transmit power to a particular target direction of interest (θ_T, ϕ_T) must be constrained in some sense to provide a main beam gain in the desired direction.

To solve this problem, a minimum variance distortionless response (MVDR) solution to the problem could be used

$$\min_{\underline{\bar{w}}_s} \ \underline{\bar{w}}_s^H \mathbb{R}_s \underline{\bar{w}}_s \tag{2.51}$$

$$\text{s.t.} \ \underline{\bar{w}}_s^H \underline{S}(\theta_T, \phi_T) = N \tag{2.52}$$

where

$$\mathbb{R}_s = \mathbb{S}_s(\underline{\theta}_a, \underline{\phi}_a)\mathbb{S}_s^H(\underline{\theta}_a, \underline{\phi}_a) + k\mathbb{I}_N. \tag{2.53}$$

The scalar k is a diagonal loading factor to guarantee the invertability of $\mathbb{R}_s$. The well-known solution is formed as

$$\underline{\bar{w}}_s = \mathbb{R}_s^{-1} \underline{S}(\theta_T, \phi_T). \tag{2.54}$$

Although offering a closed-form solution, this approach requires normalisation of the transmit weights so that the largest magnitude is no greater than unity to avoid distortion of the beampattern due to saturation of the transmitters. In doing this normalisation, the total transmit power is reduced and, in addition, the transmitters will run below saturation (see [12] for further information of this issue).

A more desirable condition for transmit element weights is for them to be constrained to the unit circle (for every i, i.e. for every clutter patch in the range ring) to avoid signal amplification, that is

$$\|\underline{\bar{w}}_s(i)\| = 1; \forall i. \tag{2.55}$$

This quadratic equality constraint makes the problem non-convex, and therefore cannot be solved efficiently and has no known closed-form solution. However, a convex relaxation of the constraint can be introduced by restricting these variables to be within the unit circle, though this could result in lower mainlobe gain. By introducing a second constraint to maximise the transmit power in the target direction, a reduction in the transmit amplitudes will conflict with this requirement. The two constraints can be weighted so that in practice the transmit element amplitudes fall on the unit circle. These constraints and the minimisation of transmit energy toward the directions specified in $\mathbb{S}(\underline{\theta}_a, \underline{\phi}_a)$ can be cast in a second order cone program (SOCP) as

$$\min_{\underline{\bar{w}}_s, \delta_1, \delta_2} \left\{ [\mu_1, \ -\mu_2] \begin{bmatrix} \delta_1 \\ \delta_2 \end{bmatrix} \right\} \tag{2.56}$$

$$\text{s.t.} \quad \begin{cases} \|\underline{\bar{w}}_s^H \mathbb{S}(\underline{\theta}_a, \underline{\phi}_a)\|_2^2 \leq \delta_1 \\ \underline{\bar{w}}_s^H \underline{S}(\theta_T, \phi_T) = \delta_2 \\ \|\underline{\bar{w}}_s\|_2^2 \leq 1. \end{cases} \tag{2.57}$$

Here, δ_1 and δ_2 are auxiliary variables and μ_1 and μ_2 are weights describing the Pareto-optimal solution trading-off mainlobe gain against beampattern null depth as described above, and are set to 1 and 0.1 in the simulations conducted here. The main-beam constraint not only maximises power to a prescribed target direction, but also constrains the resultant beam amplitude at this point to be real. Thus the imaginary part of the beampattern crosses from negative to positive at this point, which helps to maintain the symmetry of the beam about the target direction. This is advantageous for estimation procedures using monopulse-based techniques, since matching of the channels is necessary for accurate estimation. This optimisation was found in practice to outperform the closed form solution given in Eq. (2.54) due to the difference in transmit power post-normalisation. Therefore, this method is employed for the remainder of the chapter to demonstrate the results of the transmit beampattern design.

The elevation beampattern angles which are nulled need not be angles that are ambiguous to the target location. The training data used to adapt the STAP filter will also contain energy from ambiguous range rings, thus these regions can be nulled in the same manner by adding more constraints to the optimisation problem described by Eq. (2.57). This is of particular interest, as the training data can span a large range of angles which would not be covered by a single clutter notch at the target-ambiguous range.

2.4.3   *Simulation results*

In this section, the effects of transmit elevation nulling are studied for a STAP rectangular planar array radar of 80 elements of half-wavelength interelement spacing with the physical parameters specified in Table 2.1. The example used is a low-PRF ($f_r = 2000$) long-range target scenario, where a short-range ambiguous clutter ring is illuminated by a transmit sidelobe of the array.

Table 2.1   STAP radar.

Parameter	Symbol	Value
Vertical array elements	N_V	10
Horizontal array elements	N_H	8
Pulses	N_P	8
Platform height (m)	h_r	3000
Platform velocity (ms^{-1})	v_r	140
Carrier wavelength (m)	λ	0.03
System bandwidth (MHz)	BW	10
Range resolution (m)	ΔR	15
Clutter-to-noise ratio (dB)	CNR	30

Signal-to-noise ratios (SNRs) and CNRs are defined as the ratio of powers per element at the receiver of the conventional STAP transmitter. Clutter in other scenarios is normalised such that the simulated clutter patches and target are of the same strength to make the comparisons fair, though this may result in different actual SNRs and CNRs at the receiver due to the different transmit beamforming approaches in each scenario.

In the proposed scenario the unambiguous range is $R_U = 75\,\text{km}$ and the range to horizon is $R_h = 225.9\,\text{km}$, thus there are $N_{rings} = 3$ ambiguous

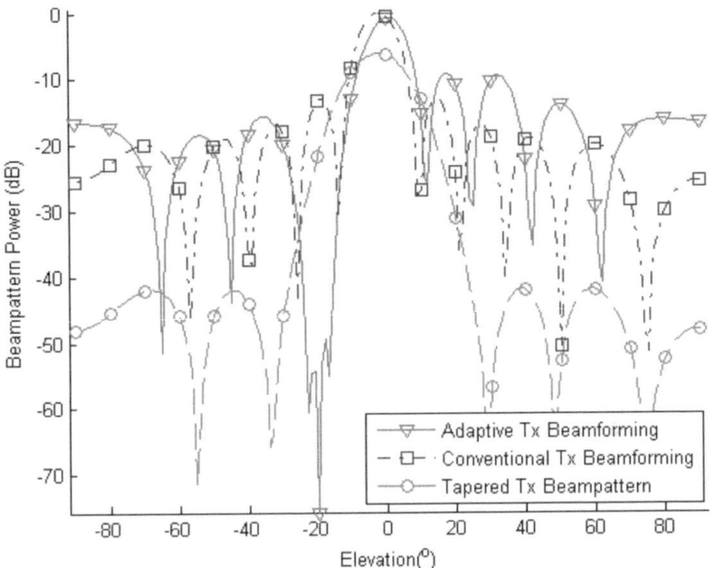

Fig. 2.13 Conventional, adaptive and Hamming weighted transmit elevation beampatterns.

range rings. It is assumed that there is a target of interest at the second range ring around $-2.1°$ elevation. The first range ring at around $-20°$ elevation is illuminated by the transmit sidelobe and therefore contributes significant clutter energy to the signal. The elevation pattern produced from solving Eq. (2.56) to null the range-ambiguous training data is shown in Fig. 2.13, with the conventional illumination pattern and a Hamming tapered transmit beampattern. The illumination of the ambiguous sidelobe clutter is greatly reduced at around 40 dB below that of the conventional transmit pattern. Note that the signal power delivered to the target direction is not significantly affected in the case of the adaptive beampattern. This is because the numerical solution allows the target power to be maintained, in contrast to the closed-form solution which would lose mainlobe signal power, as in the case of the Hamming weighted transmit pattern. In general, it is practical only to null ranges which are illuminated in the sidelobes of the beampattern, as mainlobe nulling on transmit severely degrades detection performance. The third ambiguous range is very close to the target elevation and far below the resolution of the vertical array dimension, so cannot be nulled effectively. However, due to the greater distance, it has significantly lower power than the other range rings. The 2D

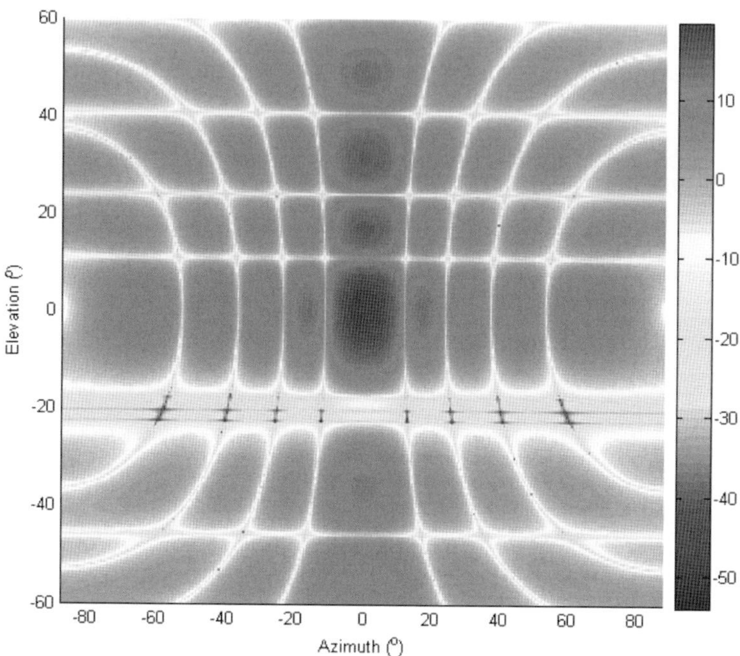

Fig. 2.14 Planar array 2D transmit beampattern with elevation nulling.

azimuth-elevation transmit beampattern is shown in Fig. 2.14. The null is shown around $-20°$ elevation and persists across all azimuth angles in the beampattern.

The eigenvalues of the clutter sample covariance matrix are shown in Fig. 2.15. From this the large reduction in clutter energy, which is contributed to the signal, can be seen. Analysis of the received target and clutter signal powers in this example shows a 7 dB reduction in clutter power and a 1.2 dB reduction in signal power caused by augmenting the beampattern, indicating an almost 6 dB gain in SINR.

Figure 2.16 compares the ideal detection performance of the adaptive nulling on transmit and on receive, i.e. the adaptive weight set can be used at the receiver. However, the receiver loses the elevation information by beamforming on this plane and this becomes detrimental to the performance of the STAP. The transmit nulling approach does not illuminate the ambiguous clutter region and therefore can perform 3D STAP at the receiver.

In this section, it has been shown that when a short-range ambiguous clutter ring is illuminated through a transmit sidelobe, a significant level of

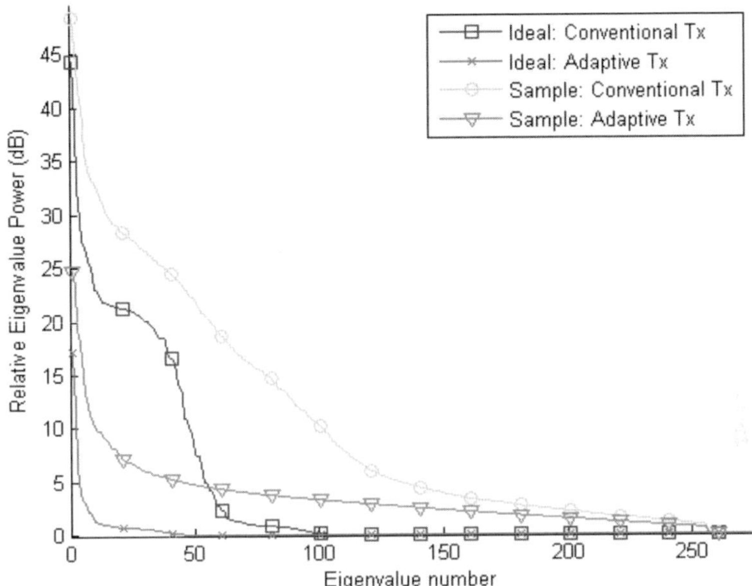

Fig. 2.15   Ordered eigenvalues of the ideal and sample clutter covariance matrices, for conventional and adaptive transmit patterns.

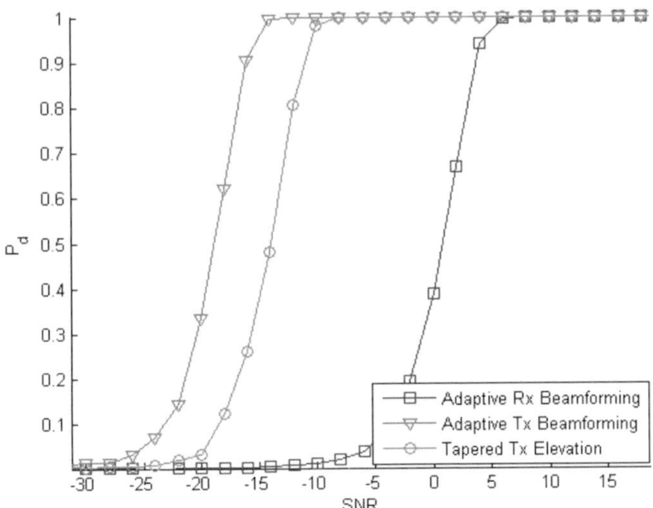

Fig. 2.16   Detection performance of STAP with ideal covariance. 1000 Monte-Carlo runs with probability of false alarm $P_{fa} = 0.002$ and CNR=30 dB per range-azimuth bin for $N_{AB} = 180$ individual clutter patches making up the range ring.

clutter energy impinges on the receiver, and the reduction in clutter energy afforded by transmit beamforming becomes significant. This technique is most effective when ambiguous clutter is illuminated by a sidelobe of the transmit pattern, as this contributes significant energy to the signal and is effectively nulled with little loss of transmit power to the target. The technique is not effective if mainlobe nulling is required, as this will affect the energy delivered to the target of interest.

2.5 Summary

In this chapter, the possibilities presented by utilising transmit adaptivity for STAP clutter nulling have been investigated. First, in Section 2.3.2, space-time illumination patterns were shown to display the potential for nulling clutter on transmit. However, additional constraints, such as per element power restrictions, which do not appear on the receiver side, degrade the performance of the transmit clutter filter and should be considered when employing any transmit adaptivity. Doppler compensation through space-time beampattern phase control, discussed in Section 2.3.3, was found to minimise the rank of the clutter covariance matrix, but also to be impractical for real-world implementation. However, MIMO and frequency diverse arrays could provide some avenues for further investigation. Furthermore, splitting the processing burden between the transmitter and receiver to achieve identically distributed spatial-temporal statistics across different ranges offers another way to improve STAP performance. Finally, in Section 2.4 the performance of elevation-nulling was investigated to overcome the diversification of clutter statistics caused by range-ambiguous clutter returns. Solutions based on the geometry of the radar scene for estimating range-ambiguous elevation angles were presented and an efficient beampattern design procedure was developed to null interfering ambiguous ranges while achieving per element power constraints on the beampattern. The results show a large improvement in detection performance due to a reduction in clutter energy, non-homogeneity and sample support requirements.

A further area of research in which transmit diversity could be greatly beneficial is that of bistatic airborne radars. In these scenarios the clutter angle-Doppler characteristics follow Lissajous curves, and the complexity in detecting targets under these conditions is a major challenge for signal processing. The effects of array errors on performance and robust beamforming techniques also warrant investigation. In addition, the transmit beampattern is only one form of transmit adaptivity that may be used.

The other major facet of the transmitter which may be adapted is the transmitted waveform. The waveform, coupled with its pulse compression filter, forms a range-Doppler filter through the ambiguity function. A significant open problem is how to use this in STAP scenarios to simplify the signal processing and improve target detection performance.

References

[1] R. Klemm, *Principles of Space-Time Adaptive Processing.* Institution of Electrical Engineers, 2002.

[2] J. Ward, "Space-time adaptive processing for airborne radar," MIT Lincoln Laboratory, Lexington, MA, Tech. Rep., 1994.

[3] A. De Maio, S. De Nicola, Y. Huang, D. Palomar, S. Zhang, and A. Farina, "Code design for radar STAP via optimization theory," *IEEE Transactions on Signal Processing*, vol. 58, no. 2, pp. 679–694, Feb. 2010.

[4] P. Corbell, M. Temple, and T. Hale, "Forward-looking planar array 3D-STAP using space time illumination patterns (STIP)," in *40th IEEE Workshop on Sensor Array and Multichannel Processing*, Jul. 2006, pp. 602–606.

[5] P. Corbell, M. Temple, T. Hale, W. Baker, and M. Rangaswamy, "Performance improvement using interpulse pattern diversity with space-time adaptive processing," in *IEEE International Radar Conference*, May 2005, pp. 55–60.

[6] D. C. Wilcox and M. Sellathurai, "Transmit beamforming for range-ambiguous clutter mitigation in forward-looking STAP radar," in *Sensor Signal Processing for Defense (SSPD)*, Sep. 2011, pp. 1–4.

[7] I. Reed, J. Mallett, and L. Brennan, "Rapid convergence rate in adaptive arrays," *IEEE Transactions on Aerospace and Electronic Systems*, vol. 10, no. 6, pp. 853–863, Nov. 1974.

[8] P. Corbell, J. Perez, and M. Rangaswamy, "Enhancing GMTI performance in non-stationary clutter using 3D STAP," in *IEEE Radar Conference*, Apr. 2007, pp. 647–652.

[9] O. Kreyenkamp and R. Klemm, "Doppler compensation in forward-looking STAP radar," *IEE Proceedings - Radar, Sonar and Navigation*, vol. 148, no. 5, pp. 253–258, Oct. 2001.

[10] M. Li and G. Liao, "Robust short-range clutter suppression algorithm for forward looking airborne radar," in *IEEE Radar Conference*, May 2010, pp. 559–562.

[11] X. Meng, T. Wang, J. Wu, and Z. Bao, "Short-range clutter suppression for airborne radar by utilizing prefiltering in elevation," *IEEE Geoscience and Remote Sensing Letters*, vol. 6, no. 2, pp. 268–272, Apr. 2009.

[12] L. Patton and B. Rigling, "Modulus constraints in adaptive radar waveform design," in *IEEE Radar Conference*, May 2008, pp. 1–6.

Chapter 3

Digital Beamforming for Synthetic Aperture Radar

Karen Mak and Athanassios Manikas

Communications and Array Processing,
Department of Electrical and Electronic Engineering,
Imperial College London

Synthetic aperture radar (SAR) has all weather capabilities, which gives it an advantage over optical instruments for remote sensing during the night or cloudy weather. SAR also has the ability to detect targets hidden below foliage. An effectively long aperture, which would otherwise be impractical to build, is created due to SAR's method of moving along a flight path during data collection. This allows the creation of higher resolution images when compared to real aperture radar (RAR), where the radar system is stationary during data collection. Therefore, SAR has been used in many remote sensing applications including geographic mapping, ocean surveillance and ground moving target indication (GMTI), to name but a few. However, with conventional SAR data collection using a single transmitter and receiver beam[1] there is a trade-off between resolution and the dimensions of the imaged area, in particular between cross-range resolution and the swath-width[2]. In order to overcome this contradiction and to provide greater flexibility in meeting specific application requirements, different transmitter and receiver configurations have been proposed in the

---

[1] It will be assumed that these directional beams are created using planar arrays, where beamforming is performed by applying weights to the elements of the array such that the desired beam is created. However, other systems can be used, such as a reflector phased array. Also directional antennas such as a horn antenna can be utilised to form the directional beam.

[2] The "cross-range" is defined along the flight path of the SAR. The "swath width" of the imaged area is the width of the imaged area perpendicular to the flight path.

literature using multiple beamformers. The use of these systems allow the employment of additional beamforming algorithms such that ambiguity, jammer and clutter suppression can be achieved.

This chapter is structured as follows. In Section 3.1 the main SAR parameters are defined which then are used in Section 3.2 to provide a mathematical model of the signals in a general single-input single-output (SISO) SAR[3] system. In particular, the signals in three common SAR operational modes, namely Stripmap, ScanSAR and Spotlight are described. Then Section 3.3 extends the SISO SAR mathematical modelling to single-input multiple-output (SIMO) SAR[4] systems. Section 3.4 is concerned with beamforming in both the elevation and cross-range directions including beamforming for ambiguity suppression in the cross-range direction for overcoming SISO SAR's wide swath, high resolution contradiction. Furthermore SIMO SAR parameter design and representative examples are given. Then target parameter estimation, namely round trip delay estimation and joint direction of arrival (DOA) and target power estimation is given in Section 3.5. The chapter is concluded in Section 3.6.

3.1 SAR radar main parameters

In general, RAR images an area by transmitting a signal to illuminate the area of interest and receives the echoes. During data collection the radar system is stationary and therefore only one set of returns from the area of interest is obtained. However, with SAR, the transmit and receive cycle is repeated at specific time intervals, given by the pulse repetition interval T_r, while travelling along a specific flight path. It is assumed that a chirp pulse of duration T is transmitted every T_r seconds with $T < T_r$. If N_p chirp pulses are transmitted along the flight path then N_p sets of echoes are obtained. This results in the creation of an effectively larger aperture, which would otherwise be impractical to build. Depending on the flight path and the steering of the transmit and receive beams, images with a higher resolution and/or images of a wider area can be obtained.

With reference to Fig. 3.1 consider a SAR system consisting of a single planar grid array with N elements, which are weighted to form a single transmitter and receiver beam moving on a flight path along the positive x-axis, with velocity v_s. At one particular receive time t_p, with

---

[3]SISO SAR: single beam transmitter and single beam receiver SAR.
[4]SIMO SAR: single beam transmitter and multiple beam receiver SAR.

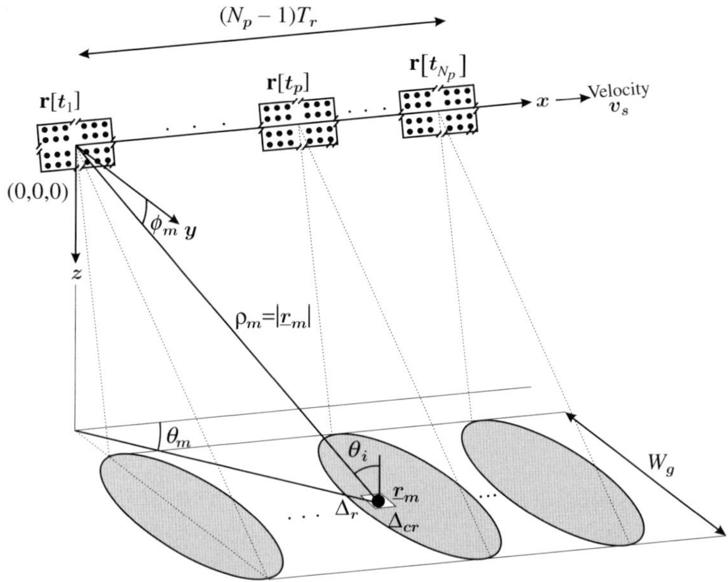

Fig. 3.1   Geometry of a SAR system with a single beamformer (SISO SAR).

$p = 1, 2, \ldots, N_p$, the Cartesian coordinates of the array elements are given by the matrix $\mathbf{r}[t_p] \in R^{3 \times N}$. The N elements are located on the same platform with Cartesian coordinates described by the matrix

$$\mathbf{r} = [\underline{r}_1, \underline{r}_2, \ldots, \underline{r}_N] \tag{3.1a}$$

$$= \left[\underline{r}_x, \underline{r}_y, \underline{r}_z\right]^T \in R^{3 \times N} \tag{3.1b}$$

with $\underline{r}_x$, $\underline{r}_y$ and $\underline{r}_z$ being column vectors containing the x, y and z coordinates of each element respectively and $\underline{r}_i \in R^{3 \times 1}$ denotes the location of the i-th element. Furthermore, the Cartesian coordinates of a single scatterer, say the m-th scatterer, located on the ground within the beam footprint are denoted by the (3×1) vector $\underline{r}_m$ as shown in Fig. 3.1, where

$$\underline{r}_m = \left[r_{m,x}, r_{m,y}, r_{m,z}\right]^T \tag{3.2a}$$

$$= \rho_m \begin{bmatrix} \cos\theta_m \cos\phi_m \\ \sin\theta_m \cos\phi_m \\ \sin\phi_m \end{bmatrix} \tag{3.2b}$$

where $(\rho_m, \theta_m, \phi_m)$ are the range, azimuth and elevation angles of the m-th scatterer with respect to the SAR system's reference point $(0, 0, 0)$, i.e. with respect to the origin of the Cartesian coordinate system. In addition, the angle the slant range makes with the perpendicular of the Earth's surface is defined as the incident angle θ_i shown in Figure 3.1.

Two important parameters of radar systems are the slant range resolution and cross range resolution which are defined as follows for SAR and RAR systems:

$$\text{Slant Range Resolution (RAR or SAR)} : \Delta_r \triangleq \frac{cT}{2\sin\theta_i} \qquad (3.3)$$

and

$$\text{Cross Range Resolution} : \Delta_{cr} \triangleq \begin{cases} \frac{\lambda\rho_m}{L_a}, & \text{RAR} \\ \frac{L_a}{2}, & \text{SAR} \end{cases} \qquad (3.4)$$

where L_a is the physical length (array aperture) of the planar grid array along the x-axis, c is the speed of light and λ is the wavelength in metres[5].

To get an idea of the increased resolution capabilities of SAR compared to RAR, consider an example [1] of a spaceborne SAR system with parameters, $\lambda = 0.057\,\text{m}$, $\rho_m = 850\,\text{km}$ and $L_a = 10\,\text{m}$ with a half-wavelength element spacing grid array and $T = 40\,\mu\text{s}$. In this case, the cross-range resolution, using Eq. (3.4), is 5 m. If a RAR system was used in order to obtain the same cross-range resolution, the length of the required beamformer would be $L_a = 9690\,\text{m}$, which would be impractical to physically build. Therefore, by moving the radar system along a flight path and transmitting a signal at intervals determined by the pulse repetition frequency (PRF) $f_r = \frac{1}{T_r}$, a larger aperture array is formed and higher resolution images can be obtained.

Note that the value of f_r plays an important role in preventing ambiguities in the received signals. In particular, to prevent cross-range ambiguities, f_r must be larger than the Doppler bandwidth B_D in order to sample the Doppler spectrum without aliasing, where the Doppler bandwidth is the range of Doppler frequencies in the cross-range direction in the antenna footprint [2]. However, f_r cannot be made arbitrarily large, as range ambiguities will occur due to the returns from different transmitted signals overlapping and being received within the same time frame $2T_r > \frac{W_g\sin\theta_i}{c}$, where W_g and θ_i are shown in Fig. 3.1. Therefore the following condition can

---

[5]For SAR $\Delta_r = \frac{cT}{2\sin\theta_i}$ before range compression and $\Delta_r = \frac{c}{2B\sin\theta_i}$ after range compression, where B is the bandwidth of the transmitted signal and θ_i is the incident angle.

be made on the choice of f_r for a given SAR system to ensure unambiguous returns are received [2]:

$$B_D < f_r < \frac{c}{2W_g \sin \theta_i} \tag{3.5}$$

where c is the speed of light, W_g is the swath-width and θ_i is the incident angle. For non-zero squint angle θ_{sq}, the Doppler bandwidth is given by

$$B_D = \frac{2v}{\lambda} \left(\sin \left(\theta_{sq} + \frac{\lambda}{2L_a} \right) - \sin \left(\theta_{sq} - \frac{\lambda}{2L_a} \right) \right) \tag{3.6}$$

while in the case of zero squint (i.e. $\theta_{sq} = 0$; boresight direction), B_D is given as [4]:

$$B_D \approx \frac{2v_s}{L_a} \approx \frac{v_s}{\Delta_{cr}}. \tag{3.7}$$

Using Eq. (3.7), Eq. (3.5) becomes

$$\frac{v_s}{\Delta_{cr}} < f_r < \frac{c}{2W_g \sin \theta_i} \tag{3.8}$$

thus relating the parameters Δ_{cr}, f_r and W_g. In particular Eq. (3.8) indicates that for a fixed value of f_r, an increase in W_g will lead to coarser resolution, and a decrease in W_g will lead to a finer resolution. Therefore, it is *not* possible to increase both W_g and the resolution Δ_{cr} simultaneously[6]. This is proven in practice with single transmitter and receiver beamformer operational modes, described in Section 3.2.

In order to overcome the contradiction between the imaging of a large swath-width W_g and imaging with a high cross-range resolution Δ_{cr} and, furthermore, to provide greater flexibility in meeting specific application requirements, different transmitter and receiver configurations have been proposed in the literature using multiple beamformers, which can be grouped as:

(a) SIMO SAR systems which consist of a single transmitter beamformer and multiple $K > 1$ receiver beamformers, with each beamformer using an array of N elements. In this chapter, it will be assumed that the transmitter beamformer is the reference beamformer of the SAR system.

---

[6]The trade-off between the swath-width and cross-range resolution has a greater importance in spaceborne SISO SAR systems, compared to airborne SAR, due to the larger velocities and smaller incidence angles in the spaceborne case.

(b) MIMO SAR systems consisting of multiple $K_{Tx} > 1$ transmitter beamformers, each formed from N_{Tx} elements, and multiple $K_{Rx} > 1$ receiver beamformers, each formed from N_{Rx} elements, resulting in a total of

$$N = K_{Tx}N_{Tx} + K_{Rx}N_{Rx} \qquad (3.9)$$

elements in the SAR system.

The use of these multiple beamformers allows additional array signal processing techniques to be applied for a number of applications such as ambiguity, jammer and clutter suppression resulting in an increase in performance. In this chapter only SISO and SIMO SAR will be studied.

Furthermore, to distinguish between the time a chirp signal is transmitted and the time when its echo is received, the following notation will be used: t_n, with $n = 1, 2, \ldots, N_p$ denoting the time when a chirp is transmitted and t_p, with $p = 1, 2, \ldots, N_p$ the time when echos are received. Therefore the chirp pulse transmitted at t_n is received at t_p.

3.2   SISO SAR

In this section, the following three common SAR operational modes will be described for a SISO SAR system:

- Stripmap mode;
- ScanSAR mode;
- Spotlight mode.

where a SAR system consisting of a single grid planar array of $N = N_x \cdot N_z$ elements forming a transmitter/receiver beamformer that travels on a straight path along the positive x-axis is considered, with N_x being the number of elements along the length of the beamformer and N_z the number of elements in the height of the beamformer where half-wavelength interelement spacing is assumed. The array forms a transmitter beamformer that transmits a pulse signal of duration T and then switches to a receiver beamformer.

3.2.1   *Stripmap SAR*

A SAR system in Stripmap SAR mode travelling with a velocity v_s along the x-axis, shown in Fig. 3.2, results in the imaging of a strip of the Earth's surface parallel to the SAR's flight path. The SAR transmits a chirp signal

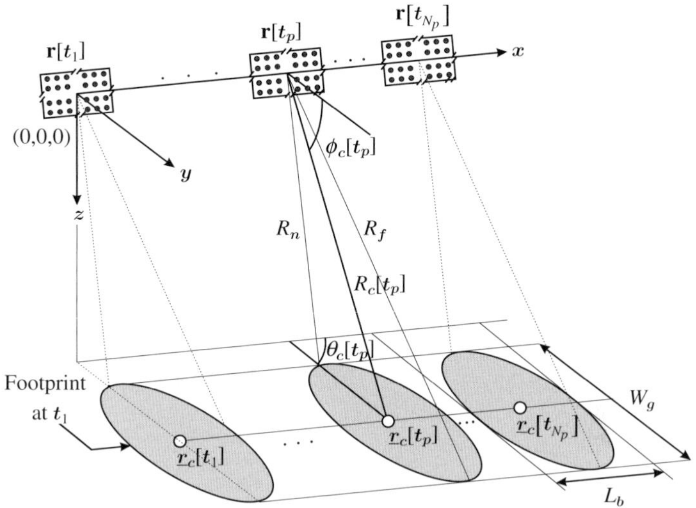

Fig. 3.2   Stripmap data collection, at instances of t_p for $p = 1, 2, \ldots, N_p$.

$m(t)$ of duration T seconds every T_r seconds and can be expressed as

$$m(t) = \exp\left(j\pi\frac{B}{T}t^2\right) \tag{3.10}$$

$$\text{with } t_n < t < t_n + T \quad \text{for } n = 1, 2, ..., N_p$$

where B is the chirp bandwidth and T the chirp pulse duration (see Fig. 3.3) with $\frac{B}{T}$ defining the chirp rate. Usually a duty cycle D is defined during transmission to give the ratio between the transmit time and pulse repetition interval (PRI) T_r.

The signal $m(t)$ is transmitted using a transmitter beamformer described by the complex $(N \times 1)$ vector $\underline{w}_{Tx}$ producing a single beam, whose dimensions on the ground are determined by the array geometry and the design of the weight vector $\underline{w}_{Tx}$. Particular beamformer parameters of interest are:

- the two 3 dB beamwidths of the mainlobe in the flight direction, i.e. the cross-range direction along the x-axis $\theta_{cr,3\mathrm{dB}}$, and elevation direction $\theta_{el,3\mathrm{dB}}$ in degrees [1]:

$$\theta_{cr,3\mathrm{dB}} = 0.886 \times \frac{180}{\pi}\frac{\lambda}{L_a} \tag{3.11}$$

$$\theta_{el,3\mathrm{dB}} = 0.886 \times \frac{180}{\pi}\frac{\lambda}{H_a} \tag{3.12}$$

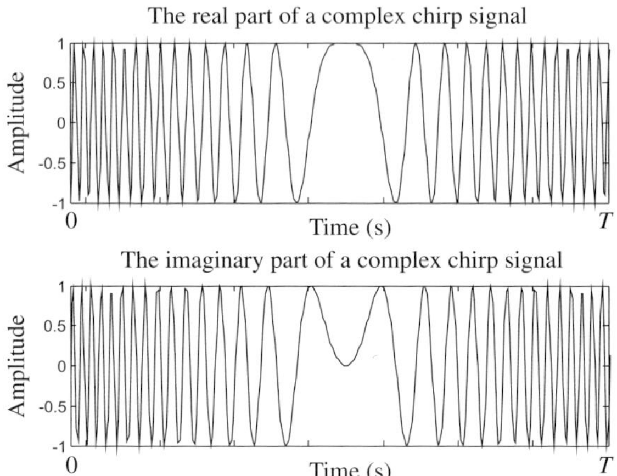

Fig. 3.3   The real and imaginary parts of a complex chirp signal.

$$\text{where} \begin{cases} L_a = \text{Beamformer length} = N_x \frac{\lambda}{2} \\ H_a = \text{Beamformer height} = N_z \frac{\lambda}{2} \\ \lambda = \text{Wavelength of carrier} \end{cases}$$

- the peak directivity D_{peak} of the beam:

$$D_{peak} = 10 \log_{10} \left(\frac{4\pi L_a H_a}{\lambda^2} \right) \tag{3.13}$$

As a result with reference to Fig. 3.2, a beam footprint of length L_b along the cross-range direction on the ground is created illuminating a swath-width W_g on the ground, which is given by the following approximate expressions $\forall t_p$ [1]

$$L_b \approx 0.886 \frac{\lambda R_c}{L_a} \tag{3.14}$$

$$W_g \approx 0.886 \frac{\lambda R_c}{H_a \cos(\theta_i)} \tag{3.15}$$

$$\text{where} \begin{cases} \theta_i = \text{Incident angle, } \forall t_p \\ R_c = \frac{R_f + R_n}{2} \\ \quad = \text{Slant range to centre of beam footprint, } \forall t_p \\ R_f = \text{Far slant range, } \forall t_p \\ R_n = \text{Near slant range, } \forall t_p. \end{cases}$$

Note that each scatterer within the beam footprint is illuminated for T seconds, starting at the near range and sweeping out towards the far range. If a chirp transmitted at t_n is received at t_p and a total of M scatterers are assumed, then the array received signal $\underline{x}(t)$ vector at a particular data collection point at time t_p at all N array elements can then be modelled as the $(N \times 1)$ vector $\underline{x}(t)$ given as[7]

$$\underline{x}(t) = \sum_{m=1}^{M} \begin{pmatrix} \beta_m \exp\left(j2\pi f_{d,m} t_p\right) \\ \cdot \underline{S}_{Rx,m} \underline{S}_{Tx,m}^{H} \underline{w}_{Tx}[t_n] \\ \cdot m\left(t - \tau_m[t_p]\right) \end{pmatrix} + \underline{n}(t) \tag{3.16}$$

with the time variable t satisfying

$$t_p \leq t \leq t_p + \frac{2\left(R_f - R_n\right)}{c} + T \tag{3.17}$$

where

$$
\begin{aligned}
\beta_m &= \text{Complex path gain associated with the } m\text{-th scatterer} \\
(\theta_m, \phi_m) &= \text{(azimuth, elevation) angle associated with the } m\text{-th scatterer with reference to } (0,0,0) \\
\tau_m[t_p] &= \text{Round trip delay associated with the } m\text{-th scatterer} \\
\rho_m &= \text{Slant range between } m\text{-th scatterer and } (0,0,0) \\
n(t) &= \text{Noise} \\
\underline{w}_{Tx} &= (N \times 1) \text{ weight vector of the } N \text{ elements of the single beamformer forming the transmit beam} \\
f_{d,m} &= \text{Doppler frequency shift associated with the } m\text{-th scatterer} \\
\underline{S}_{Rx,m} &\triangleq \underline{S}_{Rx}\left(\mathbf{r}[t_p], \theta_m, \phi_m, \rho_m\right) \\
&= \text{Receiver array manifold vector associated with the } m\text{-th scatterer} \\
\underline{S}_{Tx,m} &\triangleq \underline{S}_{Tx}\left(\mathbf{r}[t_n], \theta_m, \phi_m, \rho_m\right) \\
&= \text{Transmitter array manifold vector associated with the } m\text{-th scatterer.}
\end{aligned}
$$

---

[7]The Doppler frequency term is only required in a satellite SAR case where Doppler shifts are greater than in an airborne case and therefore create a significant phase shift. This term is also required when "stop and receive" data collection is not assumed, i.e. when it cannot be assumed that the SAR system transmits a chirp AND receives its echo when the system is still in approximately the same location in space.

It is important to note that β_m is a complex number containing the round trip attenuation from the reference element of the beamformer when the SAR system is at a reference position along the flight path to the m-th scatterer and back, as well as various constants related to the m-th scatterer, for example, scatterer reflectivity.

The $(N \times 1)$ complex vectors $\underline{S}_{Tx,m}$ and $\underline{S}_{Rx,m}$ are the transmitter and receiver array manifold vectors associated with the m-th scatterer, respectively. Throughout this chapter it is assumed that "stop and receive" data collection occurs and therefore the Cartesian coordinates satisfy $\mathbf{r}[t_p] = \mathbf{r}[t_n]$ and $\underline{r}_c[t_p] = \underline{r}_c[t_n]$, where $\underline{r}_c[t_p]$ is shown in Fig. 3.2. Thus $\underline{S}_{Tx,m}$ is given as

$$\underline{S}_{Tx,m} \triangleq \underline{S}_{Tx}\left(\mathbf{r}[t_p], \theta_m, \phi_m, \rho_m\right) \tag{3.18a}$$

$$= \rho_m^a \odot \underline{R}_{Tx,m}^{-a}[t_p]$$

$$\odot \exp\left(+j\frac{2\pi F_c}{c}\left(\rho_m \cdot \underline{1}_N - \underline{R}_{Tx,m}[t_p]\right)\right) \tag{3.18b}$$

with

$$\underline{R}_{Tx,m}[t_p] = \sqrt{\rho_m^2 \cdot \underline{1}_N + \underline{r}_x^2[t_p] + \underline{r}_y^2[t_p] + \underline{r}_z^2[t_p] - \frac{\rho_m c}{\pi F_c}\mathbf{r}^T[t_p]\underline{k}\left(\theta_m, \phi_m\right)} \tag{3.19}$$

where ρ_m is the reference slant range between the SAR system and the m-th scatterer, a denotes the path loss exponent and $\underline{k}\left(\theta_m, \phi_m\right)$ is the wavenumber vector

$$\underline{k}(\theta_m, \phi_m) = \frac{2\pi F_c}{c}\left[\cos(\theta_m)\cos(\phi_m), \sin(\theta_m)\cos(\phi_m), \sin(\phi_m)\right]^T \tag{3.20a}$$

$$= \frac{2\pi F_c}{c}\underline{u}\left(\theta_m, \phi_m\right) \tag{3.20b}$$

with $\underline{u}(\theta_m, \phi_m)$ being a (3×1) unit vector pointing in the direction (θ_m, ϕ_m) of the m-th scatterer. As it is assumed that the single array is used for both the transmitter and receiver beamformers, the transmitter and receiver manifold vectors $\underline{S}_{Tx,m}$ and $\underline{S}_{Rx,m}$ are related as follows

$$\underline{S}_{Rx,m} = \underline{S}_{Tx,m}^* \tag{3.21}$$

due to the opposite flow of energy between the transmitter and receiver, where $\underline{S}_{Rx,m} \triangleq \underline{S}_{Rx}(\mathbf{r}[t_p], \theta_m, \phi_m, \rho_m)$. Note that it is assumed that a single beamformer operates as a transmitter and receiver, that all N elements of the single beamformer are collocated and that the m-th scatterer is in the far field of the beamformer, i.e. ρ_m is much greater than the array's real aperture. Under these assumptions Eq. (3.17) reduces to the plane wave propagation manifold vector expressed as

$$\underline{S}_{Tx,m} = \exp\left(+j\mathbf{r}^T[t_p]\underline{k}\left(\theta_m,\phi_m\right)\right). \tag{3.22}$$

Correspondingly $\underline{S}_{Rx,m}$, when plane wave propagation occurs, can be expressed as

$$\underline{S}_{Rx,m} = \underline{S}_{Tx,m}^* \tag{3.23a}$$
$$= \exp\left(-j\mathbf{r}^T[t_p]\underline{k}\left(\theta_m,\phi_m\right)\right). \tag{3.23b}$$

The $(N \times 1)$ weight vector $\underline{w}_{Tx}$ is designed such that at each t_p, with $p = 1, 2, \ldots, N_p$, a single transmit beam is produced whose mainlobe points in the direction $(\theta_c[t_p], \phi_c[t_p])$, where the centre of the mainlobe has Cartesian coordinates $\underline{r}_c[t_p]$ on the ground. After t_{N_p} seconds a strip on the Earth's surface is imaged with dimensions related to the flight path distance, the beam squint and the swath-width. With reference to Fig. 3.2, the Cartesian coordinates of the centre of the formed mainlobe on the ground, $\underline{r}_c[t_p]$, change for every t_p, with $p = 1, 2, \ldots, N_p$, however the azimuth and elevation angles of the formed beam, θ_c and ϕ_c respectively, stay constant for all t_p. Therefore $\underline{w}_{Tx}$ can be described as

$$\underline{w}_{Tx}[t_p] = \underline{S}_{Tx}\left(\mathbf{r}[t_p], \theta_c, \phi_c\right). \tag{3.24}$$

The received signals at all N elements of the SAR system are then combined by applying weights such that a single beam is formed whose mainlobe covers the same area as the transmit beam for all t_p

$$\underline{w}_{Rx}[t_p] = \underline{S}_{Rx}\left(\mathbf{r}[t_p], \theta_c, \phi_c\right). \tag{3.25}$$

Therefore, the received signal due to the M scatterers at the output of the receiver beamformer can be modelled as

$$y(t) = \underline{w}_{Rx}^H[t_p]\underline{x}(t) \tag{3.26a}$$
$$= \sum_{m=1}^{M}\left(\begin{array}{c}\beta_m \exp\left(j2\pi f_{d,m}t_p\right)\\ \cdot\underline{w}_{Rx}^H[t_p]\underline{S}_{Rx,m}\underline{S}_{Tx,m}^H\underline{w}_{Tx}[t_p]\\ \cdot m\left(t-\tau_m[t_p]\right)\end{array}\right) + n(t). \tag{3.26b}$$

3.2.2   *ScanSAR*

In ScanSAR [5], the mainlobe of the beam is steered at specific time intervals such that different subswaths are illuminated periodically as the SAR system travels along the flight path. This is illustrated in Fig. 3.4 where two subswaths are shown. By assuming N_{sub} subswaths are illuminated, the n-th subswath is illuminated every $T_{p,n}$ seconds for a period of $T_{b,n}$ seconds, where $T_{p,n}$ is the burst length in seconds corresponding to the n-th subswath and $T_{b,n}$ is the burst period in seconds corresponding to the n-th subswath

$$
\text{with} \begin{cases}
\underline{r}_{c,n}[t_p] & = \ \text{Cartesian coordinates of the centre of the} \\
& \quad \ \text{formed mainlobe on the ground when the} \\
& \quad \ n\text{-th subswath is illuminated at time } t_p \\
R_{c,n}[t_p] & = \ \text{Slant range between the SAR system at} \\
& \quad \ \text{time } t_p \text{ and } \underline{r}_{c,n}[t_p] \text{ when the } n\text{-th sub-} \\
& \quad \ \text{swath is illuminated} \\
(\theta_{c,n}[t_p], \phi_{c,n}[t_p]) & = \ \text{(azimuth, elevation) angle between the} \\
& \quad \ \text{SAR system at time } t_p \text{ and } \underline{r}_{c,n}[t_p] \text{ when} \\
& \quad \ \text{the } n\text{-th subswath is illuminated.}
\end{cases}
$$

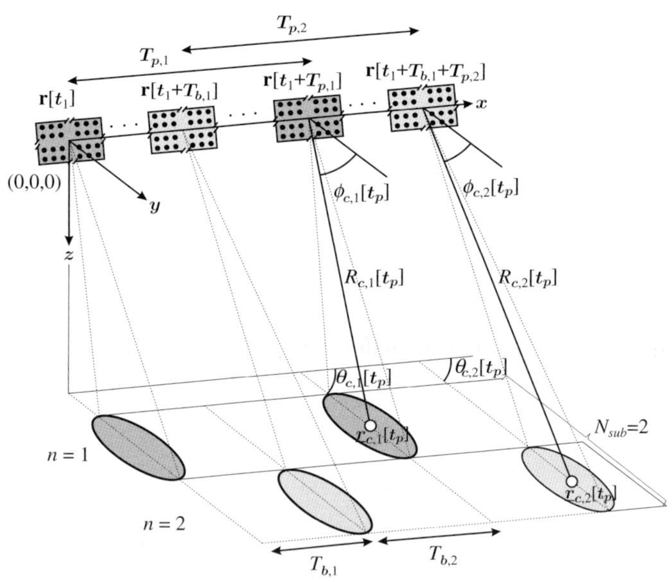

Fig. 3.4   ScanSAR data collection, at instances of t_p.

The signal $m(t)$ is transmitted using a transmitter beamformer described by the complex $(N \times 1)$ vector $\underline{w}_{Tx}[t_p]$, producing a single beam, as

$$\underline{w}_{Tx}[t_p] = \underline{S}_{Tx}(\mathbf{r}[t_p], \theta_{c,n}[t_p], \phi_{c,n}[t_p]) \tag{3.27}$$

where it has been assumed that the subswaths are illuminated one after the other. The receiver beamformer output can be modelled as

$$y(t) = \underline{w}_{Rx}^{H}[t_p]\underline{x}(t) \tag{3.28}$$

with $t_p \leq t \leq t_p + \frac{2(R_f - R_n)}{c} + T$ for $p = 1, 2, \ldots, N_p$ and where

$$\underline{w}_{Rx}[t_p] = \underline{S}_{Rx}(\mathbf{r}[t_p], \theta_{c,n}[t_p], \phi_{c,n}[t_p]) \tag{3.29}$$

forms a single beam whose mainlobe covers the same area as the transmit beam for all t_p.

If N_{sub} is the number of subswaths, ScanSAR approximately results in an N_{sub} times larger swath-width compared to that in Stripmap mode. However images formed from the collected raw data have an approximately $N_{sub} + 1$ reduction in the cross-range resolution.

3.2.3 *Spotlight SAR*

With ScanSAR, the beam of the single beamformer is steered to illuminate multiple swaths at the expense of a lower cross-range resolution. However, with Spotlight SAR the single beam is steered during data collection such that the same area is constantly illuminated [6], as illustrated in Fig. 3.5 with $t_1 < t_2 < \ldots < t_{N_p}$.

The weights of the N elements forming the transmit beam are now designed such that the location of the centre of the formed mainlobe on the ground, $\underline{r}_c$, remains constant for all t_p. Therefore the weights required to form the beam *change* with t_p, where $p = 1, 2, \ldots, N_p$, and can be modelled as

$$\underline{w}_{Tx}[t_p] = \underline{S}_{Tx}(\mathbf{r}[t_p], \theta_c[t_p], \phi_c[t_p]). \tag{3.30}$$

The received signals at all N elements of the SAR system are then combined by applying weights such that a single beam is formed whose mainlobe covers the same area as the transmit beam for all t_p and therefore

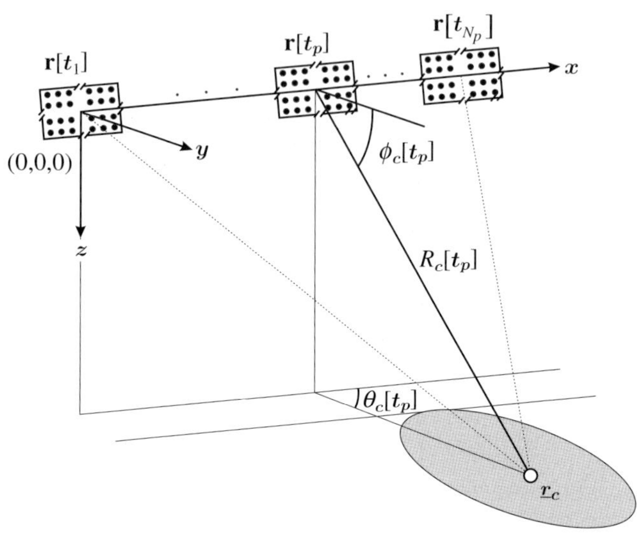

Fig. 3.5　Spotlight data collection, at instances of t_p.

$\underline{r}_c[t_p] = \underline{r}_c[t_n]$. Thus, at the receiver, the steering vector beamformer is designed as follows

$$\underline{w}_{Rx}[t_p] = \underline{S}_{Rx}(\mathbf{r}[t_p], \theta_c[t_p], \phi_c[t_p]) \qquad (3.31)$$

and the signal at the output of this beamformer is

$$y(t) = \underline{w}_{Rx}^{H}[t_p]\underline{x}(t) \qquad (3.32)$$

for $t_p \leq t \leq t_p + \frac{2(R_f - R_n)}{c} + T$ with $p = 1, 2, \ldots, N_p$.

Due to the single beam constantly illuminating the same area, images formed from the collected data after processing have a higher resolution when compared to images formed from data in Stripmap mode. However, there is also a reduction in the swath coverage.

3.2.4　*Discrete time modelling*

The received signal $\underline{x}(t)$, using the array $\mathbf{r}[t_p]$ with $p = 1, 2, \ldots, N_p$, is sampled at a sampling rate of F_s to obtain L snapshots forming an $(N \times L)$ matrix $\mathbb{X}[t_p]$. This matrix describes the data received at all N antenna elements due to the chirp received at time t_p and can be given as

$$\mathbb{X}[t_p] = [\underline{x}[1, t_p], \underline{x}[2, t_p], \ldots, \underline{x}[L, t_p]] \ (N \times L) \qquad (3.33)$$

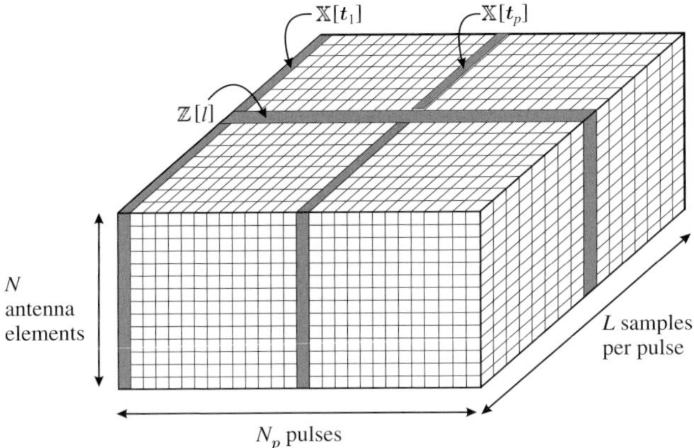

Fig. 3.6   3D datacube of the received signals at all N elements of the receiver beamformer after discretisation.

with $\underline{x}[l, t_p]$ being the l-th snapshot of the $N \times 1$ received signal vector $\underline{x}(t)$ with t satisfying Eq. (3.17). Figure 3.6 illustrates the 3D datacube formed by $\mathbb{X}[t_p]$ for $p = 1, 2, \ldots, N_p$. Furthermore, in Fig. 3.6 the matrix $\mathbb{Z}[l]$ is shown, which is an $(N \times N_p)$ matrix describing the data received at the N antenna elements corresponding to the l-th sample of all N_p pulses.

Weight vectors (i.e. beamformers) $\underline{w}[t_p]$ are then designed and applied to $\mathbb{X}[t_p]$, $\forall p$, forming a single output, described by the $(L \times N_p)$ matrix $\mathbb{Y}$

$$
\mathbb{Y} = \begin{bmatrix} \underline{w}[t_1]^H \, \mathbb{X}[t_1] \\ \underline{w}[t_2]^H \, \mathbb{X}[t_2] \\ \vdots \\ \underline{w}[t_{N_p}]^H \, \mathbb{X}[t_{N_p}] \end{bmatrix}^T .
\tag{3.34}
$$

In Fig. 3.7 a raw data image is presented by the absolute value of the elements of matrix $\mathbb{Y}$ due to a single scatterer during Stripmap data collection. From this "grey" image it can be seen that the trajectory of the scatterer changes with each transmitted pulse. Furthermore, the scatterer is only illuminated for a specific length of time during data collection related to the synthetic aperture length, defined as

$$
L_{synth} = 0.886 \frac{R_c \lambda}{L_a} \frac{v_s}{v_g}
\tag{3.35}
$$

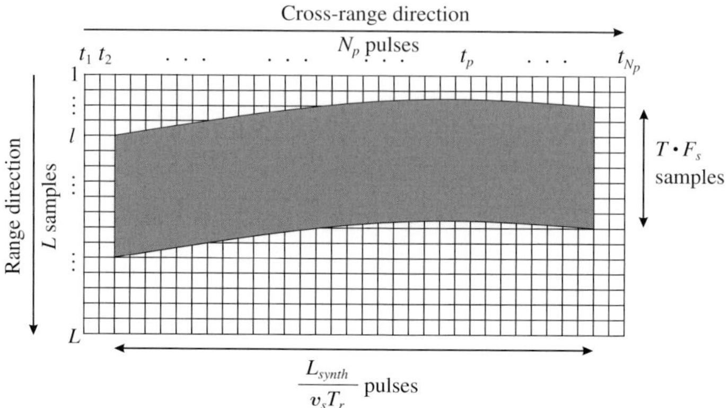

Fig. 3.7 A representation of the magnitude of the elements of the matrix $\mathbb{Y}$ (Eq. (3.34)), which represents the output of the receiver beamformer after discretising in the direction perpendicular to the cross-range direction, i.e. the range direction.

where v_s is the velocity of the SAR platform and v_g is the velocity of the beam on the ground. In airborne platforms, $v_s \approx v_g$; however, in satellite platforms the difference in velocities needs to be taken into account.

When comparing the ScanSAR and Spotlight operational modes there appears to be a trade-off between swath-width and resolution. With ScanSAR, a wider swath-width compared to Stripmap data collection is imaged but at the expense of a lower resolution. However, with Spotlight data collection a higher resolution compared to Stripmap can be achieved, although a smaller swath-width is imaged. One method to overcome this is the use of multiple receiver beamformers resulting in additional samples along the synthetic aperture. In order to understand this further, SIMO SAR systems will be examined where K receiver beamformers and a single transmitter beamformer are assumed.

3.3 SIMO SAR

In the previous section, operational modes have been described for the case of SISO SAR, in which the SAR system consists of a single planar array of N elements, forming a single transmitter beamformer that transmits a chirp signal of duration T every T_r, and then switches to a receiver beamformer. In this section, the number of receiver beamformers will be increased to K (with $K > 1$), with each beamformer being formed from N elements. Furthermore, beamforming techniques that allow the wide swath, high cross-range resolution trade-off to be overcome, along with ambiguity suppression, will be examined.

3.3.1 *SIMO SAR system mathematical modelling*

Consider a SAR system consisting of a single planar array of N elements forming a transmitter beamformer that travels along a straight path along the x-axis. The N elements are located on the same platform with Cartesian coordinates described by the matrix

$$\mathbf{r}_{Tx} = \left[\underline{r}_{Tx,1}, \underline{r}_{Tx,2}, \dots, \underline{r}_{Tx,N}\right] \tag{3.36a}$$

$$= \left[\underline{r}_{Tx,x}, \underline{r}_{Tx,y}, \underline{r}_{Tx,z}\right]^T \in R^{3\times N} \tag{3.36b}$$

with $\underline{r}_{Tx,x}$, $\underline{r}_{Tx,y}$ and $\underline{r}_{Tx,z}$ being $(N \times 1)$ column vectors containing the x, y and z coordinates of each transmitter element respectively and $\underline{r}_{Tx,i} \in R^{3\times 1}$ denoting the location of the i-th element of the array. Additionally, there are K receiver beamformers that travel along the same flight path as the transmitter beamformer with each receiver beamformer using an array of N elements with geometries identical to the Tx-array. The N elements of each Rx-array are located on the same platform; however, the K Rx-arrays can be either on separate platforms, or on the same platform. The total KN elements of the K receiver beamformers have Cartesian coordinates described by the matrix

$$\mathbf{r} = [\mathbf{r}_1, \mathbf{r}_2, \dots, \mathbf{r}_K] \in R^{3\times KN} \tag{3.37}$$

where $\mathbf{r}_k$ describes the Cartesian coordinates of the N elements of the k-th beamformer and is defined as

$$\mathbf{r}_k = \left[\underline{r}_{k,x}, \underline{r}_{k,y}, \underline{r}_{k,z}\right]^T \in R^{3\times N} \tag{3.38}$$

with $\underline{r}_{k,x}$, $\underline{r}_{k,y}$ and $\underline{r}_{k,z}$ being column vectors containing the x, y and z coordinates of each element of the k-th beamformer, respectively.

An example configuration of a SIMO SAR system in Stripmap data collection is illustrated in Fig. 3.8 showing the beam footprint of the transmitter beamformer, where it has been assumed that the transmitter beamformer switches to a receiver beamformer after transmission of a chirp pulse at t_p for $p = 1, 2, \dots, N_p$ and becomes one of the K receiver beamformers.

A chirp pulse $m(t)$, shown in Eq. (3.10), is transmitted by the transmitter beamformer, described by the complex $(N \times 1)$ vector $\underline{w}_{Tx}$, producing a single beam with dimensions given by Eqs. (3.14) and (3.15). Each scatterer within this beam footprint is illuminated for T seconds and their echoes are

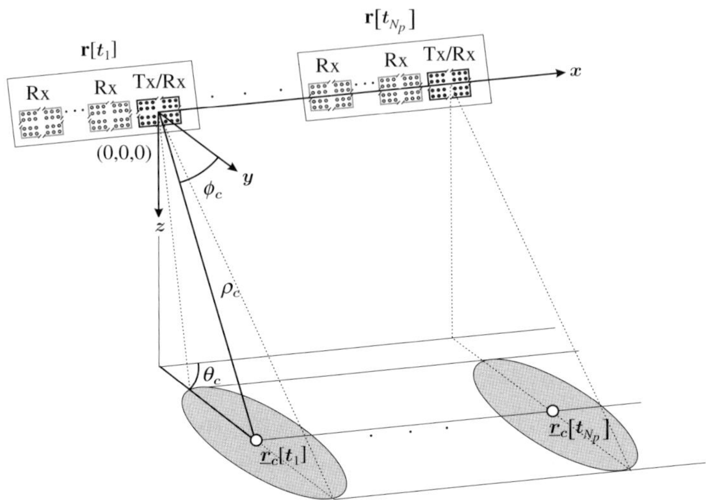

Fig. 3.8   An example of a SIMO SAR system.

received by all KN elements of the SIMO SAR system. Assuming there are
a total of M scatterers, the received signal at a particular data collection
point at time t_p can then be modelled as the $(KN \times 1)$ vector $\underline{x}(t)$ given as

$$\underline{x}(t) = \sum_{m=1}^{M} \begin{pmatrix} \beta_m \exp\left(j2\pi f_{d,m} t_p\right) \\ \cdot \underline{S}_{Rx,m} \underline{S}_{Tx,m}^{H} \underline{w}_{Tx}[t_p] \\ \cdot m\left(t - \underline{\tau}_m[t_p]\right) \end{pmatrix} + \underline{n}(t) \qquad (3.39)$$

where $\underline{\tau}_m[t_p]$ is the $(KN \times 1)$ vector with elements the round trip delays
associated with the m-th scatterer and the dependency of $\underline{w}_{Tx}$ on t_p will
depend on the required operational mode. The transmitter array manifold
vector $\underline{S}_{Tx,m}$ associated with the m-th scatterer is an $(N \times 1)$ vector and
$\underline{S}_{Rx,m}$ is now a $(KN \times 1)$ complex vector, representing the receiver array
manifold vector of the overall KN element array formed by the K individual
arrays.

All KN elements of the system are then weighted such that K beams
are formed and can be represented by the K columns of the matrix $\mathbb{W}[t_p]$
given by

$$\mathbb{W}[t_p] = \begin{bmatrix} \underline{w}_1[t_p], & \underline{0}_{N \times 1} & ,\cdots, & \underline{0}_{N \times 1} \\ \underline{0}_{N \times 1}, & \underline{w}_2[t_p] & ,\cdots, & \underline{0}_{N \times 1} \\ \vdots & \vdots & \ddots & \vdots \\ \underline{0}_{N \times 1}, & \underline{0}_{N \times 1} & ,\cdots, & \underline{w}_K[t_p] \end{bmatrix} \qquad (3.40)$$

where $\underline{w}_k[t_p]$ is an $(N \times 1)$ vector of the weights of the N elements of the k-th beamformer at time t_p. The time dependency on t_p is dictated by the required operational mode described in Section 3.2.

The output of the K receiver beamformers $\underline{y}(t) \in \mathcal{C}^{K \times 1}$ can then be modelled as

$$
\underline{y}(t) = \sum_{m=1}^{M} \left(\begin{array}{c} \beta_m \exp\left(j2\pi \underline{f}_{d,m} t_p\right) \\ \odot \left(\mathbb{W}[t_p]^H \underline{S}_{Rx,m}\right) \cdot \underline{S}_{Tx,m}^H \underline{w}_{Tx}[t_p] \\ \cdot m\left(t - \underline{\tau}_m[t_p]\right) \end{array} \right) + \underline{n}(t) \qquad (3.41)
$$

$$
\text{where} \begin{cases}
\mathbb{W}[t_p] & = & (KN \times K) \text{ matrix of the weights of the } K \\
& & \text{receiver beamformers} \\[4pt]
\underline{f}_{d,m} & = & (K \times 1) \text{ vector of the Doppler frequency shifts} \\
& & \text{associated with the } m\text{-th scatterer and } K \\
& & \text{receiver beamformers} \\[4pt]
\underline{S}_{Rx,m} & \triangleq & \underline{S}_{Rx}\left(\mathbf{r}[t_p], \theta_m, \phi_m, \rho_m\right) \\
& = & (KN \times 1) \text{ receiver array manifold vector asso-} \\
& & \text{ciated with the } m\text{-th scatterer} \\[4pt]
\underline{S}_{Tx,m} & \triangleq & \underline{S}_{Tx}\left(\mathbf{r}[t_p], \theta_m, \phi_m, \rho_m\right) \\
& = & (N \times 1) \text{ transmitter array manifold vector} \\
& & \text{associated with the } m\text{-th scatterer}
\end{cases}
$$

where the k-th element of $\underline{y}(t)$ corresponds to the signals received by the k-th receiver beamformer.

In the case where all K beamformers are on the same platform and collocated, planewave propagation can be assumed. However, in the case where all K beamformers are located on different platforms, the spherical wave propagation manifold vector should be utilised. However, the elements of the manifold vector that are on the same platform will have the same range, collapsing to planewave propagation.

3.3.2 *Discrete time modelling*

The received signals at the KN elements of the SIMO SAR system are sampled at a sampling rate of F_s to obtain L samples. A 3D datacube can then be illustrated as shown in Fig. 3.9 where $\mathbb{X}[t_p]$ is a $(KN \times L)$ matrix describing the data received at all KN elements of the SIMO SAR system

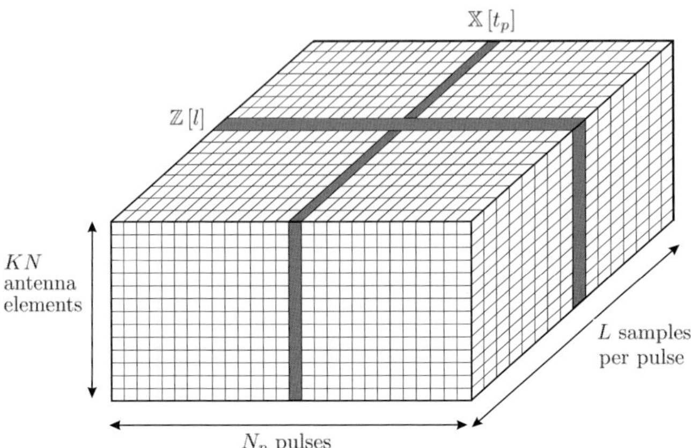

Fig. 3.9   3D datacube of the received signals at all KN elements of the SIMO SAR system after discretisation.

due to the chirp transmitted at time t_p and can be given as

$$\mathbb{X}[t_p] = [\underline{x}\,[1, t_p]\,, \underline{x}\,[2, t_p]\,, \ldots, \underline{x}[L, t_p]] \tag{3.42}$$

with $\underline{x}[l, t_p]$ being the l-th snapshot of the vector $\underline{x}(t)$ at time t_p. With reference to Fig. 3.9, $\mathbb{Z}[l]$ is a $(KN \times N_p)$ matrix describing the data received at all KN antenna elements at a particular sample at l.

Weights are then designed and applied to all KN elements such that K beamformer outputs are created, giving a 3D datacube with dimensions of $(K \times N_p \times L)$. With reference to Fig. 3.10, $\mathbb{Y}[k]$ is an $(L \times N_p)$ matrix containing all the data received by the k-th beamformer from all N_p transmitted chirp pulses and is illustrated in Fig. 3.7. Furthermore, $\bar{\mathbb{Z}}[l]$ is a $(K \times N_p)$ matrix describing the data received by the K beamformers at a particular sample at l. Therefore $\bar{\mathbb{Z}}[l]$ can be seen as a range sample space-time snapshot and the $(K \times L)$ matrix $\bar{\mathbb{X}}[t_p]$ can be seen as a cross-range sample space-time snapshot.

3.4   Beamforming in the elevation and cross-range direction using SIMO SAR

In this section, the use of beamforming for SIMO SAR systems in both the elevation and cross-range directions will be examined. In particular, the use of beamformers for ambiguity suppression will be investigated. The requirements of different applications can be met using SIMO SAR when

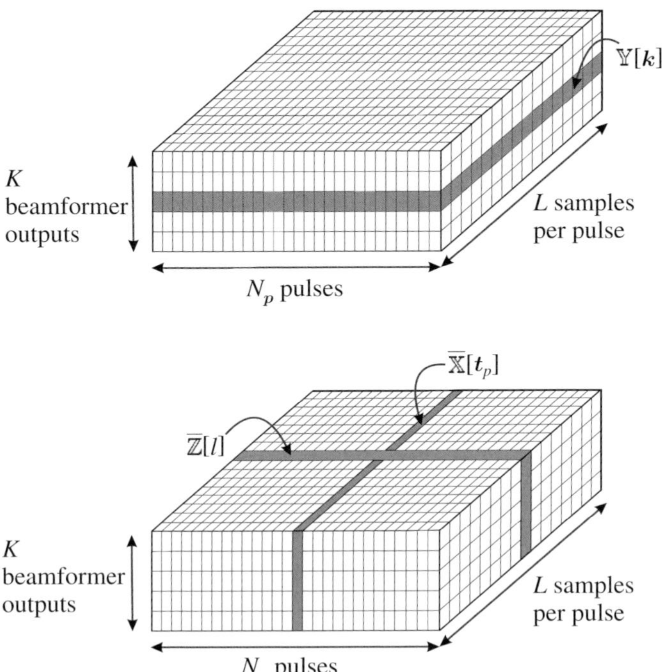

Fig. 3.10 The output 3D datacube of the K receiver beamformers after discretisation in the range direction.

the parameters corresponding to the single transmitter beamformer and K receiver beamformers are designed for optimum performance. Therefore the parameter design of a SIMO SAR system will first be given, before beamforming in both the elevation and cross-range directions is described. Finally, two SIMO SAR system examples will be presented.

3.4.1 *SIMO SAR parameter design*

The parameters of interest corresponding to the SIMO SAR system are as follows:

Beamformer height: As given by Eq. (3.15), the height of a beamformer, H_a, i.e. the dimensions along the z axis, is inversely proportional to the imaged swath-width W_g. Therefore, by decreasing the beamformer's height, the swath-width can be increased. However, as directive gain is related to the dimensions of the beamformer, the smaller the height of the beamformer, the lower the directive gain. Therefore H_a cannot be made arbitrarily small in order to increase W_g. Moreover, the resulting value of W_g needs to be taken into account due to its limitation by range ambiguities as shown by the right-hand side of the inequality in Eq. (3.5).

Beamformer length: The length of a beamformer L_a, i.e. its dimension along the x-axis, is related to the cross-range resolution of the image(s) formed from the collected raw data, where the cross-range resolution is described by

$$\Delta_{cr} \approx \frac{L_a}{2}. \tag{3.43}$$

Therefore, by decreasing L_a, higher resolution in the cross-range direction can be achieved. However, L_a cannot be made arbitrarily small, as it limits the swath-width that can be imaged unambiguously.

Number of beamformers in the z direction: By increasing the number of beamformers in the z direction from 1 to K_z, beamsteering in elevation can be achieved by steering a single beam to follow the radar returns on the ground. Due to an increased total height in the z direction, a narrower, higher gain beam is produced.

Number of beamformers in the x direction: The imaging of a wide swath is limited by ambiguities. In order to prevent these ambiguities the PRF f_r must satisfy Eq. (3.5). By increasing the number of beamformers in the x or cross-range direction from one to K_x, an optimum f_r can be reduced by a factor K_x without an increase in cross-range ambiguities. Therefore the minimum f_r that allows unambiguous imaging is decreased, allowing an increase in the possible swath-width that can be imaged unambiguously. By increasing the number of beamformers in the x direction, the cross-range resolution can also be increased due to the resulting increased number of sets of received data.

Beamformer separation: It is assumed that all N elements of a single beamformer are collocated. However, care has to be taken when choosing the separation between the K beamformers in the system because grating lobes may form due to the beamformer array geometry, resulting in ambiguities along either the elevation or cross-range directions.

By using the K beamformers as a uniform linear array (ULA) along the cross-range direction, each beamformer receives the echoes of the chirp signals transmitted every T_r. This allows f_r to be decreased further compared to a single beamformer case due to the total extra samples received by the multiple beamformers. If f_r is decreased to the extent that it does not meet the requirement given in Eq. (3.5), each beamformer will receive aliased signals. However, due to the use of K beamformers, these aliased signals can be reconstructed as long as it is no more than K times less than an f_r value that allows unaliased signals to be received. Depending on the distance between each beamformer in the cross-range direction, the sampling of the received signals in the cross-range direction, i.e. along the synthetic aperture length, will either be uniform or non-uniform.

Optimum performance of the SIMO SAR system can be obtained if there is uniform sampling of the received signal in the cross-range direction. This is achieved when there are equally spaced phase centres, which occurs when the system travels a distance equal to half the total receiver beamformer array length in T_r seconds, resulting in the condition [4]

$$f_r = \frac{2v}{L_{a,total}} \tag{3.44}$$

where $L_{a,total}$ is the total array length of the receiver beamformer. If this condition is not met, non-uniform sampling of the received signal will occur.

A summary of the above parameter design of a SIMO SAR system is presented in Fig. 3.11, showing the main parameters of interest (given in

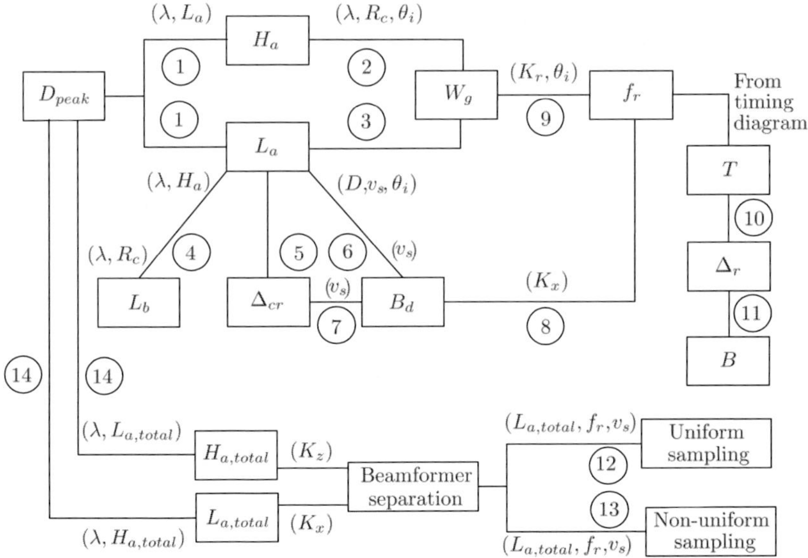

Fig. 3.11 An illustrative diagram of parameter design of a SIMO SAR system, with the numbers ①–⑭ corresponding to the rows/equations of Table 3.1.

Table 3.1   Equations related to the parameter design of a SIMO SAR system.

1	$D_{peak} = 10\log_{10}\left(\frac{4\pi L_a H_a}{\lambda^2}\right)$	dB
2	$W_g \approx \frac{0.886\lambda R_c}{H_a \cos\theta_i}$	m
3	$W_g < (1-D)\frac{cL_a}{4v_s \sin\theta_i}$	m
4	$L_b \approx 0.886\frac{\lambda R_c}{L_a}$	m
5	$\Delta_{cr} \approx \frac{L_a}{2}$	m
6	$B_d = \frac{2v_s}{L_a}$ (in terms of L_a)	Hz
7	$B_d = \frac{v_s}{\Delta_{cr}}$ (in terms of Δ_{cr})	Hz
8	$K_x f_r > B_d$ (to prevent cross-range ambiguities)	Hz
9	$K_x f_r < \frac{c}{2W_g \sin\theta_i}$ (to prevent slant-range ambiguities)	Hz
10	$\Delta_r = \frac{cT}{2\sin\theta_i}$ (before range compression)	m
11	$\Delta_r = \frac{c}{2B\sin\theta_i}$ (after range compression)	m
12	$f_r = \frac{2v_s}{L_{a,total}}$ (for uniform sampling)	Hz
13	$f_r \neq \frac{2v_s}{L_{a,total}}$ (for non-uniform sampling)	Hz
14	$D_{peak} = 10\log_{10}\left(\frac{4\pi L_{a,total} H_{a,total}}{\lambda^2}\right)$	dB

boxes). The relationships between these parameters are given by the circled
equation numbers corresponding to the rows of Table 3.1.

In a SIMO SAR system, by arranging the K beamformers in both the
elevation and cross-range direction, more possibilities can be achieved compared to the use of a single beamformer. Examples of these possibilities will
next be shown in both the elevation and cross-range directions.

3.4.2   *Beamforming in the elevation direction*

In order to describe beamforming concepts in the elevation direction it
will be assumed that all K beamformers are arranged in the elevation

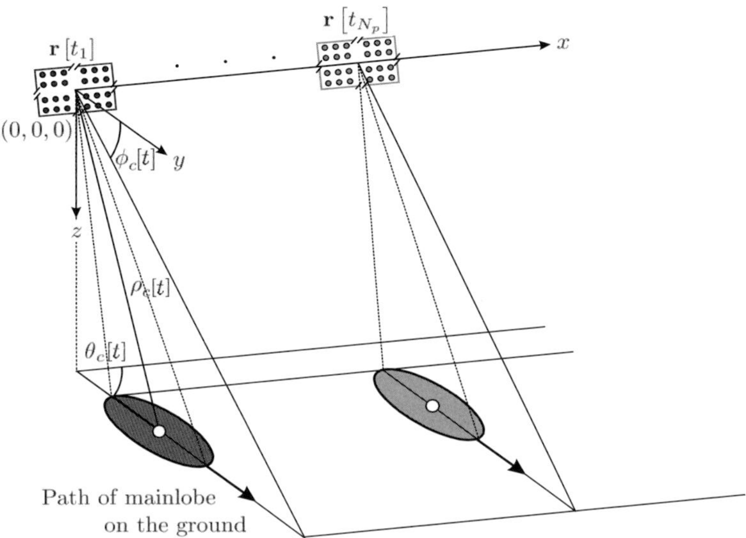

Fig. 3.12   SCORE operation.

direction, $K = K_z$, i.e.

$$\mathbf{r} = [\underline{0}_{K \times 1}, \underline{0}_{K \times 1}, \underline{r}_z]^T \tag{3.45}$$

where $\underline{r}_z$ is a $(K \times 1)$ column vector containing the z coordinates of the centre element of the K beamformers.

If K beamformers are arranged in elevation, an overall array that has a height K times larger than the height of the single beamformer is created, resulting in a narrower, higher gain beam to be formed. Beamforming can be utilised to steer this beam to "follow" the radar echoes of a transmitted chirp on the ground. A technique that uses this idea is Scan-On-Receive (SCORE), where the single beam is steered to the estimated direction of arrival of the echoes before data collection. These values are calculated from the spherical Earth model where no topographic height is assumed.

It will be assumed that Stripmap data collection is used. From Eq. (3.15), an increase in the height of a SAR system results in a smaller imaged swath-width. All the elements of the receiver beamformers are weighted such that the single narrow beam is created. With reference to Fig. 3.12, this beam follows the echoes of the transmitted signals on the ground and therefore steers from the near range to the far range.

By letting $\theta_c[t]$ and $\phi_c[t]$ be the azimuth and elevation angles between the SAR system and the centre of the mainlobe on the ground at time t, the weights of the KN elements for steering of the beam can be represented by

$$\underline{w}[t] = \text{rep}_{T_r}\left(\underline{S}\left(\theta_c[t], \phi_c[t]\right)\right) \tag{3.46}$$

for $t_p \leq t \leq t_p + \frac{2(R_f - R_n)}{c} + T$, with $p = 1, 2, \ldots, .N_p$, where $\text{rep}_{T_r}(\cdot)$ indicates a repetition every T_r seconds. Note that $\theta_c[t]$ and $\phi_c[t]$ are in terms of t not t_p.

This technique has the advantage of covering a large swath-width with a high gain. However, the performance of SCORE decreases in the presence of topographic height and surface variation, due to a difference in the expected value of $(\theta_c[t], \phi_c[t])$ before data collection and the actual values during data collection.

An extension to deal with this problem of performance degradation due to topographic height is to apply adaptive beamforming [7]. In this case, the values of $(\theta_c[t], \phi_c[t])$ of the received echoes are estimated from the received signals at the K beamformers. First, range compression is applied to the received signals using a matched filter formed from the complex conjugate of the transmitted chirp signal. Then coregistration is applied to register the signals to those received by the reference beamformer of the array of K beamformers. Afterwards $(\theta_c[t], \phi_c[t])$ can be estimated using estimation methods such as Multiple Signal Classification (MUSIC) and Capon. Then the elements of the K beamformers can be weighted such that a narrow, high gain beam is steered in the direction given by the estimated values $(\widehat{\theta}_c[t], \widehat{\phi}_c[t])$.

Beamsteering in the elevation direction can also be utilised to suppress ambiguities by forming nulls in the antenna pattern at specific angles corresponding to ambiguities. In this case the elevation angle is of interest. From the received signals and knowledge of the manifold vector, corresponding to the ambiguous elevation angles ϕ_{amb}, a $(K \times 1)$ vector of weights can be calculated to form nulls corresponding to the values of ϕ_{amb} in the beampattern. One example is Capon's MVDR beamformer, whose weights are formed from the solution to the following constrained optimisation problem

$$\underline{w}\left(\phi_{amb}, t_p\right) = \min_{\underline{w}[t_p]}\left(\underline{w}[t_p]^H \mathbb{R}_{xx}[t_p]\underline{w}[t_p]\right) \tag{3.47}$$

subject to

$$\underline{w}\left(\phi_{amb}, t_p\right)^H \underline{S}\left(\phi_{amb}, t_p\right) = 1 \tag{3.48}$$

where $\mathbb{R}_{xx}[t_p]$ is defined as the covariance matrix of the received signals at the elements of the K beamformers and $\underline{w}\left(\phi_{amb}, t_p\right)$ contains the required weights to create nulls at the ambiguous elevation angles. The solution can be found using the method of Lagrange multipliers resulting in

$$\underline{w}\left(\phi_{amb}, t_p\right) = \frac{\mathbb{R}_{xx}^{-1}[t_p]\underline{S}\left(\phi_{amb}, t_p\right)}{\underline{S}^{H}\left(\phi_{amb}, t_p\right)\mathbb{R}_{xx}^{-1}[t_p]\underline{S}\left(\phi_{amb}, t_p\right)}. \tag{3.49}$$

3.4.3 *Beamforming in the cross-range direction*

In order to describe the beamforming concepts in the cross-range direction it will be assumed that all K beamformers are arranged in the cross-range direction, $K = K_x$, i.e.

$$\mathbf{r} = \left[\underline{r}_x, \underline{0}_{K\times 1}, \underline{0}_{K\times 1}\right]^{T} \tag{3.50}$$

where $\underline{r}_x$ is a $(K \times 1)$ column vector containing the x coordinates of the centre element of each of the K beamformers.

Due to the use of K beamformers along the cross-range direction, additional samples along the synthetic aperture during data collection are obtained. It is assumed that the single transmitter beamformer transmits a chirp signal every T_r, and that the chirp signal is transmitted a total of N_p times along the flight path. Therefore, due to the use of a SIMO SAR system with a single transmitter and K receiver beamformers, where K received sets of signals for every transmitted chirp are received, there will be a total of KN_p samples in the cross-range direction compared to only N_p samples in the case where a SISO SAR system is used. The effective positions of these samples are determined by half the separation between the single transmitter beamformer and K receiver beamformers and therefore, depending on the locations of the beamformers in space, the distance between each sample will change. Three main cases can be considered. These are when:

(a) The beamformers are located sufficiently close to each other such that each receiver beamformer receives the echoes from the same transmitted chirp signal at approximately similar times. This will usually occur when the beamformers are located on the same platform. Also, the distance between the centre element of each beamformer satisfies [4]

$$\Delta_x = \frac{2v_s}{K \cdot f_r} \tag{3.51}$$

where v_s is the velocity of the SAR system, resulting in the distance between the total KN_p samples in the cross-range direction to be uniform and therefore uniform sampling occurs.

(b) The beamformers are located sufficiently close to each other, as in case (a) above, but here Eq. (3.51) is not met, thus resulting in non-uniform sampling.

(c) The beamformers are widely located forming a sparse array, resulting in all receiver beamformers receiving the echoes from the same transmitted chirp signal at significantly different times and the range histories of the imaged scatterers in the received data to be different between beamformers.

Due to these additional samples, additional beamforming compared to SISO SAR systems can be applied. This allows a wider range of application specifications to be met by allowing Eq. (3.5) to be rewritten for a SIMO SAR system as

$$B_D < K f_r < \frac{c}{2 W_g \sin \theta_i} \qquad (3.52)$$

where it can be seen that compared to a SISO SAR system

- in a SIMO SAR system, f_r can be decreased K times while still satisfying the condition and therefore allowing a wider swath-width to be imaged, and
- in a SIMO SAR system for the same value of f_r, a larger Doppler bandwidth can be used while still satisfying the condition. As a result from Eq. (3.7) it can be seen that the cross-range resolution will be improved.

If f_r is decreased such that it does not satisfy Eq. (3.5), undersampling and therefore cross-range ambiguities will occur. However, if K beamformers are used in the cross-range direction and satisfy Eq. (3.51), resulting in uniform sampling, the received signals at the K beamformers can be simply interleaved resulting in an effective higher pulse repetition frequency of $K f_r$ and therefore reducing cross-range ambiguities. However, satisfying Eq. (3.51) results in less flexibility in the locations of the K beamformers in space and specification requirements may not be met. If there is non-uniform sampling along the cross-range direction, additional processing needs to be applied. The main aim of this processing is to suppress ambiguous returns due to the undersampling and to then combine the processed signals. It is also assumed that beamforming has already been applied to the KN elements of the SAR system to form K beams.

A technique given in [8] allows the combination of these signals. As mentioned in [4], this technique is equivalent to null-steering in the case of

K collocated beamformers. In particular, it is equivalent to steering vector beamforming. As each beamformer receives a section of the Doppler bandwidth, the technique allows reconstruction of the complete Doppler bandwidth from the aliased subsampled K received signals, as well as suppression of the aliased frequency components. For this, weighting coefficients are applied to the signals at each receiver beamformer.

This can be achieved by forming a manifold vector for each return from ambiguous azimuth angles, $\underline{\theta}_{amb}$. However, in the cross-range direction it is the spectrum of the received signals that are aliased due to undersampling and therefore suppression of these ambiguities needs to be performed, which are spread across a frequency range. Doppler frequency is related to the squint angle θ_{sq}, which in turn is related to the azimuth and elevation angles θ and ϕ, allowing Doppler frequency to be given as

$$f_d = \frac{2v_s}{\lambda} \sin(\theta_{sq}) \tag{3.53a}$$

$$= \frac{2v_s}{\lambda} \cos(\theta) \cos(\phi) \tag{3.53b}$$

where the Doppler frequency is the frequency along the cross-range direction. As it is assumed that the K receiver beamformers form a linear array along the positive x-axis, the wavenumber vector $\underline{k}(\theta, \phi)$ can be reduced to a scalar and written in terms of f_d, allowing the manifold vector $\underline{S}(\theta, \phi)$ to be written in terms of Doppler frequency to give

$$\underline{S}(f_d) = \begin{cases} \exp\left(-j\frac{\pi}{v_s}\underline{r}_x[0]f_d\right) \\ \text{for plane wave propagation} \\ \rho_c^a(f_d) \odot \underline{R}^{-a}(f_d) \odot \exp\left(-j\frac{2\pi}{\lambda}\left(\rho_c(f_d)\cdot\underline{1}_K - \underline{R}(f_d)\right)\right) \\ \text{for spherical wave propagation} \end{cases}$$

$$\tag{3.54}$$

where $\underline{r}_x[0]$ is a $(K \times 1)$ vector of the x coordinates of the centre element of the K beamformers in the system at time $t_p = 0$ and ρ_c is the slant range to the centre of the beam footprint and

$$\underline{R}(f_d) = \sqrt{\rho_c^2(f_d)\cdot\underline{1}_K + \underline{r}_x^2[0] - \frac{\rho_c(f_d)\lambda}{v_s}\underline{r}_x[0]f_d}. \tag{3.55}$$

By forming a manifold vector for each Doppler frequency in the range [4]

$$f_{d,k} = \left[\left(-\frac{K}{2}+k-1\right)f_r, \left(-\frac{K}{2}+k\right)f_r\right] \tag{3.56}$$

for $k = 1, 2, \ldots, K$, where $f_{d,k}$ gives the range of Doppler frequencies within the k-th subband, the matrix $\mathbb{S}(f_d)$ is formed

$$\mathbb{S}(f_d) = [\underline{S}(f_{d,1}), \underline{S}(f_{d,2}), \ldots, \underline{S}(f_{d,K})] \tag{3.57}$$

where the k-th column of $\mathbb{S}(f_d)$, $\underline{S}(f_{d,k})$, is the manifold vector corresponding to the K beamformers for the k-th frequency range. From $\mathbb{S}(f_d)$ a weight matrix $\mathbb{W}$ can be formed from

$$\mathbb{W} = \mathbb{S}^{-1}(f_d) \tag{3.58a}$$

$$= [\underline{w}_1, \underline{w}_2, \ldots, \underline{w}_K]^T \tag{3.58b}$$

such that the vector $\underline{w}_k$ is a $(K \times 1)$ weight vector which recovers the Doppler frequencies within the Doppler frequency range $f_{d,k}$ while suppressing the frequencies within the other $K - 1$ frequency ranges. The k-th row of $\mathbb{W}$ is then applied to the received signals at the k-th receiver beamformer in the Doppler domain[8] for suppression of the ambiguous Doppler frequencies, resulting in K sets of unambiguous data for each Doppler frequency range, which are combined to form a single set of data for each Doppler frequency range. In order to obtain the full Doppler spectrum, each Doppler frequency range is concatenated, resulting in unambiguous data in the cross-range direction.

By representing the weights for all f_d, the matrix $\mathbb{W}_k$ is defined as the weights applied to recover the k-th subband for all f_d within the subband

$$\mathbb{W}_k = \begin{bmatrix} \underline{w}_k(f_{d,1}), & \underline{0}_{K \times 1} & , \cdots, & \underline{0}_{K \times 1} \\ \underline{0}_{K \times 1}, & \underline{w}_k(f_{d,2}), & \cdots, & \underline{0}_{K \times 1} \\ \vdots & \vdots & \ddots & \vdots \\ \underline{0}_{K \times 1}, & \underline{0}_{K \times 1} & , \cdots, & \underline{w}_k(f_{d,N_p}) \end{bmatrix} \tag{3.59}$$

and the reconstructed data can be given as

$$\mathbb{Y}_{recon}(f_d) = \left[\mathbb{W}_1^T \begin{bmatrix} \bar{\mathbb{X}}[f_{d,1}] \\ \bar{\mathbb{X}}[f_{d,2}] \\ \vdots \\ \bar{\mathbb{X}}[f_{d,N_p}] \end{bmatrix}, \mathbb{W}_2^T \begin{bmatrix} \bar{\mathbb{X}}[f_{d,1}] \\ \bar{\mathbb{X}}[f_{d,2}] \\ \vdots \\ \bar{\mathbb{X}}[f_{d,N_p}] \end{bmatrix}, \ldots, \mathbb{W}_K^T \begin{bmatrix} \bar{\mathbb{X}}[f_{d,1}] \\ \bar{\mathbb{X}}[f_{d,2}] \\ \vdots \\ \bar{\mathbb{X}}[f_{d,N_p}] \end{bmatrix} \right]^T. \tag{3.60}$$

So far it has been assumed that the K beamformers are collocated, however in the case when the K beamformers are sparsely located, the

---

[8] A Fourier transform on the received signals along the cross-range direction transforms the signals into the Doppler domain.

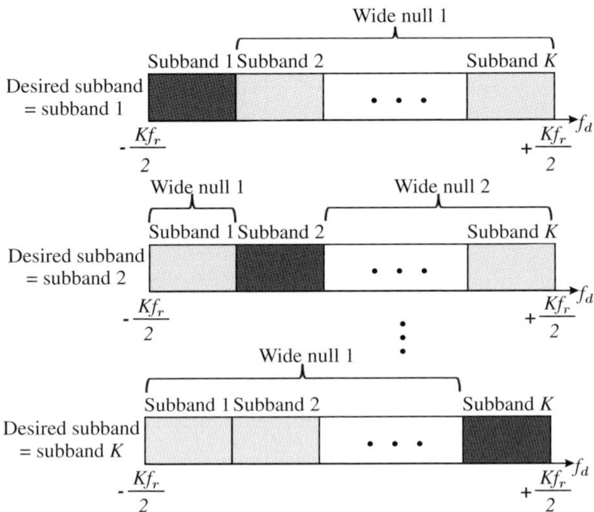

Fig. 3.13 Representation of suppression of subbands, where the "dark grey" subbands are desired.

extra phase terms of the reconstruction filters given in [4] may be required to take into account the separation between the transmitter beamformer and receiver beamformers. Also care must be taken when the K receiver beamformers are widely spaced, i.e. a sparse array is used, as the range history of the same imaged scatterer will differ between each beamformer and additional correction of the range history is required between beamformers. In addition, unambiguous reconstruction can only be achieved if the signal spectrum is within the limits of $-\frac{K}{2}f_r$ and $+\frac{K}{2}f_r$. As the signal spectrum outside these limits is not suppressed, aliasing will occur resulting in cross-range ambiguities [4].

By considering the full Doppler frequency range in the reconstructed data from $-\frac{K}{2}f_r$ to $+\frac{K}{2}f_r$, with reference to Fig. 3.13, it can be noticed that rather than suppressing the ambiguous Doppler frequencies one at a time, ranges of frequencies could be suppressed if a wide null was applied over these frequencies. In this case, at most only two wide nulls need to be formed at a time, for $K > 2$, to suppress the range of Doppler frequencies on either side of the desired subband. For $K = 2$ at most only one wide null is required.

Based on the superresolution beamformer in [10], by identifying the ambiguous Doppler frequencies from Eq. 3.56, where for the k-th receiver's data the ambiguous Doppler frequency ranges for all $f_{d,i}$ for $i \neq k$, with $k = 1, 2, \ldots, K$, wide nulls in the array pattern can be formed to suppress these range of frequencies to a level smaller than or equal to a defined threshold γ in dBs. By first considering the formation of a single wide null in the array pattern, the manifold vector $\underline{S}_{I,b}(f_d)$ is used to form the $(K \times K)$ matrix $\mathbb{Q}_I$ for the ambiguous Doppler frequency range $f_{d,I}$, calculated as

$$\mathbb{Q}_I = \int \underline{S}_{I,b}(f_d) \underline{S}_{I,b}^H(f_d) \, df_{d,I} \tag{3.61}$$

where $\underline{S}_{I,b}(f_d)$ is formed using Eq. 3.54.

By forming the complement projection of the $(K \times \tau_I)$ matrix $\mathbb{E}_I = \left[\underline{E}_{I,1}, \underline{E}_{I,2}, \ldots, \underline{E}_{I,\tau_I}\right]$, where $\underline{E}_{I,j}$ is the j-th eigenvector of $\mathbb{Q}_I$

$$\mathbb{P}_{\mathbb{E}_I}^{\perp} = \mathbb{I}_K - \left(\mathbb{E}_I \left(\mathbb{E}_I^H \mathbb{E}_I \right)^{-1} \mathbb{E}_I^H \right) \tag{3.62}$$

a $(K \times 1)$ weight vector can be formed such that a wide null is formed over the Doppler frequency range $f_{d,I}$, by performing $\mathbb{P}_{\mathbb{E}_I}^{\perp} \underline{S}_d(f_d)$, where $\underline{S}_d(f_d)$ is the manifold vector at the desired Doppler frequency. The desired Doppler frequency is assumed to be at the centre of the desired subband, and for the k-th beamformer is calculated as

$$f_{d,c} = \left(-\frac{K}{2} + k - \frac{1}{2} \right) f_r. \tag{3.63}$$

The parameter τ_I is chosen such that the condition

$$\sum_{k=\tau_I+1}^{K} \lambda_{I,k} \leq \gamma \tag{3.64}$$

is satisfied, where $\lambda_{I,k}$ is the k-th eigenvalue of $\mathbb{Q}_I$.

A $(K \times 1)$ weight vector can then be formed for the general case where M_b wide nulls are to be formed by extending the matrix $\mathbb{S}_I$ to

$$\mathbb{S}_I = \left[\mathbb{E}_{I,1}, \mathbb{E}_{I,2}, \ldots, \mathbb{E}_{I,M_b} \right] \tag{3.65}$$

on which the complement projection $\mathbb{P}_{\mathbb{S}_I}^{\perp}$ is formed to give, when normalised

$$\underline{w} = \frac{\mathbb{P}_{\mathbb{S}_I}^{\perp} \cdot \underline{S}_d}{\sqrt{\underline{S}_d^H \cdot \mathbb{P}_{\mathbb{S}_I}^{\perp} \cdot \underline{S}_d}}. \tag{3.66}$$

With reference to Fig. 3.13, M_b is either 1 or 2.

By calculating $\underline{w}$ for all K subbands and then defining $\underline{w}_k$ to be a $(K \times 1)$ vector, where the desired subband is k and wide nulls are formed in all other subbands, the reconstructed data can be given as

$$\mathbb{Y}_{recon} = [(\mathbb{I}_{N_p} \otimes \underline{w}_1)^T \mathbb{X}_{concat}, (\mathbb{I}_{N_p} \otimes \underline{w}_2)^T \mathbb{X}_{concat}, \ldots, (\mathbb{I}_{N_p} \otimes \underline{w}_K)^T \mathbb{X}_{concat}]^T \tag{3.67}$$

where

$$\mathbb{X}_{concat} = \begin{bmatrix} \bar{\mathbb{X}}[1] \\ \bar{\mathbb{X}}[2] \\ \vdots \\ \bar{\mathbb{X}}[N_p] \end{bmatrix} \tag{3.68}$$

and $\bar{\mathbb{X}}[t_p]$ is the data at the t_p-th range line in the range Doppler domain.

As an example, let $K = 2$, where there are existing SAR systems operating in such a case, e.g. TerraSAR-X in the dual receive antenna mode, as well as TerraSAR-X with TanDEM-X, and assume that there is only one imaged scatterer. Although this will not be true in practice, it allows the capabilities of both steering vector beamforming and wide null beamforming to be better seen. If f_r is decreased such that Eq. (3.5) is not met, cross-range ambiguities will occur, as shown in Fig. 3.14, which is a slant range cut of the focused image of the undersampled raw data received at a single beamformer, and it can be seen that although a single target was imaged, there are two distinct peaks, where one corresponds to the actual target response, and the other is it's ambiguous replica. The ratio of the actual target response and its ambiguous replica (peak-to-ambiguity ratio) is 15.4 dB.

By applying the steering vector beamformer on the $K = 2$ sets of undersampled data, the weights are designed to suppress the ambiguous Doppler frequencies and combine the K sets of data. The result of a log slant range cut of the reconstructed image using the steering vector beamformer is shown in Fig. 3.15. Although the ambiguous replica is still present, its response has been suppressed resulting in a peak-to-ambiguity ratio of 25.4 dB, which is an improvement from the original 15.4 dB.

When wide null beamforming is applied to the $K = 2$ sets of undersampled data, the peak-to-ambiguity ratio increases to 33.8 dB compared to when the steering vector beamformer is used, as shown in Fig. 3.16.

There is also an improvement in the 3 dB width of the target's IRF in both cases of the steering vector beamformer and wide null beamformer. From the undersampled data, the 3 dB width is 1.30 m, which is improved

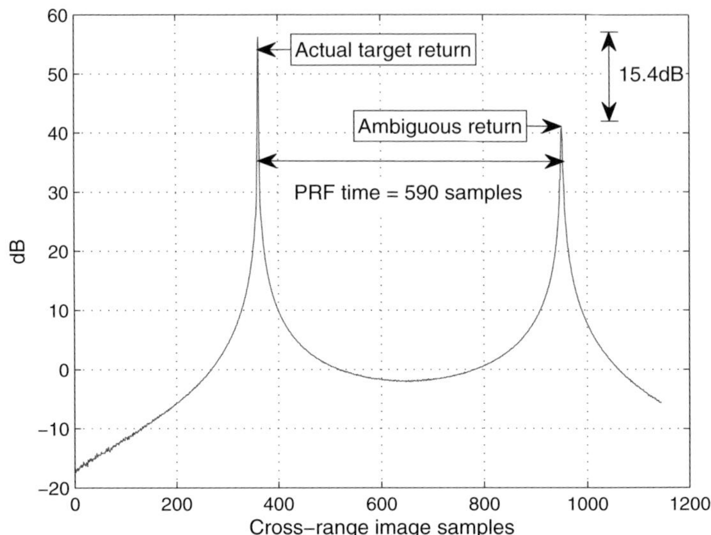

Fig. 3.14   Slant range cut across all cross-range image samples on image formed from undersampled data received by the first beamformer of an array of two.

to $0.32\,\mathrm{m}$ when the steering vector beamformer is used. When the proposed beamformer is used, the 3 dB width is $0.35\,\mathrm{m}$ and therefore slight widening occurs, however, the peak-to-ambiguity ratio is improved compared to when the steering vector beamformer is used.

As mentioned in [4], in the case when the K beamformers are sparsely located, extra phase terms may be required to take into account the separation between the transmitter beamformer and receiver beamformers. In this example extra phase terms can be added to the calculated weights as follows

$$\mathbb{W} = \begin{bmatrix} \exp\left(+j\frac{\pi r_{x,1}^2[0]}{2\lambda R_o}\right), & 0 \\ 0, & \exp\left(+j\frac{\pi r_{x,2}^2[0]}{2\lambda R_o}\right) \end{bmatrix} \mathbb{W}_K \qquad (3.69)$$

where $\mathbb{W}_K = [\underline{w}_1, \underline{w}_2]$ and R_o is defined as the slant range of the closest approach.

Beamsteering in the cross-range direction is not limited to being performed after beamforming has been applied to the KN elements of the SAR system to form K beams. It can also be performed on all the elements in order to suppress ambiguities, by forming nulls in the antenna pattern at specific corresponding angles in the cross-range direction.

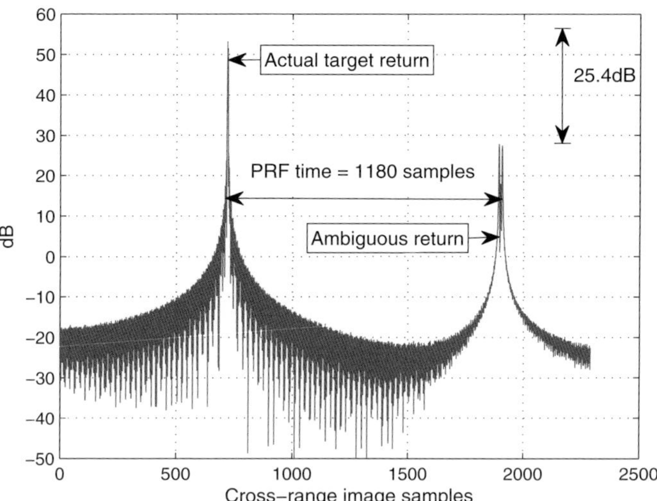

Fig. 3.15   Steering vector beamforming: slant range cut across all cross-range image samples on image after steering vector beamforming $K = 2$ sets of undersampled data.

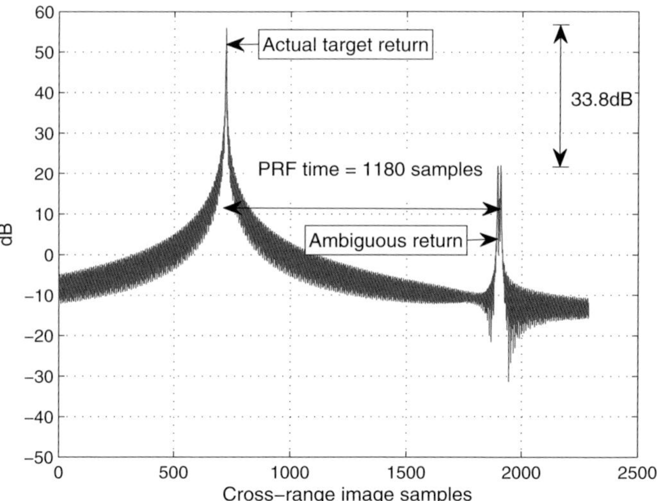

Fig. 3.16   Wide-null beamforming: slant range cut across all cross-range image samples on image after wide null beamforming $K = 2$ sets of undersampled data.

From the received signals and knowledge of the manifold vector corresponding to θ_{amb}, an $(N \times 1)$ weight vector of weights can be calculated to form nulls corresponding to these values in the beampattern of the beamformer array in the cross-range direction. By using Capon's MVDR beamformer, the required weights $\underline{w}(\theta_{amb}, t_p)$ are formed from the solution to

$$\underline{w}\left(\theta_{amb}, t_p\right) = \min_{\underline{w}[t_p]} \left(\underline{w}[t_p]^H \mathbb{R}_{xx}[t_p]\underline{w}[t_p]\right) \tag{3.70}$$

subject to $\underline{w}\left(\theta_{amb}, t_p\right)^H \underline{S}\left(\theta_{amb}, t_p\right) = 1$

where $\mathbb{R}_{xx}$ is defined as the covariance matrix of the received signals at each receiver beamformer. The solution can be found using the method of Lagrange multipliers to get

$$\underline{w}\left(\theta_{amb}, t_p\right) = \frac{\mathbb{R}_{xx}^{-1}[t_p]\underline{S}\left(\theta_{amb}, t_p\right)}{\underline{S}^H\left(\theta_{amb}, t_p\right)\mathbb{R}_{xx}^{-1}[t_p]\underline{S}\left(\theta_{amb}, t_p\right)}. \tag{3.71}$$

3.4.3.1 *SIMO SAR examples*

One example of a SIMO SAR system concept that uses beamforming is the quad-element rectangular array, which has four beamformers arranged in a two-by-two grid [11] as shown in Fig. 3.17.

Single transmitter beamformer: Three of the four beamformers are receivers only, while one beamformer is a transmitter and receiver. The

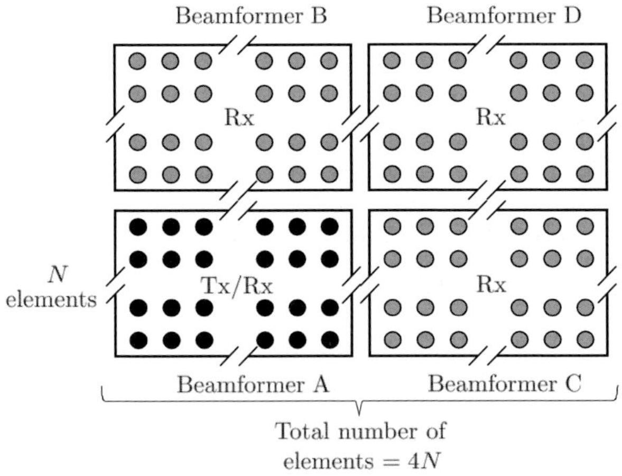

Fig. 3.17   The four beamformers of the quad-element rectangular array (adapted from [11]).

transmitter beamformer height is therefore smaller than the overall receiver beamformer array formed from all four beamformers. Due to the decrease in height, the transmitter beamformer can illuminate a wider swath and any ambiguities that arise from the decrease in f_r to illuminate the required swath are nulled during reception.

Four receiver beamformers: The four receiver beamformers forming a two-by-two-grid beamformer array are weighted such that ambiguities in both the range and cross-range directions can be nulled.

The single transmitter beamformer transmits chirp signals of length T at intervals defined by f_r to illuminate a wide swath area resulting in a need to use a decreased value of f_r. The returns are then received by four receiver beamformers in a two-by-two grid array. Any ambiguities that result due to the wide swath illumination can be removed by steering nulls in both the range and cross-range directions. This can be achieved by applying weights to each element of the four beamformers to produce trench nulls at angles related to ambiguous returns. As the ambiguities in the received signals are related to the location of each illuminated scatterer, and as the slant ranges between the system and the scatterers vary during data collection, the weights will change with each data collection point. The two beamformers in elevation allow the steering of a null in the direction of range ambiguities and can thus null one ambiguity. However, this nulling results in discontinuous imaging of the swath. In order to null ambiguities in the cross-range direction, the two beamformers along the cross-range are used to steer a null to suppress the ambiguity. These two beamformers also result in twice the number of samples along the synthetic aperture, resulting in the minimum f_r to be reduced by two for unambiguous imaging of a swath-width that is approximately twice as long, while still meeting the requirement of Eq. (3.51).

With reference to Fig. 3.17, the weights of each beamformer at a particular data collection time t_p, labelled as A, B, C and D, can be given as

$$\underline{w}_A[t_p] = \underline{1}_{N \times 1} \tag{3.72}$$

$$\underline{w}_B[t_p] = \underline{S}\left(\phi_{amb}\right) \tag{3.73}$$

$$\underline{w}_C[t_p] = \underline{S}\left(\theta_{amb}\right) \tag{3.74}$$

$$\underline{w}_D = \underline{S}\left(\phi_{amb}\right) \odot \underline{S}\left(\theta_{amb}\right) \tag{3.75}$$

where θ_{amb} and ϕ_{amb} are the azimuth and elevation angles of the ambiguities and where beamformer B controls the null in elevation and beamformer C controls the null in the cross-range direction.

When using the quad-element rectangular array, the resulting SAR image formed from the raw data at all four beamformers will have blind zones. By increasing the number of beamformers in both the elevation and cross-range directions, better suppression of ambiguities leading to the imaging of wider swaths with a high cross-range resolution can be achieved but at the cost of a larger physical array of beamformers.

Another example of a SIMO SAR system is the high-resolution wide swath (HRWS) SAR [12] which has a bistatic architecture, where the single transmitter beamformer and K receiver beamformers are either on the same platform, or on separate platforms.

Single transmitter beamformer: The transmitter beamformer height is reduced in order to illuminate a large swath, but at the expense of a decrease in the directivity gain.

K receiver beamformers: To counteract the decrease in the directivity gain of the transmitter beam, multiple receiver beamformers are arranged in the z or elevation direction, where each beamformer covers the same area as the transmitter beamformer. Therefore each receiver beamformer must have a height equal to or less than the height of the transmitter beamformer and a length equal to that of the transmitter beamformer. To allow the correct sampling of the received signals along the synthetic aperture, multiple receiver beamformers are arranged in the x or cross-range direction.

The single transmitter beamformer forms a beam to transmit a chirp pulse of length T at intervals defined by f_r to illuminate a wide swath area. All the returns are received by the KN receiver elements, which are weighted to form an array of K_z beamformers in the elevation direction and K_x beamformers in the cross-range direction. Due to the multiple receiver beamformers in elevation, the overall height of the receiver array is increased, allowing a narrow, high gain beam to be steered to "follow" the echoes of the transmitted chirp signals received from the near slant range to the far slant range using SCORE. Due to the imaging of a wide swath, f_r needs to be decreased. However, a minimum value of f_r exists for unambiguous imaging and is a constraint on the maximum swath-width that can be imaged unambiguously. By using the K_x received beamformers along the cross-range direction for data collection, more samples along the synthetic aperture are obtained, allowing the minimum value of f_r to be further decreased by a factor of K_x and therefore allowing a swath-width that is approximately K_x times longer to be unambiguously imaged. By applying beamforming techniques, the K_x sets of aliased received data at each data collection point are combined to form a single unaliased output. However, for every 100 km increase in the imaged swath-width, the required receiver beamformer array's length needs to be increased by about 10 m [13]. The HRWS SAR system uses Stripmap data collection. Therefore in order to avoid such a large increase in length, similar concepts of the HRWS SAR system can be extended to ScanSAR or TOPSAR (Terrain Observation with Progressive Scans SAR) mode, which allows the illumination of multiple subswaths during data collection.

3.5 Target parameter estimation using SIMO SAR

In this section the estimation of key parameters from SIMO SAR data is presented. In particular these parameters are the round trip delay, direction of arrival and power of imaged targets, where the term target will now be used to define a scatterer with a power above the clutter level.

3.5.1 *Round trip delay estimation*

The KN elements of a SIMO SAR system are weighted such that K receive beam outputs are created, forming the 3D datacube shown in Fig. 3.10. Thus, every $(L \times N_p)$ matrix $\mathbb{Y}[k]$, with $k = 1, 2, \ldots, K$, provides a similar image to that shown in Fig. 3.7. In order to achieve a high resolution in the slant range direction, a matched filter is applied along each range line. This results in an approximate resolution of $\frac{c}{2B \sin \theta_i}$ metres in the range direction in the range direction and focuses the data in range so that a sharp peak corresponding to the target is formed at sample number

- In collocated arrays[9]: $\left(\tau_m[t_p] - \frac{2R_n}{c}\right) \cdot F_s = \left(\frac{2R_m[t_p]}{c} - \frac{2R_n}{c}\right) \cdot F_s$
- In sparse arrays:
$$\left(\underline{\tau}_m[t_p] - \frac{2R_n}{c}\right) \cdot F_s = \left(R_{m,Tx}[t_n] \cdot \underline{1}_K + \underline{R}_m[t_p] - 2R_n \cdot \underline{1}_K\right) \cdot \frac{F_s}{c}$$

where the minimum time required for a complete echo of length T seconds to be received is $\frac{2R_n}{c}$ seconds, which corresponds to a target located in the near slant range, and $R_m[t_p]$ is the slant range between the SAR system and

---

[9] As "stop and receive" data collection and a collocated array is assumed, $R_m[t_n] = R_m[t_p]$, hence the round trip delay in samples given by $\left(\frac{R_{m,Tx}[t_n] + R_m[t_p]}{c} - \frac{2R_n}{c}\right) \cdot F_s$ can be simplified to $\left(\frac{2R_m[t_p]}{c} - \frac{2R_n}{c}\right) \cdot F_s$.

m-th target at time t_p. The parameters $R_{m,Tx}[t_n]$ and $\underline{R}_m[t_p]$ are the slant ranges between the transmitter beamformer and the m-th target at time t_n and the slant ranges between the K receiver beamformers and m-th target at time t_p respectively. In terms of samples, the first sample of a single range line will correspond to a round trip delay of $\frac{2R_n}{c}$ seconds or $\frac{2R_n}{c}F_s$ samples. Therefore the sample number relating to the sharp peak of a target is not the true round trip delay, but rather the difference between the true round trip delay and the round trip delay of a target at the near range.

By performing range matched filtering, not only is the data focused in slant range, but also the round trip delay of a particular target at a particular range line can be estimated. However, as the response of a target after matched filtering is a sinc function, the sidelobes of two or more closely located targets may interfere with each other, reducing their individual peak-to-sidelobe ratios.

Mathematically, by using the K sets of received data, a cost function for round trip delay estimation along a single range line at a data collection time t_p using the matched filter method can be given as [14]

$$\xi_\tau[t_p] = \frac{1}{KL}\left\|\left(\mathbb{I}_K \otimes \mathbb{J}^\tau \underline{m}_L^*\right)^T \underline{y}_{aug}[t_p]\right\|^2 \qquad (3.76)$$

$$\text{where}\begin{cases} L &= \text{Number of samples in a range line} \\ \underline{m}_L &= \text{Reference chirp signal of length } T \text{ seconds, sampled at } F_s \text{ and zero padded with } (L - TF_s) \text{ samples.} \end{cases}$$

with $\mathbb{J}$ being the shift matrix defined as

$$\mathbb{J} = \begin{bmatrix} \underline{0}_{L-1}^T & 0 \\ \mathbb{I}_{L-1} & \underline{0}_{L-1} \end{bmatrix} \quad (L \times L) \qquad (3.77)$$

which results in $\mathbb{J}^\tau \underline{m}_L^*$ being the complex conjugate of the reference chirp signal vector $\underline{m}_L$ shifted by τ samples. This is then applied to $\underline{y}_{aug}[t_p]$ formed by stacking the columns of the matrix $\bar{\mathbb{X}}^T[t_p]$

$$\underline{y}_{aug}[t_p] = \text{vec}\left(\bar{\mathbb{X}}^T[t_p]\right) \quad (KL \times 1) \qquad (3.78)$$

This can be seen as a convolution between the reference chirp signal and $\underline{y}_{aug}[t_p]$, which can be equivalently performed with a multiplication in frequency which is less computationally demanding to give the cost function

$$\xi_\tau[t_p] = \frac{1}{KL}\left\|\text{FT}^{-1}\left\{\text{FT}\{\mathbb{Y}_{concat}[t_p]\} \odot \text{FT}\left\{\underline{1}_K^T \otimes \underline{m}_L^*\right\}\right\}\right\|_{\text{row}}^2 \qquad (3.79)$$

where $\begin{cases} \text{FT}\{\mathbb{A}\} & = & \text{Fourier transform performed down each column of} \\ & & \text{matrix } \mathbb{A} \\ \text{FT}^{-1}\{\mathbb{A}\} & = & \text{Inverse Fourier transform performed down each} \\ & & \text{column of the matrix } \mathbb{A} \\ \|\mathbb{A}\|_{\text{row}} & = & \text{A norm applied to each row of the matrix } \mathbb{A} \end{cases}$

with $\mathbb{Y}_{concat}[t_p]$ formed by concatenating the $(L \times 1)$ vectors $\underline{x}_k[t_p]$ for $k = 1, 2, \ldots, K$.

$$\mathbb{Y}_{concat}[t_p] = [\underline{x}_1[t_p], \underline{x}_2[t_p], \ldots, \underline{x}_K[t_p]] \quad (L \times K) \tag{3.80}$$

where $\underline{x}_k[t_p]$ is the transpose of the k-th row of $\bar{\mathbb{X}}[t_p]$

In order to apply subspace partitioning, first consider the matrix $\mathbb{C}$, which contains in its columns all possible delays of the complex conjugate of the reference chirp signal and can be defined as [15]

$$\mathbb{C} = \left[\mathbb{J}^0 \underline{m}^*, \mathbb{J}^1 \underline{m}^*, \ldots, \mathbb{J}^{2T \cdot F_s - 1} \underline{m}^*\right] \quad (L \times 2TF_s) \tag{3.81}$$

By defining the matrix

$$\mathbb{P}_\tau = \mathbb{I}_K \otimes \mathbb{P}_{\mathbb{C}_\tau}^\perp \quad (KL \times KL) \tag{3.82}$$

where

$$\mathbb{P}_{\mathbb{C}_\tau}^\perp = \mathbb{I}_L - \mathbb{C}_\tau \left(\mathbb{C}_\tau^H \mathbb{C}_\tau\right)^{-1} \mathbb{C}_\tau^H \quad (L \times L) \tag{3.83}$$

with $\mathbb{C}_\tau$ being the matrix $\mathbb{C}$ with its τ-th column removed, it can be shown that multiplication between $\mathbb{P}_\tau$ and $\underline{y}_{aug}[t_p]$ will result in contributions only from imaged scatterers with a round trip delay equal to the delay shift τ. The signals from all other undesired scatterers, e.g. clutter and noise, with a round trip delay not equal to the delay shift τ will be projected onto the null subspace. By combining this concept with Eq. (3.76), the following cost function can then be defined

$$\xi_\tau[t_p] = \frac{1}{KL} \left\| \left(\mathbb{I}_K \otimes \mathbb{P}_{\mathbb{C}_\tau}^\perp \mathbb{J}^\tau \underline{m}_L^*\right)^T \underline{y}_{aug}[t_p] \right\|^2 \tag{3.84}$$

which suppresses interference returns while minimising the attenuation of the returns from desired imaged targets. Using Eq. (3.84)) the round trip delay of the M_d targets of interest, $\widehat{\underline{\tau}}[t_p] = [\tau_1[t_p], \tau_2[t_p], \ldots, \tau_{M_d}[t_p]]^T$, can be estimated. However, Eq. (3.84) can be equivalently written by a multiplication in the frequency domain as

$$\xi_\tau[t_p] = \frac{1}{KL} \left\| \text{FT}^{-1}\left\{ \text{FT}\left\{\mathbb{P}_{\mathbb{C}_\tau}^\perp \mathbb{Y}_{concat}[t_p]\right\} \odot \text{FT}\left\{\underline{1}_K^T \otimes \underline{m}_L^*\right\} \right\} \right\|_{\text{row}}^2 \tag{3.85}$$

Note that in a collocated array case, the round trip delay of a particular target will be approximately equal between all K sets of data and therefore $\mathbb{P}^{\perp}_{\mathbb{C}_\tau}$ will be equal for all K beamformers. However, this may not be the case in a sparse array case, and therefore "coregistration" of the K sets of data will need to be applied first to ensure that the range history of the same target is approximately equal between all K sets of data.

The matrix $\mathbb{C}$ has been defined as being an $(L \times 2TF_s)$ matrix such that all possible round trip delays are covered. However, depending on the SAR design parameters, conventional range matched filtering can be used to obtain an initial estimate of a range of round trip delays containing imaged targets. Therefore $\mathbb{C}$ can be reduced to $(L \times \Delta_\eta)$

$$\mathbb{C} = \left[\mathbb{J}^\eta \underline{m}^*, \mathbb{J}^{\eta+1} \underline{m}^*, \ldots, \mathbb{J}^{\Delta_\eta - 1} \underline{m}^* \right] \quad (L \times \Delta_\eta) \qquad (3.86)$$

which in turn reduces computational requirements, where η is the smallest round trip delay in the range Δ_η.

As an example of the use of subspace partitioning with range matched filtering, consider the smallest possible target separation, which corresponds to a round trip delay difference of two samples in slant range, sampled at F_s. By comparing Fig. 3.18, when only a matched filter is applied, with Fig. 3.19, when subspace partitioning is applied, it can be seen that although in both cases the target responses are distinguishable, in the case

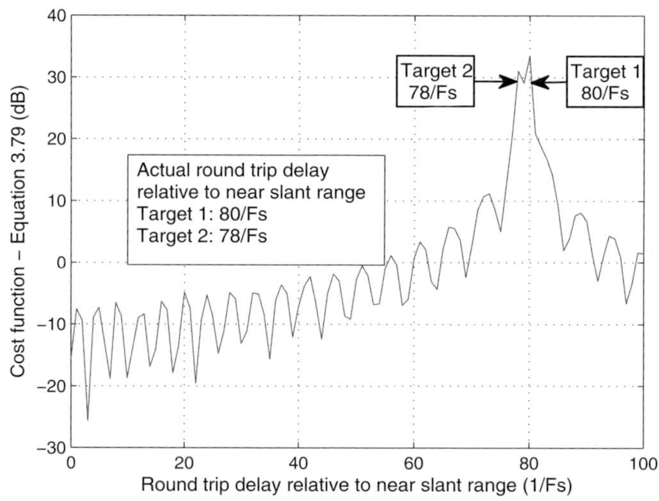

Fig. 3.18   Round trip delay estimation using matched filtering.

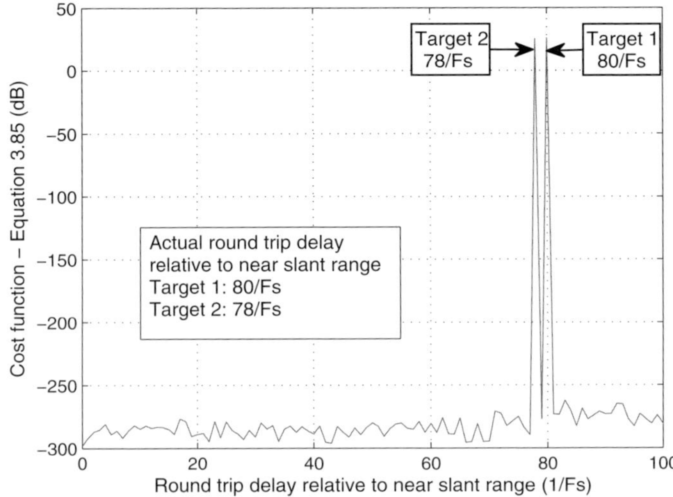

Fig. 3.19   Round trip delay estimation using subspace partitioning.

when only a matched filter is applied, the difference in dB between the
target peaks at sample number 78 and 80 with the response at sample
number 79 is only about 3 dBs (see Fig. 3.18). In the case where there is
a low SNR or if the clutter levels are high, or if a "strong, bright" target
is located nearby, the target responses at sample number 78 and 80 may
become undetectable. However, when subspace partitioning is applied, it
can be seen that the average difference in dBs between the peaks at sample
number 78 and 80 with the response at sample number 79 is about 300
dBs (see Fig. 3.19), despite the slight attenuation of the target responses.
This is more than sufficient for detection purposes. This technique can
also be applied to $K = 1$ sets of data during the range compression stage
for sidelobe suppression by appling subspace partitioning for all t_p, with
$p = 1, 2, \ldots, N_p$.

3.5.2   *Joint direction of arrival and slant range estimation*

The MUSIC algorithm is a superresolution algorithm which can be used
for signal parameter estimation, such as direction of arrival (DOA) and
frequency estimation. This is achieved by finding the intersection of the
signal subspace of the received signals $\mathcal{L}[\mathbb{E}_s]$ (i.e. the subspace spanned by
the columns of the signal eigenvector matrix $\mathbb{E}_s$) and the array manifold.

At a particular parameter of interest value $\widehat{p}$, the MUSIC algorithm can be given as [16]

$$\xi_{music}\left(\widehat{p}\right) = (\underline{S}(\widehat{p})^{H}\mathbb{E}_{n}\mathbb{E}_{n}^{H}\underline{S}(\widehat{p}))^{-1} \tag{3.87}$$

where $\mathbb{E}_{n}$ are the eigenvectors that span the noise subspace of the covariance matrix of the received signals, and for all possible values of $\widehat{p}$, $\xi_{music}(\widehat{p})$ goes towards ∞ when $\widehat{p}$ is equal to the actual value of the parameter of interest. Therefore parameter estimation can be achieved. However, estimation accuracy decreases in the coherent signal case when the received signals are fully correlated. To overcome this a spatial smoothing is applied, which in effect performs decorrelation of the received signals [17].

In SAR applications MUSIC can be exploited to improve target resolution, i.e. target separation within a resolution cell of dimension $\Delta_{r} \times \Delta_{cr}$ (see Fig. 3.1), due to its superresolution property in both the range and cross-range directions and also for the parameter estimation of imaged targets of interest. However, there is one main factor that needs to be taken into account: MUSIC assumes that the noise is "white" but in SAR applications this "noise" also includes "clutter". Therefore for cases where the clutter consists of diffused scatterers, for example in terrain applications, MUSIC is not directly suitable without using a QR decomposition-preprocessor if the aim is SAR imaging. Also the absolute value of $\xi_{music}\left(\widehat{p}\right)$ for all $\widehat{p}$ does not correspond to the backscattering power of the imaged area [18]. However, this does not affect detection or estimation algorithms as only the responses of targets of interest are required and therefore MUSIC's superresolution and estimation properties are of particular use.

In the case where plane wave propagation occurs, the 2D MUSIC algorithm cost function for joint azimuth and elevation estimation along a particular range line is given as

$$\xi\left(\theta_{m}, \phi_{m}, t_{p}\right) = \frac{1}{\underline{S}^{H}\left(\theta_{m}, \phi_{m}\right)\mathbb{E}_{n}\left[t_{p}\right]\mathbb{E}_{n}^{H}\left[t_{p}\right]\underline{S}\left(\theta_{m}, \phi_{m}\right)} \tag{3.88}$$

where $\underline{S}(\theta_{m}, \phi_{m}) \triangleq \underline{S}(\theta_{m}, \phi_{m}, t_{p})$ is the plane wave manifold vector and $\mathbb{E}_{n}[t_{p}]$ is the matrix with columns the noise eigenvectors of the covariance matrix of the data matrix $\bar{\mathbb{X}}[t_{p}]$. There are many methods for estimating the covariance matrix of $\bar{\mathbb{X}}[t_{p}]$, including forward averaging and forward-backward averaging. However, as reported in [20], the forward averaging method yields poor results with SAR data. By additionally performing backward averaging a higher performance can be obtained. Also forward-backward averaging yields better results when the DOA differences are small compared to when only forward averaging is applied.

As range lines of data at a particular t_p are used, the reference for DOA estimation is the location of the centre element of the reference beamformer at time t_p. Therefore the following can be calculated

	Parameters with respect to $(0,0,0)$	Parameters with respect to $r_x[t_p]$
Elevation	$\phi_m = \sin^{-1}\left(\frac{h}{\rho_m}\right)$	$\phi_m[t_p] = \sin^{-1}\left(\frac{h}{R_m[t_p]}\right)$
Azimuth	$\theta_m = \cos^{-1}\left(\frac{r_{m,x}}{\rho_m \cos(\phi_m)}\right)$	$\theta_m[t_p] = \cos^{-1}\left(\frac{r_{m,x}-r_x[t_p]}{R_m[t_p]\cos\left(\phi_m[t_p]\right)}\right)$

with ρ_m being the slant range of the m-th target with respect[10] to $(0,0,0)$ and $R_m[t_p]$ being the slant range of the m-th target with respect to $r_x[t_p]$ calculated as

$$R_m[t_p] = \sqrt{\rho_m^2 + r_x^2[t_p] + r_y^2[t_p] + r_z^2[t_p] - \frac{\rho_m c}{\pi F_c}\underline{r}^T[t_p]\underline{k}\left(\theta_m,\phi_m\right)} \qquad (3.89)$$

$$\text{where} \begin{cases} \underline{r} & = & [r_x,r_y,r_z]^T \\ & = & \text{Cartesian coordinates of the centre element of the} \\ & & \text{reference beamformer, assumed to be the Tx/Rx} \\ & & \text{beamformer of the system} \\ r_{m,x} & = & x \text{ coordinate of the } m\text{-th target.} \end{cases}$$

However depending on the chosen range line, the estimated parameters may differ.

As an example with $K = 4$ arranged as a (2×2) grid array, it will be assumed that there is only one imaged target, although this will not be true in practice, and the matrix $\mathbb{X}[t_p]$, with p chosen to be where the maximum energy of the target lies, is used for the $(\theta_m[t_p], \phi_m[t_p])$ estimation of the target, and therefore the expected azimuth angle will be approximately $90°$.

With reference to Figs 3.20 and 3.21 it can be seen that this is indeed the case, however there is a $0.2°$ difference between the true and estimated elevation angles. This discrepancy is more evident when the range line index p is chosen to be $\frac{N_p}{2}$ rounded to the nearest integer, i.e. away from the target's cross-range location in samples in the signal space. This is shown

---

[10]If the slant range at the zero Doppler shift of a particular target is to be used, i.e. the slant range of closest approach, $R_{o,m}$, it can be seen that ρ_m is related to $R_{o,m}$ by
$$\rho_{,m} = \sqrt{R_{o,m}^2 + r_{m,x}^2}.$$

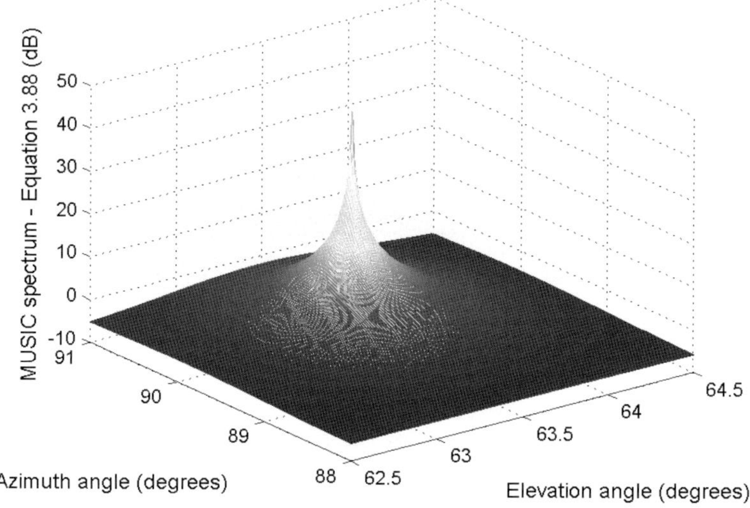

Fig. 3.20 Joint azimuth and elevation angle estimation surface plot, with p chosen to be where the maximum energy of the scatterer lies.

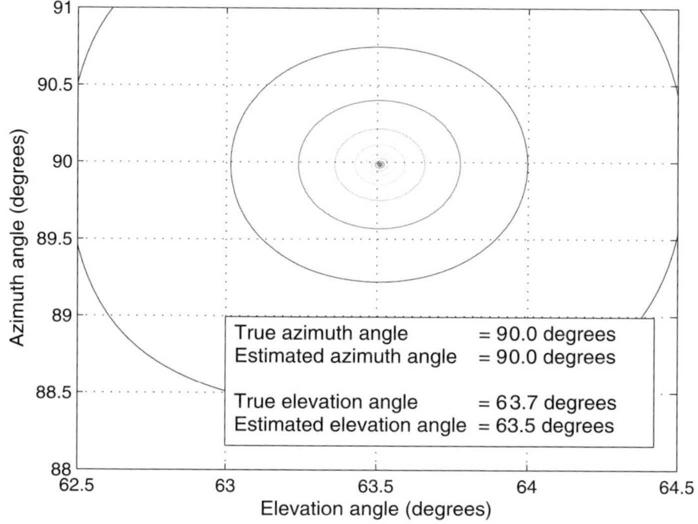

Fig. 3.21 Contour map of Fig. 3.20.

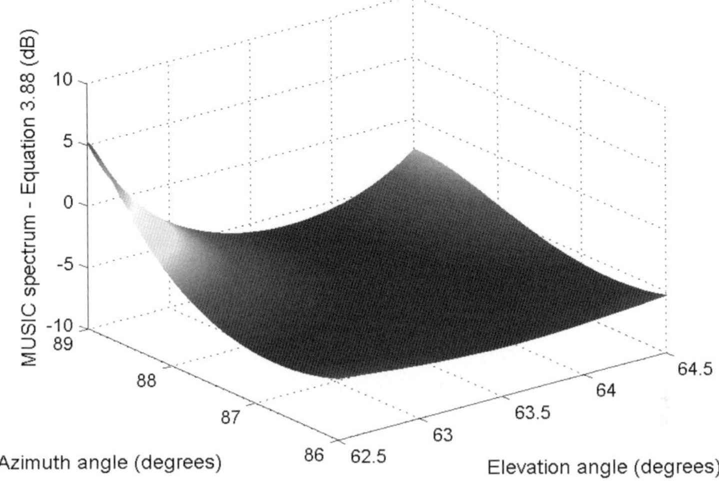

Fig. 3.22   Joint azimuth and elevation angle estimation surface plot at range line index p chosen to be $\frac{N_p}{2}$ rounded to the nearest integer.

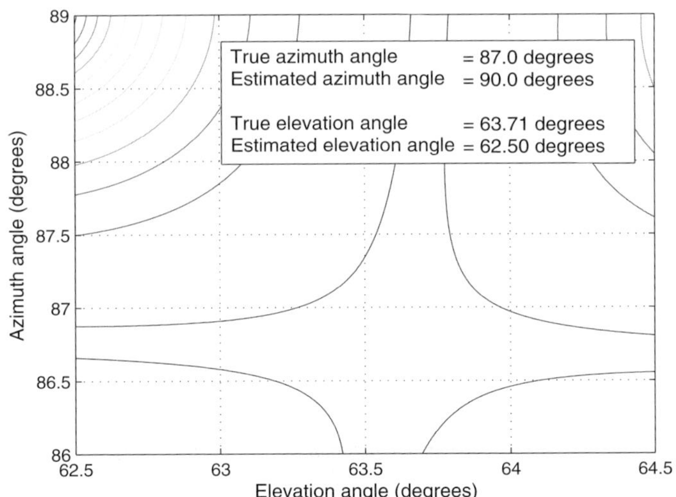

Fig. 3.23   Contour map of Fig. 3.22.

in Figs. 3.22 and 3.23, where it can clearly be seen that estimation of the azimuth and elevation angles has failed.

As only a (2×2) grid array of four beamformers has been used, one method to increase the accuracy of the estimation would be to increase the array size. However, platform space is often limited and a large array of beamformers may not always be feasible in real world applications. Therefore one method is to make use of the way SAR collects data. In total there are KN_p range lines which could be used, but so far only K of these have been used for parameter estimation. Therefore groups of adjacent range lines could be utilised, thus increasing the signal space from $(K \times L)$ given by the matrix $\bar{\mathbb{X}}[t_p]$ to $(K \cdot N_{rl} \times L)$ given by $[\bar{\mathbb{X}}[t_{p-0.5(N_{rl}-1)}]^T, \dots, \bar{\mathbb{X}}[t_p], \dots, \bar{\mathbb{X}}[t_{p+0.5(N_{rl}-1)}]^T]^T$ in the case where N_{rl} is odd and is defined as the number of range lines within a single group and will be defined as the range line block size.

By letting $N_{rl} > 1$, and performing joint azimuth and elevation estimation, it can be seen that more accurate estimates are obtained in the case when p is equal to $\frac{N_p}{2}$ rounded to the nearest integer, compared to when $N_{rl} = 1$. This is shown in Fig. 3.24 with $N_{rl} = 21$ compared to Fig. 3.23 with $N_{rl} = 1$.

However, there is a trade-off between the accuracy of the estimates and N_{rl} as illustrated in Figs. 3.25 and 3.26 for this example, where it can be seen that there is an optimum range of values N_{rl} can take.

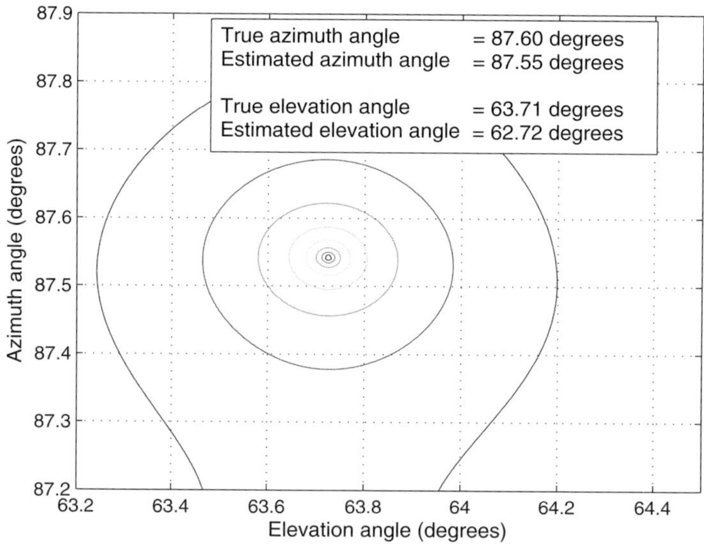

Fig. 3.24   Joint azimuth and elevation angle estimation contour map using $N_{rl} = 21$ and with $p = \frac{N_p}{2}$ rounded to the nearest integer.

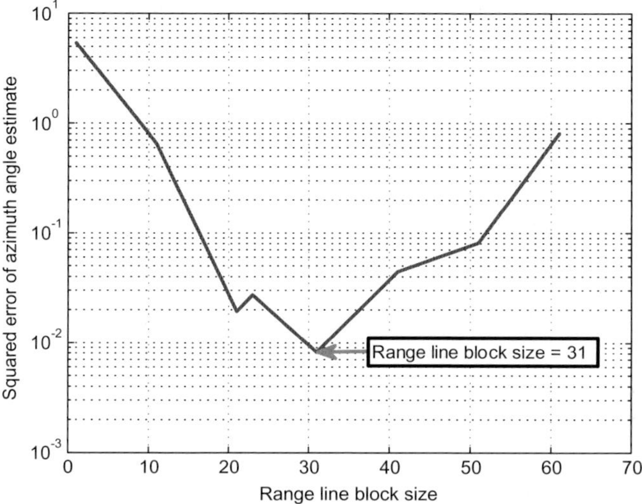

Fig. 3.25 Squared error of *azimuth* angle estimates with changes in range line block size.

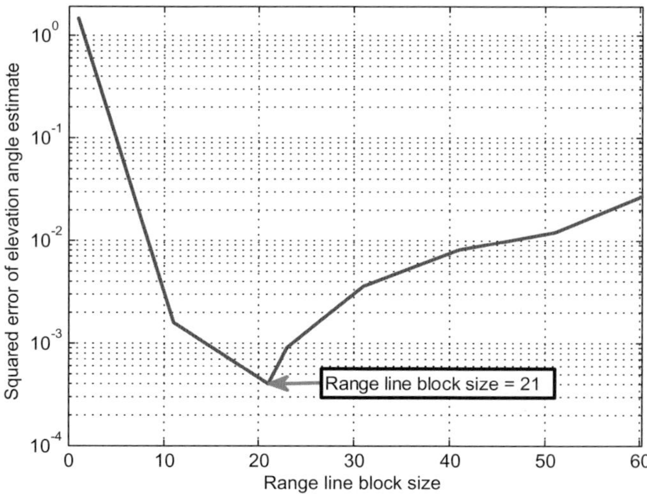

Fig. 3.26 Squared error of *elevation* angle estimates with changes in range line block size.

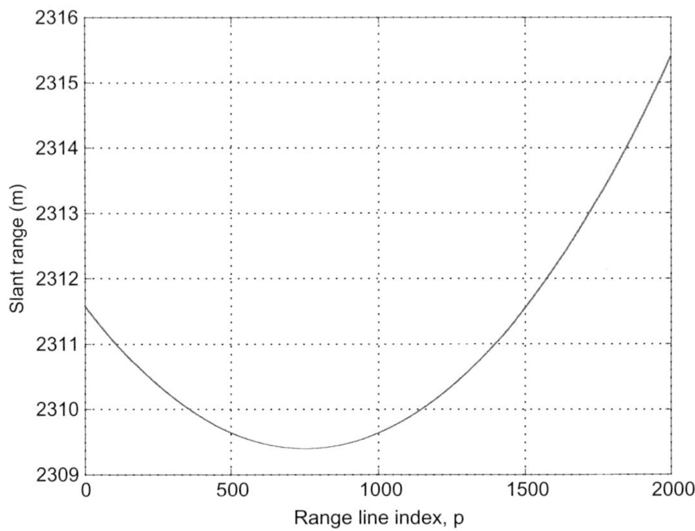

Fig. 3.27   Range history variation of a single target with range line index p for $p = 1, 2, \ldots, N_p$.

Small values of N_{rl} correspond to a small number of beamformers in the cross-range direction which may decrease the accuracy of the estimates. In terms of large values of N_{rl}, the range history of a single target varies over t_p for $p = 1, 2, \ldots, N_p$, as shown Fig. 3.27 where the range history of a single target in a collocated SAR system is given, therefore when N_{rl} is large there are significant range cell migration effects within a block of range lines, which in turn affects the accuracy of the estimates.

3.5.3   *Joint direction of arrival and power estimation*

The MUSIC provides a joint azimuth and elevation angle estimation but gives no information about the target power. Here an approach is presented that jointly estimates the direction and power of the target. Without any loss of generality consider the case where a linear array of K beamformers is used. By using the squint angle of the m-th target $\theta_{sq,m}$ with

$$\sin \theta_{sq,m} = \cos \theta_m \cos \phi_m \tag{3.90}$$

rather than azimuth θ_m and elevation ϕ_m angles, the target power P_s and DOA estimation applied to range lines of data can be jointly estimated using the following cost function inspired from [10]

$$\xi\left(P_s, \theta_{sq,m}, t_p\right) = \sum_{\substack{k=1 \\ \mathrm{eig}_k > 0}}^{K} \left(1 + \mathrm{eig}_k\left(\mathbb{R}\left(P_s, \theta_{sq,m}, t_p\right)\right)\right)$$

$$+ \sum_{\substack{k=1 \\ \mathrm{eig}_k < 0}}^{K} \left|\mathrm{eig}_k\left(\mathbb{R}\left(P_s, \theta_{sq,m}, t_p\right)\right)\right| \tag{3.91}$$

where

$$\mathbb{R}\left(P_s, \theta_{sq,m}, t_p\right) = \mathbb{R}_{yy} - \left(\frac{1}{K - M_d} \sum_{i=M_d+1}^{K} \mathrm{eig}_i\left(\mathbb{R}_{yy}\right)\right) \cdot \mathbb{I}_K \tag{3.92}$$

$$- P_s \underline{S}\left(\theta_{sq,m}, t_p\right) \underline{S}^H\left(\theta_{sq,m}, t_p\right)$$

with $\underline{S}(\theta_{sq,m}, t_p)$ representing the manifold vector of a linear array $\underline{r}_x[t_p]$ along the x-axis. More specifically, the array $\underline{r}_x[t_p]$ is now defined as the x coordinates of the centre element of the K receiver beamformers. Note that, in Eq. (3.91) M_d is the number of targets whose parameters are to be estimated.

As an example, consider a single illuminated target. Although this may not be true in practice, it allows the capabilities of the algorithm to be better seen. By performing joint squint angle and power estimation, from

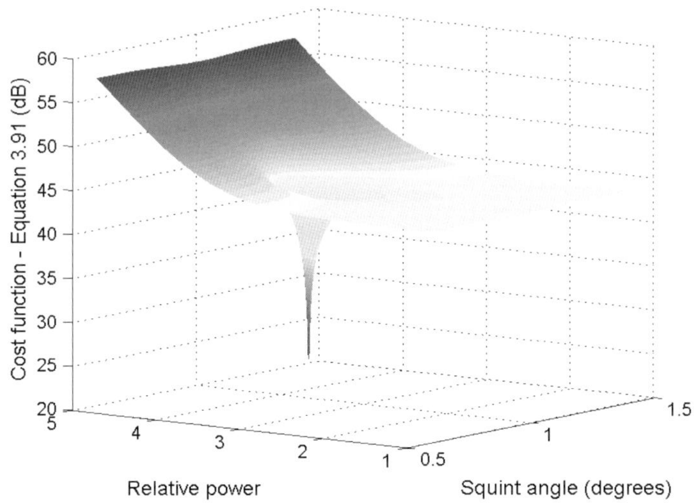

Fig. 3.28   Joint squint angle and relative power estimation using range lines of data.

Table 3.2 True and estimated parameter values (see Fig. 3.28) of imaged target.

True parameter values		
$\phi_m[t_p]$	$(^o)$	68.42
$\theta_m[t_p]$	$(^o)$	87.08
$\theta_{sq}[t_p]$	$(^o)$	1.07
P_s		3.69
Estimated parameter values		
$\theta_{sq}[t_p]$	$(^o)$	1.04
P_s		3.69

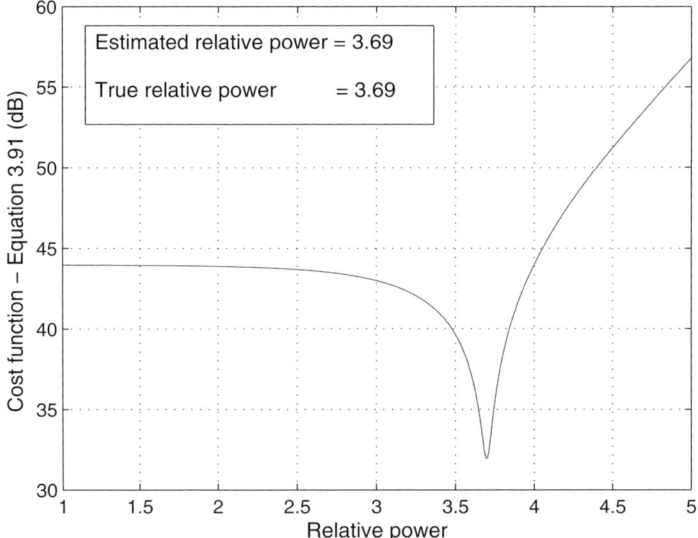

Fig. 3.29 Relative target power estimation using range lines.

Fig. 3.28 and with reference to Table 3.2, it can be seen that joint estimation has been achieved.

Joint azimuth and elevation angle estimation can also be firstly performed using 2D MUSIC and then by sustituting the estimates into Eq. 3.91, a 1D search over all values of P_s allows the estimation of the target's power along a particular range line, as shown in Fig. 3.29.

3.6 Summary and conclusions

In this chapter, the mathematical model of the signals in a general SISO SAR system in Stripmap, Spotlight and ScanSAR operational modes using a single beamformer assumed to be both a transmitter and receiver was given, indicating a trade-off between the width of the imaged swath and the cross-range resolution. Then the mathematical model was extended to SIMO SAR systems, where a single transmitter beamformer and K receiver beamformers were assumed, and examples from the literature of beamforming techniques for SIMO SAR systems were given.

In SIMO SAR systems, due to the additional samples along the synthetic aperture compared to SISO SAR systems, digital beamforming in the cross-range direction can be utilised to overcome the trade-off between wide swath and resolution. In particular, by reducing f_r by K, a wider swath can be imaged while K ambiguous sets of signals are collected by K receiver-beamformers in the cross-range direction due to undersampling. Beamforming allows the combination of these signals to reconstruct the full Doppler bandwidth signal by performing ambiguity suppression. In the case where the received signals are unambiguous, a higher cross-range resolution can be achieved compared to SISO SAR systems due to the increase in the number of phase centres in the cross-range direction. Then target parameter estimation was presented. In particular the use of subspace partitioning with the traditional matched filter using range lines of data was given for round trip delay estimation, as well as a 2D MUSIC algorithm for joint azimuth and elevation angle estimation. Then a joint DOA and target power algorithm was given.

The operational modes that were described for the case of SISO SAR systems can be applied to SIMO SAR systems for more capabilities and in order to meet different application specifications. Examples include the use of multi-channel ScanSAR [22], as well as for Spotlight data collection [23] with multiple SAR beamformers. In particular, the use of ScanSAR along with multiple SAR beamformers allow the illumination of wider swaths at a higher cross-range resolution compared to SIMO SAR systems in Stripmap mode, where in ultrawide swath cases the use of multiple beamformers using Stripmap data collection is not feasible, as the dimensions of the required SAR system would be impractical to build [24].

References

[1] I. Cumming and F. Wong, *Digital Signal Processing of Synthetic Aperture Radar Data: Algorithms and Implementation*, ser. Artech House Remote Sensing library.   Artech House, 2005.

[2] A. Freeman, "On ambiguities in SAR design," in *6th European Conference on Synthetic Aperture Radar*, 2006.

[3] A. Currie and M. A. Brown, "Wide-swath SAR," *IEE Proceedings Part F, Radar and Signal Processing*, vol. 139, no. 2, pp. 122–135, Apr. 1992.

[4] N. Gerbert, G. Krieger, and A. Moreira, "Digital beamforming on receive: Techniques and optimization strategies for high-resolution wide-swath SAR imaging," *IEEE Transactions on Aerospace and Electronic Systems*, vol. 45, no. 2, pp. 564–592, Apr. 2009.

[5] G. Franceschetti and R. Lanari, *Synthetic Aperture Radar Processing*.   CRC Press, 1999.

[6] W. G. Carrara, R. M. Majewski, and R. S. Goodman, *Spotlight Synthetic Aperture Radar: Signal Processing Algorithms*, ser. Artech House remote sensing library.   Artech House, 1995.

[7] F. Bordoni, M. Younis, E. Varona, and G. Krieger, "Adaptive scan-on-receive based on spatial spectral estimation for high-resolution, wide-swath synthetic aperture radar," in *IEEE International Geoscience and Remote Sensing Symposium*, vol. 1, Jul. 2009, pp. I-64–I-67.

[8] G. Krieger, N. Gebert, and A. Moreira, "Unambiguous SAR signal reconstruction from nonuniform displaced phase center sampling," *IEEE Geoscience and Remote Sensing Letters*, vol. 1, no. 4, pp. 260–264, Oct. 2004.

[9] J. L. Brown, "Multi-channel sampling of low-pass signals," *IEEE Transactions on Circuits and Systems*, vol. 28, no. 2, pp. 101–106, Feb. 1981.

[10] P. Karaminas and A. Manikas, "Super-resolution broad null beamforming for cochannel interference cancellation in mobile radio networks," *IEEE Transactions on Vehicular Technology*, vol. 49, no. 3, pp. 689–697, May 2000.

[11] G. D. Callaghan and I. D. Longstaff, "Wide-swath space-borne SAR using a quad-element array," *IEE Proceedings - Radar, Sonar and Navigation*, vol. 146, no. 3, pp. 159–165, Jun. 1999.

[12] M. Suess, B. Grafmueller, and R. Zahn, "A novel high resolution, wide swath SAR system," in *International Geoscience and Remote Sensing Symposium*, vol. 3, Jul. 2001, pp. 1013–1015.

[13] G. Krieger, M. Younis, S. Huber, F. Bordoni, A. Patyuchenko, J. Kim, P. Laskowski, M. Villano, T. Rommel, P. Lopez-Dekker, and A. Moreira, "Digital beamforming and MIMO SAR: review and new concepts," in *9th European Conference on Synthetic Aperture Radar*, Apr. 2012, pp. 11–14.

[14] See Chapter 4 of this book.

[15] M. Sethi and A. Manikas, "Code reuse DS-CDMA - a space time approach," in *International Conference on Acoustics, Speech, and Signal Processing*, vol. 3, May 2002, pp. 2297–2300.

[16] A. Manikas, *Differential Geometry in Array Processing*.   Imperial College Press, 2004.

[17] T. J. Shan, M. Wax, and T. Kailath, "On spatial smoothing for direction of arrival estimation of coherent signals," *IEEE Transactions on Acoustics, Speech and Signal Processing*, vol. 33, no. 4, pp. 806–811, Aug. 1985.

[18] P. Thompson, M. Nannini, and R. Scheiber, "Target separation in SAR image with the MUSIC algorithm," in *IEEE International Geoscience and Remote Sensing Symposium*, Jul. 2007, pp. 468–471.

[19] J. Odendaal, E. Barnard, and C. W. I. Pistorius, "Two-dimensional super-resolution radar imaging using the MUSIC algorithm," *IEEE Transactions on Antennas and Propagation*, vol. 42, no. 10, pp. 1386–1391, Oct. 1994.

[20] S. DeGraaf, "SAR imaging via modern 2-D spectral estimation methods," *IEEE Transactions on Image Processing*, vol. 7, no. 5, pp. 729–761, May 1998.

[21] P. Stoica and N. Arye, "MUSIC, maximum likelihood, and Cramer-Rao bound," *IEEE Transactions on Acoustics, Speech and Signal Processing*, vol. 37, no. 5, pp. 720–741, May 1989.

[22] Y. Wei, L. ChunSheng, W. Pengbo, and C. Jie, "Performance analysis and data processing of space-borne multi-channel ScanSAR mode for high-resolution wide-swath," in *IET International Radar Conference*, Apr. 2009, pp. 1–4.

[23] Q. Wu, M. Xing, H. Shi, and Z. Bao, "High azimuth resolution wide swath imaging based on the intrapulse spotlight SAR," in *2nd Asian-Pacific Conference on Synthetic Aperture Radar*, Dec. 2009, pp. 421–425.

[24] N. Gebert, G. Krieger, and A. Moreira, "Multichannel azimuth processing in ScanSAR and TOPS mode operation," *IEEE Transactions on Geoscience and Remote Sensing*, vol. 48, no. 7, pp. 2994–3008, Jul. 2010.

Chapter 4

Arrayed MIMO Radar: Multi-target Parameter Estimation for Beamforming

Harry Commin, Kai Luo and Athanassios Manikas

Communications and Array Processing,
Department of Electrical and Electronic Engineering,
Imperial College London

Arrayed multiple-input multiple-output (MIMO) radar [1,2] offers enhanced resolution and improved estimation accuracy compared to conventional radar. Consequently, significant research attention has been drawn to the field of MIMO radar. An arrayed MIMO radar is a radar system which employs two antenna arrays: one to transmit and one to receive. For each of these arrays, the antenna elements are distributed in three-dimensional real space about a common reference point. This chapter is concerned with the task of exploiting the arrays' geometries (antenna locations) in both the transmitter and receiver in order to detect, resolve and estimate the various parameters of multiple radar targets. Thus, amongst various conventional and nonconventional MIMO radar approaches presented in this chapter, an alternative representation of the arrayed MIMO radar system — namely, its equivalent virtual SIMO (single-input-multiple-output) configuration — is employed, allowing direct exploitation and analysis of the full MIMO system geometry. In particular, while the virtual array concept is heavily focused on the employment of specific algorithms (most commonly relying on the use of a bank of matched filters at the front end of the receiver) and assumes the absence of relative path delays or Doppler frequency, the formulation presented in this chapter makes no such assumptions about any specific processing applied at the receiver. Furthermore, the spatiotemporal

119

receiver architecture presented in this chapter exploits the superior performance of a unique and fresh subspace-based Doppler, delay and direction of arrival (DOA) estimation system.

4.1 Introduction

Depending on the MIMO antenna configuration, research in MIMO radar may be divided in two categories:

(1) MIMO configurations with widely separated Tx and Rx antennas;
(2) MIMO configurations with closely spaced Tx and Rx antennas.

The main objective of the first category, having widely separated transmit antennas [3], is the mitigation of target scintillations (fluctuations in amplitude). Both experimental and theoretical results demonstrate that even very small changes in the range or orientation of the target can result in a large increase or decrease in the amount of energy reflected from the target [4, 5]. When a large decrease occurs, the performance of the radar system degrades severely. However, since targets are viewed as complex bodies composed of rich scatterers radiating signal energy towards the radar's receiver antennas from multiple aspect angles, the spatial diversity of their radar cross sections (RCSs) can be exploited. This category of MIMO radar is variously referred to as "statistical" MIMO radar, or MIMO radar with widely separated antennas.

In the second category of MIMO radar, the transmit antennas are sufficiently close together in space so that each of them views the same aspect of the target. Therefore, an array of closely spaced receive antennas will receive echoes from the same target aspect, allowing a simplified point target model to be adopted. If the closely spaced transmit antennas and the closely spaced receive antennas are widely separated in space, then the system is referred to as a "bistatic" MIMO radar. Conversely, closely spaced transmit and receive antennas (for which the direction of departure (DOD) is equal to the direction of arrival (DOA)) form a "monostatic", "coherent" or "collocated" MIMO radar [6]. If the sets of transmitter (Tx) antennas and the set of receives (Rx) antennas operate as two antenna array systems then the system is defined as an arrayed MIMO radar. The main focus of the chapter is on the collocated arrayed MIMO radar.

Collocated arrayed MIMO radar is concerned with the task of exploiting the arrays' geometries (antenna locations) using array signal processing in order to *detect, resolve* and *estimate* the desired parameters of multiple radar targets. The parameters of interest can be numerous, but commonly include DOA, radial velocity, range and RCS (echo strength). In this context, *detection* performance is defined as the capability of the system to detect the presence of K targets or, equivalently, to correctly estimate the number of targets K. Then, *resolution* performance refers to the system's ability to subsequently resolve all the parameters of the K detected targets even if two or more targets are closely located in space. *Estimation* performance is then determined by the accuracy of those parameter estimates, following successful detection and resolution.

In general, the directional detection, resolution and estimation performance of an array system is a function of the array aperture, the number of sensors and array geometry. In practice, these resources are limited and so the purpose of a "superresolution" algorithm is to achieve high performance without increasing the size of the array (aperture or number of antennas), or changing the array geometry. In particular, "superresolution" refers to the algorithm's ability to achieve asymptotically infinite resolving capability for a given array geometry as the number of data snapshots L tends to infinity. Superresolution techniques have therefore been an important research topic in array signal processing for several decades [7].

In this chapter, a number of multi-target parameter estimation algorithms are presented based on two different parametric frameworks for achieving superresolution. The first is an *optimisation*-type based on a *space*-only framework which employs the array manifold vectors of both Rx and Tx of the MIMO radar. The second is a new "subspace-based" spatiotemporal framework which employs Doppler spatiotemporal array manifolds of a virtual single-input multiple-output (SIMO) system that is equivalent to the MIMO system under consideration.

Before introducing these two different frameworks, the model of the received signal (upon which the algorithms will be applied) will be described.

4.2 Arrayed MIMO radar received signal model

Consider a narrowband MIMO radar system with an array of N_{Tx} antennas at the Tx and an array of N_{Rx} antennas at the Rx. Assume that both the Tx and Rx arrays have a common reference point and are located sufficiently close together in space (compared with the distance between the common reference point and targets in the far field) such that target bearings can be considered to be the same in both the Tx and Rx arrays (i.e. monostatic MIMO radar).

A set of N_{Tx} independent known symbol-sequences $\{\underline{a}[n] \in \mathcal{C}^{N_{Tx}}; \forall n\}$, are transmitted in parallel using the Tx array of N_{Tx} antennas, with each sequence having $\mathcal{N}_s$ symbols[1] and with a symbol period T_{cs}. All the known symbols for transmission may be represented by the following matrix:

$$\mathbb{M} \triangleq [\underline{a}[1], \underline{a}[2], \dots, \underline{a}[\mathcal{N}_s]]. \tag{4.1}$$

Each vector-symbol (column of $\mathbb{M}$) is then spread using the sequence $\{\alpha[1], \alpha[2], \dots, \alpha[\mathcal{N}_c], 0, 0, \dots, 0\}$ which is an m-sequence of length (number of chips) $\mathcal{N}_c$, padded with $\mathcal{N}_c$ zeros. This m-sequence is denoted by the following $(2\mathcal{N}_c \times 1)$ vector of chips

$$\begin{aligned}
\underline{c} &\triangleq [c[1], c[2], \dots, c[2\mathcal{N}_c]]^T \\
&= [\alpha[1], \alpha[2], \dots, \alpha[\mathcal{N}_c], \underline{0}_{\mathcal{N}_c}^T]^T.
\end{aligned} \tag{4.2}$$

If the chip period is T_c, then the period of the spreading sequence $\{\underline{c}^T\}$ is $2\mathcal{N}_c T_c = T_{cs}$. This implies that the symbol $\underline{a}[n]$ after spreading with the sequence $\{\underline{c}^T\}$ will provide the matrix $\underline{a}[n]\underline{c}^T$. Furthermore, for the $(n \times 2\mathcal{N}_c + k)$-th chip interval, the transmitted vector at point A in Fig. 4.1 is $\underline{a}[n]c[k]$. Thus, the entire $(N_{Tx} \times 2\mathcal{N}_c\mathcal{N}_s)$ matrix with columns the transmitted vectors at every chip period may be written in a convenient format as:

$$[\underline{a}[1]\underline{c}^T, \underline{a}[2]\underline{c}^T, \dots, \underline{a}[\mathcal{N}_s]\underline{c}^T] = \mathbb{M} \otimes \underline{c}^T \tag{4.3}$$

where $\otimes$ denotes the Kronecker product.

Assume that there are K targets which are located at distinct directions θ_k, $k = 1, 2, \dots, K$ with respect to the radar's origin $(0, 0, 0)$ and which have different relative delays τ_k, $k = 1, 2, \dots, K$ with respect to Rx's reference clock. For the k-th moving target, the Doppler frequency is denoted by $\mathcal{F}_k$ and is a known function of radial target velocity, v_k. That is,

$$\mathcal{F}_k \triangleq -\frac{2v_k F_c}{c} \tag{4.4}$$

where F_c denotes carrier frequency and c is the speed of light. Consider that the symbols of Eq. (4.3) are transformed to the baseband waveforms $\underline{m}(t)$ which are transmitted from the Tx array, reflected by the K targets and received by the Rx array. The propagation of the signal vector transmitted from the Tx array to each of the targets and reflected to the Rx array is

---

[1]In this chapter the terms "symbols" and "pulses" are considered equivalent and can be used interchangeably.

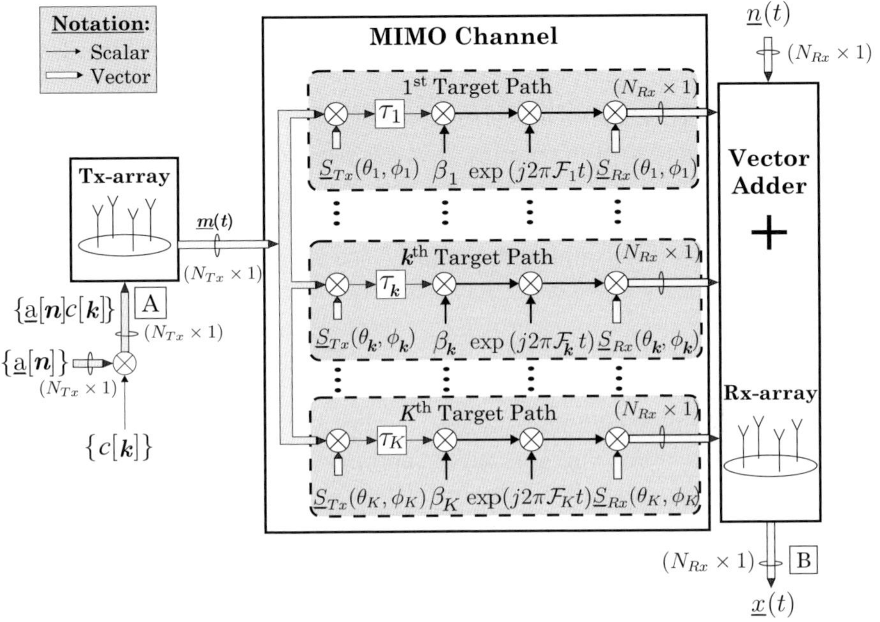

Fig. 4.1   The equivalent baseband propagation model of the MIMO radar operating in the presence of K targets and noise.

illustrated in Fig. 4.1. Therefore, the equivalent ($N_{Rx} \times 1$) baseband received signal vector (see Point B in Fig. 4.1) can be written as

$$\underline{x}(t) = \sum_{k=1}^{K} \beta_k \exp(j2\pi\mathcal{F}_k t)\underline{S}_{Rx}(\theta_k, \phi_k)\underline{S}_{Tx}^{H}(\theta_k, \phi_k)\underline{m}(t - \tau_k) + \underline{n}(t).$$

$$(4.5)$$

With reference to both Fig. 4.1 and Eq. 4.5, the vectors $\underline{S}_{Tx}(\theta_k, \phi_k) \in \mathcal{C}^{N_{Tx} \times 1}$ and $\underline{S}_{Rx}(\theta_k, \phi_k) \in \mathcal{C}^{N_{Rx} \times 1}$ denote the manifold vectors of the Tx and Rx arrays, respectively, of the k-th target and are defined as

$$\underline{S}_{Rx}(\theta_k, \phi_k) \triangleq \exp(-j\left[\underline{r}_x, \underline{r}_y, \underline{r}_z\right]_{Rx} \underline{k}(\theta_k, \phi_k)) \tag{4.6}$$

$$\underline{S}_{Tx}(\theta_k, \phi_k) \triangleq \exp(j\left[\underline{r}_x, \underline{r}_y, \underline{r}_z\right]_{Tx} \underline{k}(\theta_k, \phi_k)) \tag{4.7}$$

where (θ_k, ϕ_k) is the DOD or DOA of the k-th target. Furthermore, in Eqs. (4.6) and (4.7), the matrices

$$\left[\underline{r}_x, \underline{r}_y, \underline{r}_z\right]_{Rx} \in \mathcal{R}^{N_{Rx} \times 3} \tag{4.8}$$

$$\left[\underline{r}_x, \underline{r}_y, \underline{r}_z\right]_{Tx} \in \mathcal{R}^{N_{Tx} \times 3} \tag{4.9}$$

denote the Cartesian coordinates of the Rx and Tx antenna-array elements, respectively, and

$$\underline{k}(\theta_k, \phi_k) = \frac{2\pi F_c}{c} \left[\cos(\theta_k)\cos(\phi_k), \sin(\theta_k)\cos(\phi_k), \sin(\phi_k)\right]^T \qquad (4.10)$$

is the wavenumber vector pointing towards the target's direction (θ_k, ϕ_k). In both Fig. 4.1 and in Eq. (4.5), the parameter β_k represents the complex path gain associated with the k-th target. In addition $\underline{n}(t)$ is the spatially and temporally white complex Gaussian noise of zero mean and covariance matrix

$$\mathbb{R}_{nn} \triangleq \mathcal{E}\{\underline{n}(t)\underline{n}^H(t)\}$$
$$= \sigma_n^2 \mathbb{I}_N \qquad (4.11)$$

where σ_n^2 is the unknown noise power.

Without loss of generality, it will be assumed in this chapter that the Tx array, the Rx array and all the targets lie on the (x,y) plane (i.e. $\phi_k = 0$, for $k = 1, 2, \ldots, K$). Therefore, the unknown parameters of interest associated with the k-th target are the azimuth angle θ_k, relative delay τ_k, fading coefficient β_k and Doppler frequency $\mathcal{F}_k$.

Let l_k be the quantised/discretised relative path delay of the k-th target, which is defined as

$$l_k \triangleq \left\lceil \frac{\tau_k}{T_c} \right\rceil \bmod N_c. \qquad (4.12)$$

A very compact and effective way to represent relative delays of the targets is to use the matrix defined as

$$\mathbb{J} \triangleq \begin{bmatrix} \underline{0}_{2N_c-1}^T & 0 \\ \mathbb{I}_{2N_c-1} & \underline{0}_{2N_c-1} \end{bmatrix} = \begin{bmatrix} 0, & 0, & \cdots & 0, & 0 \\ 1, & 0 & \cdots & 0, & 0 \\ 0, & 1, & \cdots & 0, & 0 \\ \vdots & \vdots & \ddots & \vdots & \vdots \\ 0, & 0, & \cdots & 1, & 0 \end{bmatrix}. \qquad (4.13)$$

This is known as the "downshift" matrix since if $\mathbb{J}^l$ operates on a column vector, it downshifts the elements of the vector by l elements. Thus, if for the k-th target, the transmitted sequence associated with the n-th symbol, i.e.

$$\underline{a}[n]\underline{c}^T = \begin{bmatrix} a_1[n]\alpha[1], & a_1[n]\alpha[2], & \ldots, & a_1[n]\alpha[\mathcal{N}_c], & \overbrace{0,\ldots,0}^{\mathcal{N}_c} \\ a_2[n]\alpha[1], & a_2[n]\alpha[2], & \ldots, & a_2[n]\alpha[\mathcal{N}_c], & 0,\ldots,0 \\ \vdots & \vdots & \ddots & \vdots & \vdots \ddots \vdots \\ a_{N_{Tx}}[n]\alpha[1], & a_{N_{Tx}}[n]\alpha[2], & \ldots, & a_{N_{Tx}}[n]\alpha[\mathcal{N}_c], & 0,\ldots,0 \end{bmatrix}$$

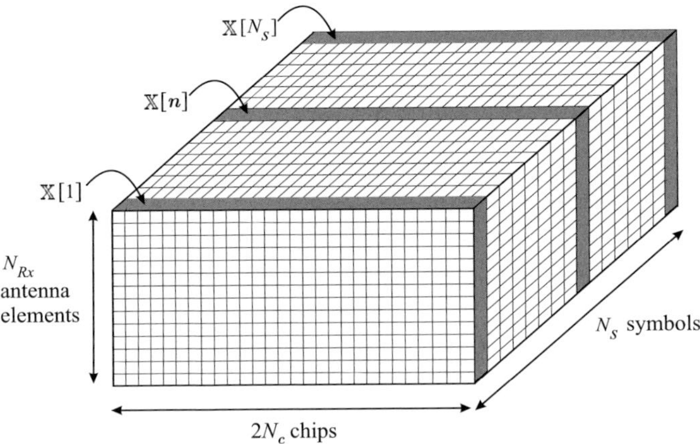

Fig. 4.2   3D datacube describing the discretized received data matrix $\mathbb{X}$.

is delayed by $l_k T_c$, then this can be represented as the $(N_{Tx} \times 2N_c)$ matrix

$$
\underline{a}[n] \left(\mathbb{J}^{l_k} \underline{c} \right)^T = \begin{bmatrix} \overbrace{0,\ldots,0,}^{l_k} & a_1[n]\alpha[1], & a_1[n]\alpha[2], & \ldots, & a_1[n]\alpha[\mathcal{N}_c], & \overbrace{0,\ldots,0}^{\mathcal{N}_c - l_k} \\ 0,\ldots,0, & a_2[n]\alpha[1], & a_2[n]\alpha[2], & \ldots, & a_2[n]\alpha[\mathcal{N}_c], & 0,\ldots,0 \\ \vdots & \vdots & \vdots & \ddots & \vdots & \vdots \\ 0,\ldots,0, & a_{N_{Tx}}[n]\alpha[1], & a_{N_{Tx}}[n]\alpha[2], & \ldots, & a_{N_{Tx}}[n]\alpha[\mathcal{N}_c], & 0,\ldots,0 \end{bmatrix}.
$$

The discretised signal of $\underline{x}(t)$ can be represented by the 3D datacube shown in Fig. 4.2 where the n-th slice, associated with the n-th data symbol interval, is an $(N_{Rx} \times 2\mathcal{N}_c)$ matrix that can be represented as

$$
\mathbb{X}[n] = \sum_{k=1}^{K} \beta_k \underline{S}_{Rx}(\theta_k) \underline{S}_{Tx}^H(\theta_k) \left(\underline{a}[n] \otimes \left(\mathbb{J}^{l_k} \underline{c} \odot \underline{\mathcal{F}}_k[n] \right)^T \right) + \mathbb{N}[n] \quad (4.14)
$$

with $\mathbb{N}[n] \in \mathcal{C}^{N_{Rx} \times 2\mathcal{N}_c}$ denoting the discretised version of the noise $\underline{n}(t)$ and the vector $\underline{\mathcal{F}}_k[n]$ capturing the Doppler effects associated with the k-th target, defined as follows

$$
\underline{\mathcal{F}}_k[n] = \underbrace{\begin{bmatrix} 1 \\ \exp\left(j2\pi\mathcal{F}_k T_c\right) \\ \vdots \\ \exp\left(j2\left(2\mathcal{N}_c - 1\right)\pi\mathcal{F}_k T_c\right) \end{bmatrix}}_{\triangleq \underline{\mathcal{F}}_{chips,k}} \exp\left(j2\pi n\mathcal{F}_k T_{cs}\right) \quad (4.15)
$$

Finally, the 3D datacube, which is the overall discretised received signal $\mathbb{X} \in \mathcal{C}^{N_{Rx} \times 2\mathcal{N}_c \mathcal{N}_s}$ may be written as

$$\mathbb{X} = \left[\mathbb{X}\left[1\right], \mathbb{X}\left[2\right], \cdots, \mathbb{X}\left[n\right], \cdots, \mathbb{X}\left[\mathcal{N}_s\right] \right] \tag{4.16}$$

$$= \sum_{k=1}^{K} \beta_k \underline{S}_{Rx}(\theta_k) \underline{S}_{Tx}^{H}(\theta_k) \left(\mathbb{M}\mathrm{diag}\{\underline{\mathcal{F}}_{s,k}\} \otimes \left(\mathbb{J}^{l_k} \underline{c} \odot \underline{\mathcal{F}}_{chips,k} \right)^{T} \right) + \mathbb{N}$$

where

$$\mathbb{N} = \left[\mathbb{N}\left[1\right], \mathbb{N}\left[2\right], \cdots, \mathbb{N}\left[n\right], \cdots, \mathbb{N}\left[\mathcal{N}_s\right] \right] \in \mathcal{C}^{N_{Rx} \times 2\mathcal{N}_c \mathcal{N}_s} \tag{4.17}$$

and $\underline{\mathcal{F}}_{s,k}$ represents the inter-symbol Doppler effects, due to the term $\exp\left(j2\pi n\mathcal{F}_k T_{cs}\right)$ in Eq. (4.15), defined as

$$\underline{\mathcal{F}}_{s,k} \triangleq \exp\left(j2\pi\mathcal{F}_k \left[0, 1, \cdots, n, \cdots, \mathcal{N}_s - 1 \right]^{T} 2\mathcal{N}_c T_c \right). \tag{4.18}$$

4.3 Space arrayed MIMO radar: Target echoes arriving with equal delays

In the MIMO radar literature, a number of parameter estimation algorithms have been proposed, which are based on the extension of existing passive radar methods to the MIMO radar model [8–14]. Here, three typical optimisation-type parameter estimation methods [8] for the estimation of DOA and path fading coefficients, will be briefly presented:

- least squares (LS),
- Capon,
- amplitude and phase estimation (APES) methods.

These three approaches were derived under the assumptions that

(1) Doppler effects are negligible and
(2) all target echoes arrive at the receiver in a synchronised manner (i.e. no relative path delays exist).

where the first assumption implies that the Doppler vector $\underline{\mathcal{F}}_k\left[n\right]$ in Eq. (4.15) becomes a vector of ones while the second assumption indicates that $l_1 = l_2 = \cdots = l_K = l$; $(l = 0,$ say$)$. Thus, the term $\underline{a}[n] \otimes (\mathbb{J}^{l_k} \underline{c} \odot \underline{\mathcal{F}}_k[n])^{T}$ in Eq. (4.14) is simplified to $\underline{a}[n] \otimes \underline{c}^{T}$ and, consequently, the term $(\mathbb{M}\mathrm{diag}\{\underline{\mathcal{F}}_{s,k}\} \otimes (\mathbb{J}^{l_k} \underline{c} \odot \underline{\mathcal{F}}_{chips,k})^{T})$ in Eq. (4.16) becomes $\mathbb{M} \otimes \underline{c}^{T}$.

In this section, under the above two assumptions, the matrix $\mathbb{M} \otimes \underline{c}^{T}$ is redefined as $\mathbb{M}$, i.e. $\mathbb{M} \triangleq \mathbb{M} \otimes \underline{c}^{T}$. That is, the columns of $\mathbb{M}$ are a known vector-sequence of symbols (spread or unspread) and the received signal data matrix $\mathbb{X}$ in Eq. (4.16) can be rewritten as:

$$\mathbb{X} = \left(\sum_{k=1}^{K} \beta_k \underline{S}_{Rx}(\theta_k) \underline{S}_{Tx}^{H}(\theta_k) \right) \mathbb{M} + \mathbb{N} \tag{4.19}$$

which can be rewritten in a more compact form as:

$$\mathbb{X} = \mathbb{S}_{Rx}(\underline{\theta}) \operatorname{diag}\{\underline{\beta}\} \mathbb{S}_{Tx}^{H}(\underline{\theta}) \mathbb{M} + \mathbb{N} \tag{4.20}$$

where

$$\mathbb{S}_{Rx}(\underline{\theta}) = [\underline{S}_{Rx}(\theta_1), \underline{S}_{Rx}(\theta_2), \ldots, \underline{S}_{Rx}(\theta_K)] \tag{4.21}$$

$$\mathbb{S}_{Tx}(\underline{\theta}) = [\underline{S}_{Tx}(\theta_1), \underline{S}_{Tx}(\theta_2), \ldots, \underline{S}_{Tx}(\theta_K)] \tag{4.22}$$

$$\underline{\theta} = [\theta_1, \theta_2, \ldots, \theta_K]^{T} \tag{4.23}$$

$$\underline{\beta} = [\beta_1, \beta_2, \ldots, \beta_K]^{T}. \tag{4.24}$$

Thus, $\underline{\theta}$ and $\underline{\beta}$ are the two unknown parameter vectors to be estimated in such a MIMO radar system. The parameter vector $\underline{\theta}$ gives the directions of the multiple targets, while the parameter vector $\underline{\beta}$ is related to the ranges and the cross section of these targets. In fact, the basic concept in these three methods is to view the signal model as the desired target under consideration in the presence of the combined effects of all other targets and noise. That is,

$$\mathbb{X} = \underbrace{\beta_d \underline{S}_{Rx}(\theta_d) \underline{S}_{Tx}^{H}(\theta_d) \mathbb{M}}_{\text{desired signal term}} + \mathbb{Z}_0 \tag{4.25}$$

where the desired signal arrives from angle θ_d with path fading coefficient β_d, and $\mathbb{Z}_0$ denotes the mutual target interference plus noise effects, i.e.

$$\mathbb{Z}_0 = \sum_{\substack{k=1 \\ k \neq d}}^{K} \beta_k \underline{S}_{Rx}(\theta_k) \underline{S}_{Tx}^{H}(\theta_k) \mathbb{M} + \mathbb{N}. \tag{4.26}$$

Thus, $\mathbb{Z}_0$ is considered an equivalent additive independent noise.

4.3.1   *Least squares*

The optimisation problem for the LS method, based on Eq. (4.25), can be formulated as follows:

$$\min_{\beta} \| \mathbb{X} - \beta \underline{S}_{Rx}(\theta) \underline{S}_{Tx}^{H}(\theta) \mathbb{M} \|_{F}^{2} \tag{4.27}$$

which aims to find the β that minimises the power of the residual term $\mathbb{Z}_0$ and the solution can be easily obtained as follows:

$$\text{LS:} \quad \beta(\theta) = \frac{\underline{S}_{Rx}(\theta)^H \mathbb{R}_{xm} \underline{S}_{Tx}(\theta)}{N_{Rx} \underline{S}_{Tx}^H(\theta) \mathbb{R}_{mm} \underline{S}_{Tx}(\theta)} \tag{4.28}$$

with

$$\mathbb{R}_{mm} \triangleq \frac{1}{L} \mathbb{M}\mathbb{M}^H \tag{4.29}$$

$$\mathbb{R}_{xm} \triangleq \frac{1}{L} \mathbb{X}\mathbb{M}^H \tag{4.30}$$

where L is the total number of snapshots. The spatial spectrum (as a function of θ) is then computed from Eq. (4.28) and parameter estimates are obtained at the peaks in this spectrum.

4.3.2   *Capon's method*

An extension of LS is the Capon method, proposed in [8]. Specifically, the optimisation problem for this method can be written as:

$$\min_{\beta} \left\| \underline{w}^H \left(\mathbb{X} - \beta \underline{S}_{Rx}(\theta) \underline{S}_{Tx}^H(\theta) \mathbb{M} \right) \right\|^2 \tag{4.31}$$

where $\underline{w} \in \mathcal{C}^{N_{Rx} \times 1}$ is the Rx beamforming vector, which uses the solution to Capon's well-known minimum variance distortionless response (MVDR) optimisation problem:

$$\min_{\underline{w}} \underline{w}^H \mathbb{R}_{xx} \underline{w} \tag{4.32a}$$

$$\text{subject to} \quad \underline{w}^H \underline{S}_{Rx}(\theta) = 1 \tag{4.32b}$$

in which

$$\mathbb{R}_{xx} \triangleq \frac{1}{L} \mathbb{X}\mathbb{X}^H. \tag{4.33}$$

In other words, this beamformer seeks to minimise its output power, subject to the constraint that the response in the look-direction θ is equal to unity. The solution to Eq. (4.32) is readily obtained as:

$$\text{Capon:} \quad \underline{w} = \frac{\mathbb{R}_{xx}^{-1} \underline{S}_{Rx}(\theta)}{\underline{S}_{Rx}^H(\theta) \mathbb{R}_{xx}^{-1} \underline{S}_{Rx}(\theta)}. \tag{4.34}$$

Finally, substituting for $\underline{w}$ and solving the original minimization problem in Eq. (4.31), the final solution is given by the following equation:

$$\text{Capon:} \quad \beta(\theta) = \frac{\underline{S}_{Rx}^H(\theta) \mathbb{R}_{xx}^{-1} \mathbb{R}_{xm} \underline{S}_{Tx}(\theta)}{\underline{S}_{Rx}(\theta)^H \mathbb{R}_{xx}^{-1} \underline{S}_{Rx}(\theta) \underline{S}_{Tx}^H(\theta) \mathbb{R}_{mm} \underline{S}_{Tx}(\theta)}. \tag{4.35}$$

4.3.3   *Amplitude and phase estimation (APES)*

The APES method was first proposed as an adaptive finite impulse response (FIR) filtering approach in [15], followed by related discussion in [16]. More recently, the APES method has been applied in MIMO radar [8] for multi-target localisation and can be formulated as the following constrained optimisation problem

$$\min_{\underline{w},\beta} \left\| \underline{w}^H \mathbb{X} - \beta \underline{S}_{Tx}^H(\theta)\mathbb{M} \right\|^2 \tag{4.36a}$$

$$\text{subject to} \quad \underline{w}^H \underline{S}_{Rx}(\theta) = 1 \tag{4.36b}$$

which seeks to find the beamformer whose output is as close as possible to the undistorted desired signal term. The solution to this constrained optimisation problem is, in fact, related to the solution of Capon's method. Let $\xi(\underline{w},\beta)$ denote the following cost function to be minimized:

$$\xi(\underline{w},\beta) \triangleq \left\| \underline{w}^H \mathbb{X} - \beta \underline{S}_{Tx}^H(\theta)\mathbb{M} \right\|^2. \tag{4.37}$$

Then, setting the derivative of $\xi(\underline{w},\beta)$ with respect to β equal to zero, the solution of the complex path gain β can be obtained as a function of $\underline{w}$ as follows:

$$\beta(\theta) = \frac{\underline{w}^H \mathbb{R}_{xm} \underline{S}_{Tx}(\theta)}{\underline{S}_{Tx}^H(\theta)\mathbb{R}_{mm}\underline{S}_{Tx}(\theta)}. \tag{4.38}$$

Insertion of Eq. (4.38) into Eq. (4.37) yields the following constrained optimisation problem for the determination of $\underline{w}$:

$$\min_{\underline{w}} \underline{w}^H \mathbb{Q}\underline{w} \tag{4.39a}$$

$$\text{subject to} \quad \underline{w}^H \underline{S}_{Rx}(\theta) = 1 \tag{4.39b}$$

in which

$$\mathbb{Q} \triangleq \mathbb{R}_{xx} - \frac{\mathbb{R}_{xm}\underline{S}_{Tx}(\theta)\underline{S}_{Tx}^H(\theta)\mathbb{R}_{xm}^H}{\underline{S}_{Tx}^H(\theta)\mathbb{R}_{mm}\underline{S}_{Tx}(\theta)}. \tag{4.40}$$

By comparing Eq. (4.39) with Eq. (4.32), it is clear to see that Eq. (4.39) has the same structure as Capon's constrained minimisation. Borrowing that result and substituting back into Eq. (4.38), the APES estimator is finally given as follows:

$$\text{APES:} \quad \beta(\theta) = \frac{\underline{S}_{Rx}^H(\theta)\mathbb{Q}^{-1}\mathbb{R}_{xm}\underline{S}_{Tx}(\theta)}{\underline{S}_{Rx}(\theta)^H\mathbb{Q}^{-1}\underline{S}_{Rx}(\theta)\underline{S}_{Tx}^H(\theta)\mathbb{R}_{mm}\underline{S}_{Tx}(\theta)}. \tag{4.41}$$

4.3.4   *Discussion*

In practice, it has been found that APES provides accurate β estimates, while the Capon approach offers superior DOA estimation performance. In order to exploit the benefits of both Capon and APES, a combined approach, the so called "CAPES" method, was proposed in [17]. Specifically, CAPES estimates the DOAs of targets by Capon's method, then provides estimates of the path gains associated with these DOAs via the APES method. Similarly, the "CAML" method was proposed in [11], which combines Capon and, instead of APES, an approximate maximum likelihood (AML) method [18], thus providing more accurate estimation of the complex path gains (βs) than LS, Capon and APES.

However, in all these methods, small angular separations among two or more targets result in high mutual coherent target interference and this significantly degrading their performance. Using an alternative signal modelling which does not combine the effects of unwanted targets and noise into the matrix $\mathbb{Z}_0$, given by Eq. (4.26), two interference cancellation-based techniques for multitarget parameter estimation were proposed in [19]. Table 4.1, provides a summary of the main techniques discussed so far, including the two techniques presented in [19].

4.3.5   *Comparative studies and computer simulation results*

Without any loss of generality consider a MIMO radar system where a uniform linear array (ULA) with $N_{Tx} = N_{Rx} = 10$ antennas with half wavelength spacing is employed for both Tx and Rx arrays. The relative delays of the targets are assumed negligible. The probing waveforms/symbols (matrix $\mathbb{M}$) employ unspread Hadamard codes, which have good orthogonality properties, and P_T is the total transmitted power. Consider the first scenario in which $K = 5$ targets located at $\theta_1 = 60°$, $\theta_2 = 75°$, $\theta_3 = 90°$, $\theta_4 = 105°$ and $\theta_5 = 120°$ with the reflection coefficients $\beta_1 = 1$, $\beta_2 = 0.6$, $\beta_3 = 0.4$, $\beta_4 = 0.2$ and $\beta_5 = 0.8$, respectively. The number of snapshots is $L = 256$. The received signal is corrupted by a spatially and temporally white complex Gaussian noise with mean zero and covariance matrix $0.1\mathbb{I}_{10}$.

The (θ_k, β_k), $k = 1, ..., K$ of the LS, Capon and APES methods are obtained by searching the peaks of the spectra of $|\beta_{LS}(\theta)|$, $|\beta_{Capon}(\theta)|$ and $|\beta_{APES}(\theta)|$, respectively, with respect to the variable θ. The results are

Table 4.1   MIMO radar multi-target (optimisation-type) parameter estimation methods.

LS	$\min_\beta \left\| \mathbb{X} - \beta \underline{S}_{Rx}(\theta)\underline{S}^H_{Tx}(\theta)\mathbb{M} \right\|^2_F$ $\beta_{LS}(\theta) = \dfrac{\underline{S}^H_{Rx}(\theta)\mathbb{R}_{xm}\underline{S}_{Tx}(\theta)}{N_{Rx}\underline{S}^H_{Tx}(\theta)\mathbb{R}_{mm}\underline{S}_{Tx}(\theta)}$
Capon	$\min_\beta \left\| \underline{w}^H\left(\mathbb{X} - \beta\underline{S}_{Rx}(\theta)\underline{S}^H_{Tx}(\theta)\mathbb{M}\right) \right\|^2$ $\beta_{Capon}(\theta) = \dfrac{\underline{S}^H_{Rx}(\theta)\mathbb{R}^{-1}_{xx}\mathbb{R}_{xm}\underline{S}_{Tx}(\theta)}{\underline{S}^H_{Rx}(\theta)\mathbb{R}^{-1}_{xx}\underline{S}_{Rx}(\theta)\underline{S}^H_{Tx}(\theta)\mathbb{R}_{mm}\underline{S}_{Tx}(\theta)}$
APES	$\min_{\underline{w},\beta} \left\| \underline{w}^H\mathbb{X} - \beta\underline{S}^H_{Tx}(\theta)\mathbb{M} \right\|^2 ; \text{s.t. } \underline{w}^H\underline{S}_{Rx}(\theta) = 1$ $\hat{\beta}_{APES} = \dfrac{\underline{S}^H_{Rx}(\theta)\mathbb{Q}^{-1}\mathbb{R}_{xm}\underline{S}_{Tx}(\theta)}{\underline{S}_{Rx}(\theta)^H\mathbb{Q}^{-1}\underline{S}_{Rx}(\theta)\underline{S}^H_{Tx}(\theta)\mathbb{R}_{mm}\underline{S}_{Tx}(\theta)}$ where $\mathbb{Q} = \mathbb{R}_{xx} - \dfrac{\mathbb{R}_{xm}\underline{S}_{Tx}(\theta)\underline{S}^H_{Tx}(\theta)\mathbb{R}^H_{xm}}{\underline{S}^H_{Tx}(\theta)\mathbb{R}_{mm}\underline{S}_{Tx}(\theta)}$
Kai's 1D iterative[19]	$\hat{\underline{\theta}}^{[i]} = \arg\max_{\underline{\theta}} \left\| \text{diag}\left\{ \left(\mathbb{W}^{[i-1]}_{Rx}\right)^H \mathbb{R}_{xm}\mathbb{S}_{Tx}(\underline{\theta}) \right\} \right\|$ $\underline{\beta}^{[i]} = \frac{1}{N_{Tx}}\underline{\text{diag}}\{\mathbb{W}^{[i]H}_{Rx}\mathbb{R}_{xm}\mathbb{S}_{Tx}(\underline{\theta})\}$ $\mathbb{W}^{[i]}_{Rx} = \mathbb{R}^{-1}_{xx}\left(\mathbb{I}_{N_{Rx}} - \mathbb{S}_{Rx}(\underline{\theta}^{[i]})\right.$ $\times (\mathbb{S}^H_{Rx}(\underline{\theta}^{[i]})\mathbb{R}^{-1}_{xx}\mathbb{S}_{Rx}(\underline{\theta}^{[i]}))^{-1}\mathbb{S}^H_{Rx}(\underline{\theta}^{[i]})\mathbb{R}^{-1}_{xx}\left.\right)$ $\cdot \mathbb{R}_{xm}\mathbb{S}_{Tx}(\underline{\theta}^{[i]})\,\text{diag}\{(\underline{\beta}^{[i]})^*\}$ $+ \mathbb{R}^{-1}_{xx}\mathbb{S}_{Rx}(\underline{\theta}^{[i]})(\mathbb{S}^H_{Rx}(\underline{\theta}^{[i]})\mathbb{R}^{-1}_{xx}\mathbb{S}_{Rx}(\underline{\theta}^{[i]}))^{-1}$
Kai's multi-dimensional[19]	$\hat{\underline{\theta}} = \arg\min_{\underline{\theta}} \xi_{\text{BBO}}(\underline{\theta})$ where $\xi_{\text{BBO}}(\underline{\theta}) \triangleq \frac{1}{L}\left\| \mathbb{X} - \mathbb{S}^H_{Rx}(\underline{\theta})\text{diag}\{\underline{\beta}(\underline{\theta})\}\mathbb{S}^H_{Tx}(\underline{\theta})\mathbb{M} \right\|^2$

shown in Figs. 4.3–4.5 respectively, where the "asterisks" indicate the true values of $\{\beta\}$ and the "dashed" lines show the associated true directions $\{\theta\}$ of these targets.

Next, consider a scenario, where there are $K = 3$ targets closely located at directions $\theta_1 = 60°, \theta_2 = 72°, \theta_3 = 79°$ and complex path gains $\beta_1 = 0.7\exp(j1.3), \beta_2 = \exp(j0.9), \beta_3 = 0.8\exp(j0.7)$, respectively, which need to be estimated. The average root-mean-square errors (RMSEs)

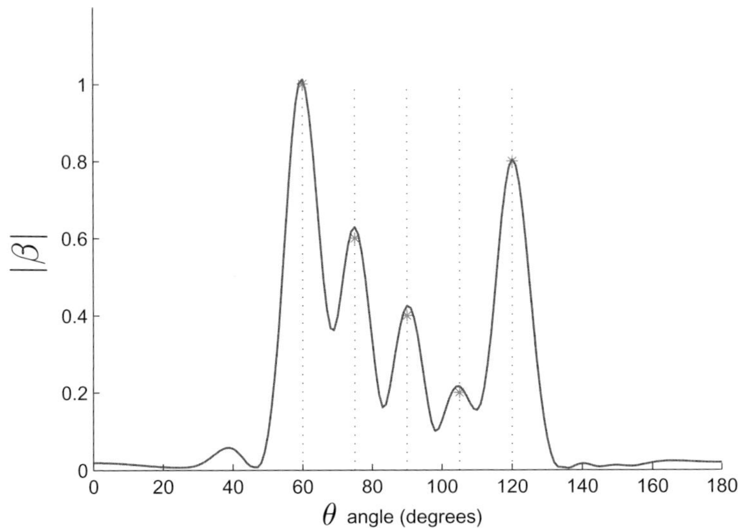

Fig. 4.3 LS DOA estimates of MIMO radar when the space between the adjacent targets is 15°.

of the estimated parameters $\hat{\theta}_k$, $\hat{\beta}_k$, $k = 1, \ldots, K$ are defined as

$$\mathrm{RMSE}_\theta = \frac{1}{K} \sum_{k=1}^{K} \sqrt{\mathcal{E}\left\{|\,\hat{\theta}_k - \theta_k\,|^2\right\}} \tag{4.42}$$

$$\mathrm{RMSE}_\beta = \frac{1}{K} \sum_{k=1}^{K} \sqrt{\mathcal{E}\left\{|\,\hat{\beta}_k - \beta_k\,|^2\right\}} \tag{4.43}$$

and are used to compare the performance of the following methods: LS, Capon and APES methods as well as Kai's one-dimensional iterative and multi-dimensional optimal methods. Moreover, here, the initial receiver's beamforming matrix $\mathbb{W}_{Rx}^{[0]}$ of the one-dimensional iterative method is formed as in [19].

First, consider a scenario in which the number of snapshots $L = 128$ and $\frac{P_T}{\sigma_n^2}$ equals[2] 0 dB. Table 4.2 provides the average estimation error of Kai's algorithms as well as Capon, CAPES, CAML and LS, obtained by 1000 Monte-Carlo simulations.

The RMSE_θ and RMSE_β values given by Kai's multi-dimensional optimal method implies that the multi-dimensional optimal method provides

---

[2]Note that the input signal-to-noise ratio (SNR) at the Rx array is smaller than 0 dB due to the path gains that attenuate the transmit signals.

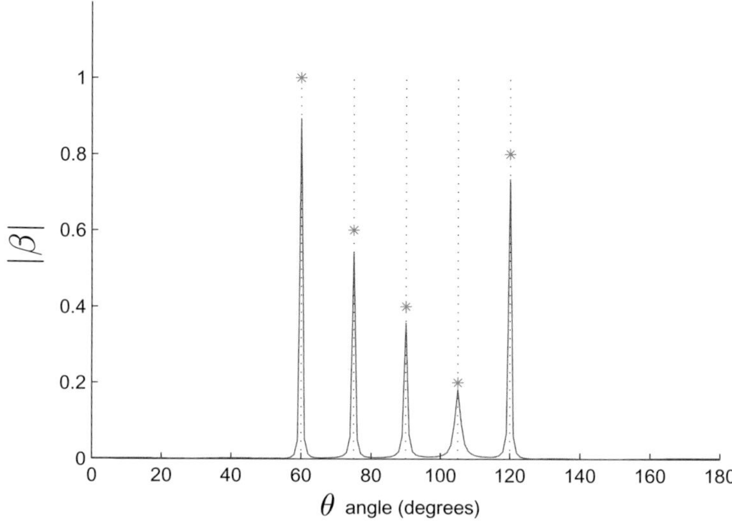

Fig. 4.4   Capon DOA estimates of MIMO radar when the space between the adjacent targets is 15°.

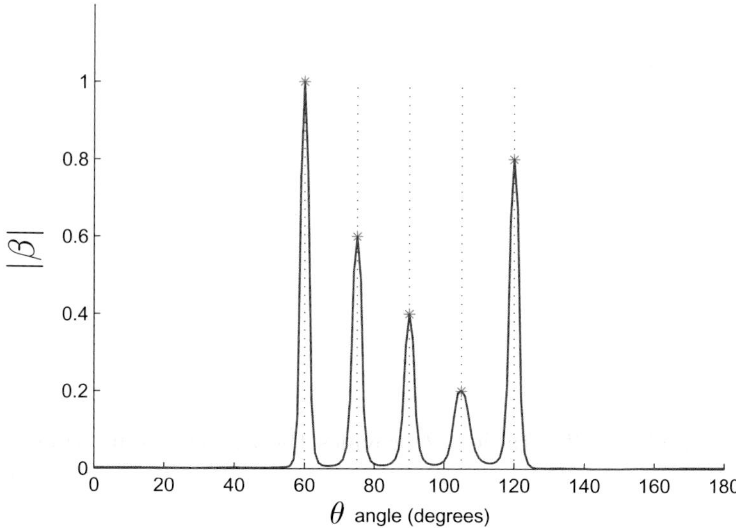

Fig. 4.5   APES DOA estimates of MIMO radar when the space between the adjacent targets is 15°.

Table 4.2   The average estimation error.

Methods	Avg. Error of $\hat{\underline{\theta}}$	Avg. Error of $\hat{\underline{\beta}}$
LS	1.35°	0.52
Capon	0.95°	0.40
CAPES	0.95°	0.39
CAML	0.95°	0.37
Kai's iterative	0.27°	0.10
Kai's multidim	0.13°	0.07

a more accurate estimation than Kai's one-dimensional iterative method. This is due to the fact that the multi-dimensional optimal method utilises the optimal solution of complex path gain vector $\underline{\beta}$ for the optimisation problem, whereas the one-dimensional iterative method utilises a suboptimal solution.

Furthermore, the performance of Capon, CAPES, CAML and LS decreases significantly when targets are close together in space since the small angular separations among multiple targets lead to high mutual target interference, as discussed previously. However, Kai's two multi-target parameter estimation approaches show superior performance against all the other methods since both are based on suppressing the mutual interference effects during the estimation process.

4.3.5.1 *Finite averaging effects*

Next, with $\frac{P_T}{\sigma_n^2}$ fixed at 0 dB, the variations of RMSE$_\theta$ and RMSE$_\beta$ as a function of the number of snapshots L are shown in Figs. 4.6 and 4.7, respectively. It is clear that, for the entire range of snapshots, Kai's multi-dimensional optimal method provides the best parameter estimation of multiple targets while LS the worst performance. These results are similar to the case of fixed $\frac{P_T}{\sigma_n^2}$ and fixed snapshots presented in Table 4.2.

4.3.5.2 *Noise effects (variable levels of $\frac{P_T}{\sigma_n^2}$)*

Finally, Figs. 4.8 and 4.9 depict the effect of $\frac{P_T}{\sigma_n^2}$ on the performance of the various methods while $\frac{P_T}{\sigma_n^2}$ is varying. The number of snapshots is fixed at $L = 128$ and all the other parameters are set as before. Again, Kai's multi-dimensional optimal method outperforms all the other methods throughout the whole range of $\frac{P_T}{\sigma_n^2}$. Furthermore, when $\frac{P_T}{\sigma_n^2}$ is high, the Capon, CAPES and CAML methods tend to reach the directional rms error

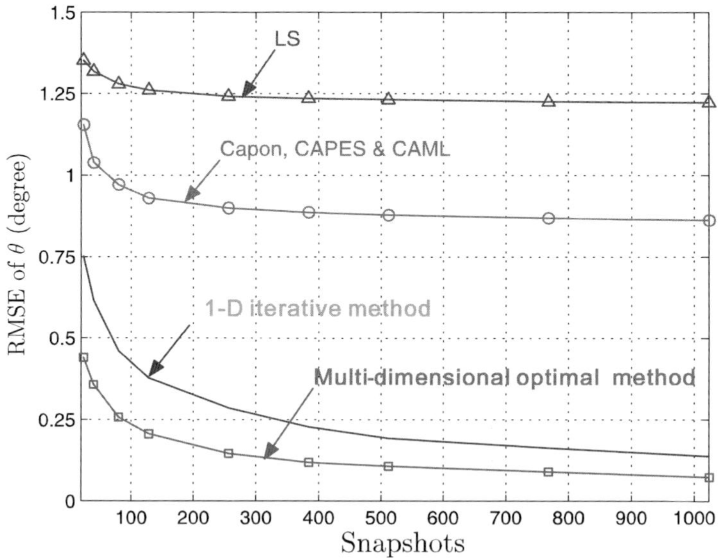

Fig. 4.6   The RMSE$_\theta$ performance versus the number of snapshots when P_T/σ_n^2 equals 0dB.

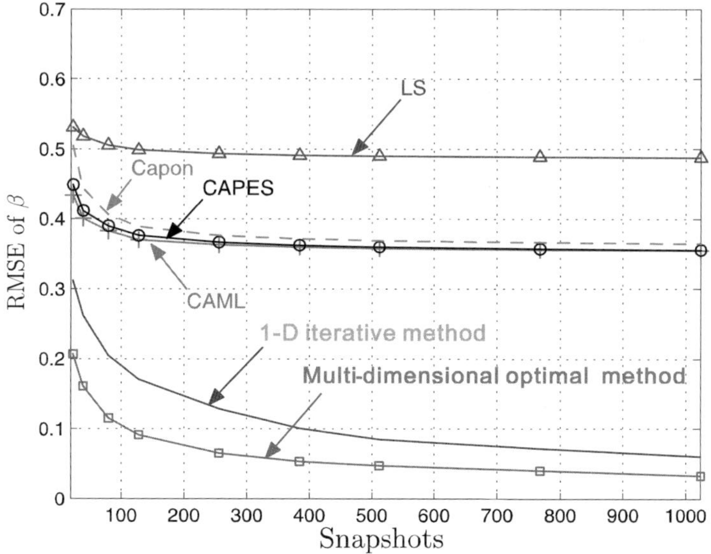

Fig. 4.7   The RMSE$_\beta$ performance versus the number of snapshots when P_T/σ_n^2 equals 0 dB.

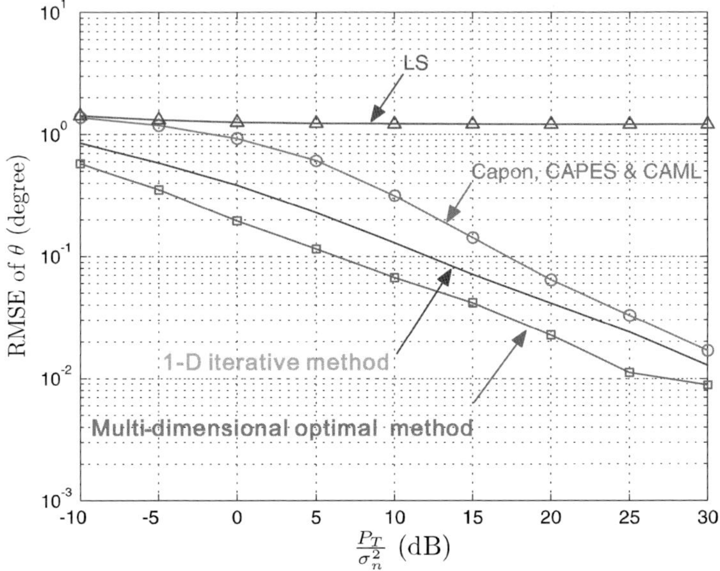

Fig. 4.8　The performance of RMSE_θ versus P_T/σ_n^2 when L equals 128 snapshots.

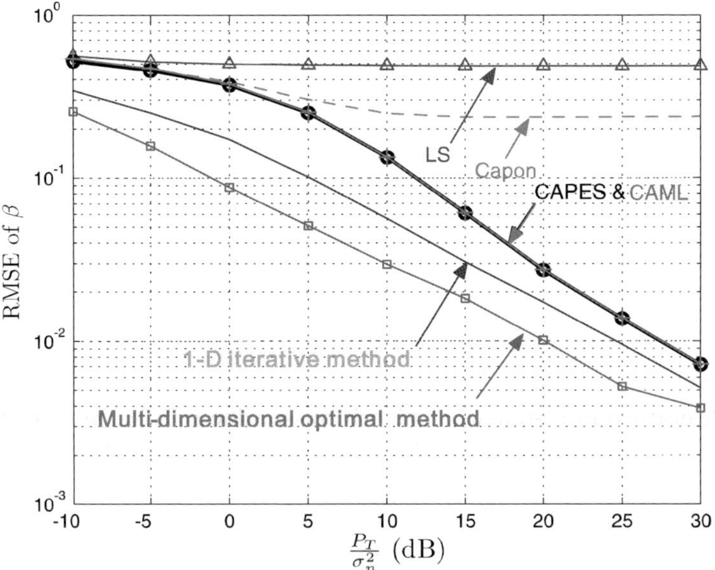

Fig. 4.9　The performance of RMSE_β versus P_T/σ_n^2 when L equals 128 snapshots.

(RMSE$_\theta$) performance of the one-dimensional iterative method, while the CAPES and CAML methods outperform both LS and Capon methods with respect to β estimation error (RMSE$_\beta$) and tend to reach the performance of the one-dimensional iterative method.

However, it is inappropriate to assume such a high $\frac{P_T}{\sigma_n^2}$ environment for a radar system since the target returns are seldom hundreds of times stronger than the noise. It is worth noting that when $\frac{P_T}{\sigma_n^2}$ increases, the RMSE$_\beta$ decreases for all described methods except the Capon method. This is because the estimate of θ is unbiased and the estimate of β is biased in Capon, while for all the other methods the estimates of both θ and β are unbiased.

4.4 Arrayed MIMO radar: Target echoes with different delays

The parameter estimation procedure developed in this section differs significantly from the optimisation-based approaches studied in Section 4.3. In particular in this section a subspace-based perspective on parameter estimation is adopted. Subspace-based parameter estimation algorithms (the MUSIC algorithm [20] is the most representative example) have been shown to exhibit superresolution capability by seeking to locate intersections between an estimated signal subspace and the array manifold. Indeed, the key concept in any array system is its manifold, which is a mathematical object (curve or surface) embedded in a multidimensional complex space that completely characterises the array. Specifically, the array manifold is defined as the locus of all the response vectors (manifold vectors) of the array over the feasible set of signal/target parameters [20].

While traditional array processing theory only takes the receiver array geometry into account, it has since been proven that the full transmit-receive MIMO radar configuration can be characterised using the concept of its "virtual" SIMO array. The virtual array first appeared in the MIMO radar literature in a somewhat algorithm-specific context most commonly relying on the use of matched filters at the front-end of the receiver. Furthermore, this was under the additional assumptions that targets are all stationary and situated within the same range bin (see, for example, [22,23]).

However, these early results here are expanded and generalised to enable geometrical and algorithm-independent analysis of MIMO radar systems operating in the presence of moving targets with different delays. In particular, the key challenge in MIMO radar is to determine how the transmit array geometry can be exploited effectively. This is not trivial, since the transmit and receive arrays are characterised by two separate entities: the Tx and Rx array manifolds, respectively. Therefore, it is not immediately obvious what level of performance the radar operator can hope to achieve when these arrays are operated in a collaborative and unified manner.

To maintain focus on resolution and estimation, it will be assumed in this section that the number of targets is known in advance, for example by using a detection algorithm such as minimum description length or Akaike's information criterion [24].

4.4.1   *Spatiotemporal arrayed MIMO radar: Doppler, delay, DOA and path gain estimation*

One way of analysing and/or exploiting the transmit array geometry is to — in some sense — "virtually" transfer the transmit antennas across to the receiver through the channel, yielding a virtual SIMO array whose response is a function of both Tx and Rx manifolds. In this case the virtual SIMO array sensor locations (virtual array geometry) are given by:

$$[\underline{r}_x, \underline{r}_y, \underline{r}_z]_{\mathrm{v}} \triangleq \left([\underline{r}_x, \underline{r}_y, \underline{r}_z]_{Tx} \otimes \underline{1}_{N_{Rx}}\right) + \left(\underline{1}_{N_{Tx}} \otimes [\underline{r}_x, \underline{r}_y, \underline{r}_z]_{Rx}\right) \quad (4.44)$$

which is the spatial convolution of the MIMO transmit and receive array geometries. Thus, the virtual SIMO array response (manifold) vector is given as follows

$$\text{Eq. } (4.44) \Leftrightarrow \underline{S}_{\mathrm{v}}(\theta) \triangleq (\underline{S}_{Tx}^* \otimes \underline{1}_{N_{Rx}}) \odot (\underline{1}_{N_{Tx}} \otimes \underline{S}_{Rx}) \quad (4.45)$$
$$= \underline{S}_{Tx}^*(\theta) \otimes \underline{S}_{Rx}(\theta) \quad (4.46)$$

Thus, the fundamental capabilities of the MIMO radar system may be characterised by the large aperture, number of sensors and array geometry of its virtual array. Furthermore, a MIMO radar system (which uses these extra resources judiciously) can achieve significant performance enhancements (compared to any receiver-only approach), particularly with respect to resolution performance and parameter identifiability (i.e. maximum number of detectable targets [10]).

In order to utilise the virtual array, the 3D datacube presented in Fig. 4.2 is employed here to create a single long ("augmented") spatiotemporal data vector

$$\underline{x}_{\mathrm{aug}} = \mathrm{vec}\left\{\text{3D-datacube}\right\} \in \mathcal{C}^{2\mathcal{N}_s \mathcal{N}_c N_{Rx} \times 1}$$

which is associated with the virtual geometry, providing a huge number of degrees of freedom and a conventional structure for space-time processing.

In particular, using the signal modelling presented in Section 4.2, and in particular Eq. (4.16), this vector can be written as follows:

$$\underline{x}_{\mathrm{aug}} \triangleq \mathrm{vec}\left\{\left[\mathbb{X}\left[1\right]^{T}, \mathbb{X}\left[2\right]^{T}, \cdots, \mathbb{X}\left[n\right]^{T}, \cdots, \mathbb{X}\left[\mathcal{N}_{s}\right]^{T}\right]\right\} \tag{4.47}$$

$$= \sum_{k=1}^{K} \beta_{k} \underbrace{\left(\mathrm{diag}\left\{\underline{\mathcal{F}}_{s,k}\right\}\mathbb{M}^{T}\underline{S}_{Tx}^{*}\left(\theta_{k}\right) \otimes \underline{S}_{Rx}\left(\theta_{k}\right) \otimes \left(\mathbb{J}^{l_{k}}\underline{c} \odot \underline{\mathcal{F}}_{chips,k}\right)\right)}_{\triangleq \underline{h}_{\mathrm{aug}}(\theta_{k},\mathcal{F}_{k},l_{k})}$$

$$+\,\underline{n}_{\mathrm{aug}}$$

where $\underline{n}_{\mathrm{aug}}$ is the vectorised discrete noise while $\underline{\mathcal{F}}_{s,k} \in \mathcal{C}^{\mathcal{N}_{s}}$ denotes the Doppler effects across the symbols (corresponding to columns of $\mathbb{M}$) and $\underline{\mathcal{F}}_{chips,k}$ models the chip-to-chip (intra-symbol) Doppler effects, defined in Eq. (4.15), both for the k-th target. That is,

$$\underline{\mathcal{F}}_{s,k} \triangleq \exp(j2\pi\mathcal{F}_{k}\left[0,1,\ldots,\mathcal{N}_{s}-1\right]^{T}2\mathcal{N}_{c}T_{c}) \tag{4.48}$$

$$\underline{\mathcal{F}}_{chips,k} = \exp(j2\pi\mathcal{F}_{k}\left[0,1,\ldots,2\mathcal{N}_{c}-1\right]^{T}T_{c}). \tag{4.49}$$

For the rest of this chapter the Doppler effects within a given symbol period (i.e. corresponding to the elements of $\underline{c}$) are negligible[3], which implies that $\underline{\mathcal{F}}_{chips} = \underline{1}_{2\mathcal{N}_{c}}$ and Eq. (4.47) is simplified to

$$\underline{x}_{\mathrm{aug}} = \sum_{k=1}^{K} \beta_{k} \underbrace{\left(\mathrm{diag}\left\{\underline{\mathcal{F}}_{s,k}\right\}\mathbb{M}^{T}\underline{S}_{Tx}^{*}\left(\theta_{k}\right) \otimes \underline{S}_{Rx}\left(\theta_{k}\right) \otimes \mathbb{J}^{l_{k}}\underline{c}\right)}_{\triangleq \underline{h}_{\mathrm{aug}}(\theta_{k},\mathcal{F}_{k},l_{k})} +\,\underline{n}_{\mathrm{aug}} \tag{4.50}$$

With reference to Fig. 4.10, the spatiotemporal parameter estimator presented in this section is based on Eq. (4.50), see point C, and provides at points D, E and F the delays, Doppler frequencies, DOAs and complex path gains for all targets.

4.4.1.1 *Subspace partitioning and delay estimation*

First of all, note that — from a subspace perspective — targets having different delays lie in (almost) orthogonal subspaces (i.e. their respective response vectors, $\underline{h}_{\mathrm{aug}}(\theta,\mathcal{F},l)$, are almost orthogonal), since the columns of:

$$\mathbb{C} \triangleq [\mathbb{J}^{0}\underline{c}, \mathbb{J}^{1}\underline{c}, \ldots, \mathbb{J}^{\mathcal{N}_{c}-1}\underline{c}] \in \mathcal{C}^{2\mathcal{N}_{c}\times\mathcal{N}_{c}} \tag{4.51}$$

are almost orthogonal, due to the optimal auto-correlation properties of the m-sequence $\underline{c}$. In other words, contrary to DOA and Doppler, there is

---

[3]The effect of neglecting intra-symbol Doppler on estimation performance is studied in Section 4.4.3.

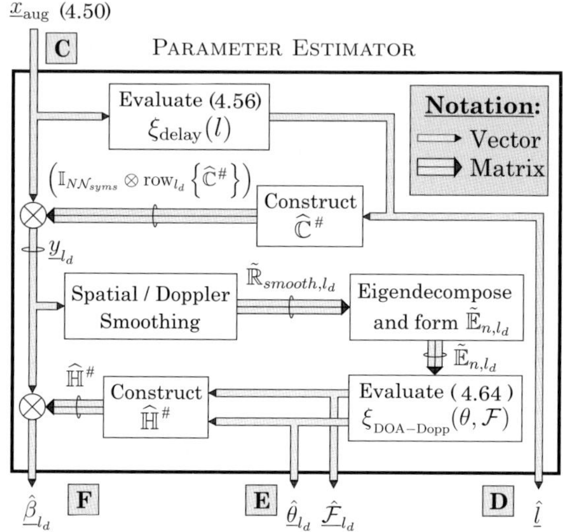

Fig. 4.10   Proposed receiver architecture for estimating delays, Doppler shifts, DOAs and fading coefficients from the vectorised received data, $\underline{x}_{\mathrm{aug}}$.

no real sense of proximity in delay; any two targets' delay subspaces are either identical or (almost) orthogonal. As a result, each of these subspaces may be isolated and processed independently. To clarify this procedure, it is useful to rewrite Eq. (4.50) as:

$$\underline{x}_{\mathrm{aug}} = \sum_{k=1}^{K} \beta_k (\underline{h}(\theta_k, \mathcal{F}_k) \otimes \mathbb{J}^{l_k}\underline{c}) + \underline{n}_{\mathrm{aug}} \tag{4.52}$$

where $\underline{h}(\theta, \mathcal{F})$, is related to the virtual array manifold vector $\underline{S}_{\mathrm{v}}(\theta)$ as follows:

$$\underline{h}(\theta, \mathcal{F}) \triangleq \mathsf{A}(\mathcal{F}) \underbrace{(\underline{S}_{Tx}^{*}(\theta) \otimes \underline{S}_{Rx}(\theta))}_{\underline{S}_{\mathrm{v}}(\theta)}. \tag{4.53}$$

The Doppler-dependent linear mapping, $\mathsf{A}(\mathcal{F}_k)$, is defined as:

$$\mathsf{A}(\mathcal{F}_k) \triangleq \mathrm{diag}\{\underline{\mathcal{F}}_{s,k}\}\mathbb{M}^{T} \otimes \mathbb{I}_{N_{Rx}}. \tag{4.54}$$

Therefore, $\underline{h}(\theta, \mathcal{F})$ can simply be viewed as the extended manifold vector [25] obtained from $\underline{S}_{\mathrm{v}}(\theta)$ via the linear mapping $\mathsf{A}(\mathcal{F})$.

So, based on Eq. (4.52), a traditional (range compression type) method for isolating each desired delay, l_d, would be to compute $(\mathbb{I}_{\mathcal{N}_s N_{Rx}} \otimes \mathbb{J}^{l_d}\underline{c})\underline{x}_{\mathrm{aug}}$. However, under this approach, residual contributions from other delays can remain. Despite being small in general, these "leakage" terms can be severely damaging to subsequent estimation stages (see Fig. 4.12 in Section 4.4.3).

As described in Appendix 4.A, an initial improvement is to first completely eliminate all unwanted contributions (while also minimising any attenuation to desired terms) by computing $(\mathbb{I}_{\mathcal{N}_s N_{Rx}} \otimes \mathbb{P}_{\mathbb{C}} \mathbb{P}^{\perp}_{\mathbb{C}_{l_d}}) \underline{x}_{\mathrm{aug}}$, where

$$\mathbb{P}_{\mathbb{C}} \triangleq \mathbb{C} \underbrace{\left(\mathbb{C}^H \mathbb{C}\right)^{-1} \mathbb{C}^H}_{\mathbb{C}^\#} \tag{4.55}$$

is the projection onto the $\mathcal{N}_c$-dimensional delay space spanned by the columns of $\mathbb{C}$ (and $\mathbb{C}^\#$ is the pseudoinverse of $\mathbb{C}$). Similarly,

$$\mathbb{P}^{\perp}_{\mathbb{C}_{l_d}} \triangleq \mathbb{I}_{2\mathcal{N}_c} - \mathbb{P}_{\mathbb{C}_{l_d}}$$

is the projection onto the subspace *orthogonal* to $\mathbb{C}_{l_d}$ (where $\mathbb{C}_{l_d}$ is defined as $\mathbb{C}$ with the column $\mathbb{J}^{l_d}\underline{c}$ removed).

A second improvement is to note that any small attenuation to the desired term (see A_{l_d} in Eq. (4.81)) can be even further reduced by removing any unnecessary columns from $\mathbb{C}$. In other words, if the $K_\tau \leq \mathcal{N}_c$ distinct delays existing in the signal environment can be estimated first, then $\mathbb{C}$ can be replaced by $\widehat{\mathbb{C}}$, whose columns correspond only to these estimated delays.

Specifically, to obtain these estimates, the following simple cost function can be evaluated (for $l_d = 0, 1, \ldots, \mathcal{N}_c - 1$):

$$\xi_{\mathrm{delay}}(l_d) \triangleq \frac{1}{N_{Rx}L} \left\| (\mathbb{I}_{\mathcal{N}_s N_{Rx}} \otimes \mathbb{J}^{l_d}\underline{c})^T \underline{x}_{\mathrm{aug}} \right\|^2 \tag{4.56}$$

which provides delay estimates, $\underline{\hat{l}} \triangleq [\hat{l}_1, \hat{l}_2, \ldots, \hat{l}_{K_\tau}]^T$, at its K_τ largest values (Point D in Fig. 4.10). Thus, $\widehat{\mathbb{C}}$ can be constructed as follows:

$$\widehat{\mathbb{C}} = [\mathbb{J}^{\hat{l}_1}\underline{c}, \mathbb{J}^{\hat{l}_2}\underline{c}, \ldots, \mathbb{J}^{\hat{l}_{K_\tau}}\underline{c}]. \tag{4.57}$$

Note that by applying the proposed projection, $(\mathbb{I}_{\mathcal{N}_s N_{Rx}} \otimes \mathbb{P}_{\widehat{\mathbb{C}}} \mathbb{P}^{\perp}_{\widehat{\mathbb{C}}_l})$, any remaining signal/noise is confined to a specific $\mathcal{N}_s N_{Rx}$-dimensional subspace (while the remainder of the $2\mathcal{N}_s \mathcal{N}_c N_{Rx}$-dimensional observation space is now redundant). In Appendix 4.A, it is shown that the relevant subspace may be extracted explicitly by computing $\underline{\breve{y}}_{l_d}$ in (4.81). For the present discussion, a more convenient normalisation of exactly the same procedure is:

$$\underline{y}_{l_d} \triangleq (\mathbb{I}_{N_{Rx}\mathcal{N}_s} \otimes \mathrm{row}_{l_d}\{\widehat{\mathbb{C}}^\#\}) \underline{x}_{\mathrm{aug}}$$

$$= \sum_{\{l_k = l_d\}} \beta_k \underline{h}(\theta_k, \mathcal{F}_k) + \underline{n}_{l_d} \tag{4.58}$$

where $\mathrm{row}_{l_d}\{\widehat{\mathbb{C}}^\#\}$ denotes the row of the pseudoinverse of $\widehat{\mathbb{C}}$ associated with delay l_d and $\{l_k = l_d\}$ corresponds to the K_{l_d} targets having delay l_d. The properties of the noise are unchanged (assuming any small amplification to be negligible).

This provides the basis for the proposed two-stage estimation procedure (summarised in Fig. 4.10), since the known structure of $\underline{h}(\theta, \mathcal{F})$ can now be exploited (as described in the next subsection) to estimate DOA and Doppler from $\underline{y}_{l_d}$. Although it is proved in Appendix 4.A that this is fundamentally equivalent to an exhaustive three-parameter search of $\underline{h}_{\text{aug}}(\theta, \mathcal{F}, l)$, there are three main advantages offered by the two-stage procedure. First, since $(\theta, \mathcal{F})$ estimation must only be computed for $K_\tau \leq K$ delays, a significant reduction in computational complexity can be attained. Second, subspace-based (MUSIC-type) parameter estimation algorithms are known to introduce more non-ideality for larger noise subspaces [26]. The proposed two-stage approach operates on noise subspaces of reduced dimension, yielding a predictable and significant performance enhancement for algorithms of this type. Finally, signals having different delays are completely separated, thus avoiding a variety of coherent sources-type problems that may otherwise arise when different targets have equal DOA and Doppler.

4.4.1.2 *Joint DOA-Doppler estimation*

Equation (4.58) provides only a single snapshot of data and, furthermore, describes a classic rank-1 "coherent sources"-type scenario (since fading coefficients $\beta_k, \forall_k$, are not time-varying over a small observation interval). As such, subspace-type parameter estimation approaches cannot be applied directly to $\underline{y}_{l_d}$.

In order to obtain multiple snapshots, one may consider "unstacking" $\underline{y}_{l_d}$ into a matrix in some structured manner. However, many such approaches may exist and managing the trade-off between number of snapshots and dimensionality of the manifold vector is not straightforward. Furthermore, such approaches would be unlikely to effectively combat the issue of rank-1 coherent type paths.

Instead, the proposed approach is to select overlapping subvectors from $\underline{y}_{l_d}$ according to two "smoothing" schemes (which are detailed below). Then, using existing results for selecting optimal subvector sizes [27], a reasonable trade-off between snapshots, manifold dimensionality and "decorrelation" will be achieved.

First, it is necessary to look in more detail at the transmitted symbol matrix, $\mathbb{M}$. In particular, it will be seen that it is useful to make the transmitted symbols repeat in a periodic sequence. Therefore, we can write:

$$\mathbb{M} = \underline{1}_{\mathcal{D}}^{T} \otimes \mathbb{M}_0 \tag{4.59}$$

where $\mathcal{D}$ is the total number of periods transmitted and the $(N_{Tx} \times N_{Tx})$ matrix $\mathbb{M}_0$ denotes the symbol sequence (such that $\mathcal{N}_s = N_{Tx}\mathcal{D}$). Therefore, $\underline{h}(\theta, \mathcal{F})$ in Eq. (4.53) can equivalently be written as:

$$\underline{h}(\theta, \mathcal{F}) = \underline{\mathcal{F}}_p \otimes (\text{diag}\{\underline{\mathcal{F}}_{ss}\}\mathbb{M}_0^T \otimes \mathbb{I}_{N_{Rx}})(\underline{S}_{Tx}^* \otimes \underline{S}_{Rx}) \tag{4.60}$$

where $\underline{\mathcal{F}}_{ss}$ and $\underline{\mathcal{F}}_p$ are, respectively, the symbol-to-symbol and period-to-period Doppler effects, defined in the obvious manner from:

$$\underline{\mathcal{F}}_s = \underline{\mathcal{F}}_p \otimes \underline{\mathcal{F}}_{ss}. \tag{4.61}$$

With reference to Eq. (4.60), the two smoothing procedures can now be more clearly described. First, for certain specific receiver array geometries (in particular, uniform linear), spatial smoothing [28] can be overlaid directly to "decorrelate" paths having different DOAs. Furthermore, if it is known *a priori* that radial velocities are sufficiently small such that

$$(\mathbb{I}_{\mathcal{D}} \otimes \mathbb{M}_0^* \otimes \mathbb{I}_{N_{Rx}})\underline{h}(\theta, \mathcal{F}) \approx (\underline{\mathcal{F}}_p \otimes \underline{S}_{Tx}^* \otimes \underline{S}_{Rx}),$$

and certain collinear transmit-receive array structures are used [29], then it is possible to apply spatial smoothing across the full extent of $\underline{S}_{Tx}^* \otimes \underline{S}_{Rx}$ (i.e. across the virtual array).

Based on the same concept as spatial smoothing, a new decorrelating technique termed "Doppler smoothing" is depicted in Fig. 4.11.

Doppler Smoothing requires the extraction of Q overlapping subvectors from $\underline{y}_{l_d}$, each of length $dN_{Tx}N_{Rx}$, where $d = \mathcal{D} - Q + 1$ is the effective

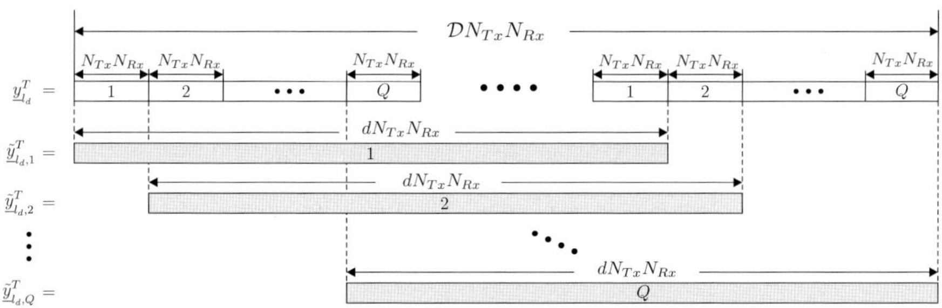

Fig. 4.11   Forming the Q overlapping subvectors for Doppler smoothing.

Doppler dimension after smoothing. Denoting the q-th subvector of $\underline{y}_{l_d}$ as $\underline{\tilde{y}}_{l_d,q}$ and collecting these for all $q = 1, 2, \ldots, Q$ yields:

$$\tilde{\mathbb{Y}}_{l_d} \triangleq [\underline{\tilde{y}}_{l_d,1}, \underline{\tilde{y}}_{l_d,2}, \ldots, \underline{\tilde{y}}_{l_d,Q}]. \tag{4.62}$$

In an equivalent manner to spatial smoothing, this acts to "decorrelate" paths at the expense of reducing the length of $\underline{\mathcal{F}}_p$ in $\underline{h}(\theta, \mathcal{F})$ from $\mathcal{D}$ to d.

Since Doppler smoothing operates only on $\underline{\mathcal{F}}_p$, spatial smoothing can now be overlaid independently on $\tilde{\mathbb{Y}}_{l_d}$ by extracting the appropriate Q_{ss} subvectors from each column. The resulting $Q_{\mathrm{ss}}Q$ total subvectors are denoted by the matrix $\tilde{\mathbb{Y}}_{\mathrm{smooth},l_d}$, allowing us to finally compute the "smoothed" sample covariance matrix:

$$\tilde{\mathbb{R}}_{\mathrm{smooth},l_d} \triangleq \frac{1}{Q_{\mathrm{ss}}Q} \tilde{\mathbb{Y}}_{\mathrm{smooth},l_d} \tilde{\mathbb{Y}}^H_{\mathrm{smooth},l_d}. \tag{4.63}$$

Thus, joint DOA-Doppler estimation is achieved (for each of the K_τ distinct delays) by evaluating the following MUSIC-like subspace-based cost function:

$$\xi_{\mathrm{DOA-Dopp}}(\theta, \mathcal{F}) \triangleq \frac{\|\underline{\tilde{h}}(\theta, \mathcal{F})\|^2}{\underline{\tilde{h}}^H(\theta, \mathcal{F}) \tilde{\mathbb{E}}_{n,l_d} \tilde{\mathbb{E}}^H_{n,l_d} \underline{\tilde{h}}(\theta, \mathcal{F})} \tag{4.64}$$

where the columns of $\tilde{\mathbb{E}}_{n,l_d}$ are the estimated noise eigenvectors of $\tilde{\mathbb{R}}_{\mathrm{smooth},l_d}$ and $\underline{\tilde{h}}(\theta, \mathcal{F})$ is the reduced-aperture space-Doppler manifold vector after smoothing. The estimated DOAs and Doppler frequencies associated with the delay l_d are therefore obtained by locating the K_{l_d} largest *maxima* in $\xi_{\mathrm{DOA-Dopp}}(\theta, \mathcal{F})$. These estimates are denoted, respectively, as $\underline{\hat{\theta}}_{l_d}$ and $\underline{\hat{\mathcal{F}}}_{l_d}$ (Point E in Fig. 4.10).

4.4.1.3 *Complex fading coefficients estimation*

Although path fading coefficients were treated as nuisance parameters in the above two-stage parameter estimation process, they can now be estimated in a straightforward manner by noting that (4.58) can be rewritten as:

$$\underline{y}_{l_d} = \mathbb{H}_{l_d} \underline{\beta}_{l_d} + \underline{n}_{l_d} \tag{4.65}$$

where $\mathbb{H}_{l_d}$ is the $(\mathcal{N}_s N_{Rx} \times K_{l_d})$ matrix of virtual space-Doppler manifold vectors associated with delay l_d and $\underline{\beta}_{l_d}$ comprises the related fading coefficients. Therefore, $\underline{\beta}_{l_d}$ may simply be estimated as follows (Point F in Fig. 4.10):

$$\underline{\hat{\beta}}_{l_d} = \widehat{\mathbb{H}}^\# \underline{y}_{l_d} \tag{4.66}$$

where $\widehat{\mathbb{H}}$ is the estimated space-Doppler channel response matrix, whose columns are constructed by inserting elements of $\underline{\hat{\theta}}_{l_d}$ and $\underline{\hat{\mathcal{F}}}_{l_d}$ (pairwise) into Eq. (4.53). Thus, if perfect $(\theta, \mathcal{F})$ estimation is achieved, Eq. (4.66) will provide the maximum likelihood estimate of $\underline{\beta}_{l_d}$.

4.4.1.4 *Algorithm summary — spatiotemporal arrayed MIMO*

(1) Pre-process:

    (a) Sample $\underline{x}(t)$ and collect snapshots to form $[\mathbb{X}[1], \mathbb{X}[2], \ldots, \mathbb{X}[\mathcal{N}_s]]$.

    (b) Vectorise received data to form $\underline{x}_{\mathrm{aug}}$ (Eq. (4.50)).

(2) Estimate delays:

    (a) Obtain delay estimates at K_τ largest values in ξ_{delay} (Eq. (4.56)).

(3) Estimate DOA and Doppler:

    (a) Construct estimated temporal response matrix $\widehat{\mathbb{C}}$ (Eq. (4.57)) and use it to compute $\underline{y}_{l_d}$ (Eq. (4.58)).

    (b) for $l_d = \hat{l}_1, \hat{l}_2, \ldots, \hat{l}_{K_\tau}$

        i. Apply Doppler smoothing and/or spatial smoothing and compute $\tilde{\mathbb{R}}_{\mathrm{smooth},l_d}$ (Eq. (4.63)).

        ii. Eigendecompose $\tilde{\mathbb{R}}_{\mathrm{smooth},l_d}$ and evaluate $\xi_{\mathrm{DOA-Dopp}}(\theta, \mathcal{F})$ (Eq. (4.64)).

        iii. Obtain joint DOA-Doppler estimates at the K_{l_d} highest spectral *maxima* in $\xi_{\mathrm{DOA-Dopp}}(\theta, \mathcal{F})$.

(4) Estimate path fading coefficients:

    (a) Construct estimated space-Doppler channel response matrix, $\widehat{\mathbb{H}}$.

    (b) Obtain fading coefficient estimates by inserting $\widehat{\mathbb{H}}$ in Eq. (4.66).

4.4.2 *Iterative adaptive approach (IAA)*

Another joint Doppler, delay and DOA estimation is the IAA method [30]. This works by iteratively updating the estimates of a traditional "delay and sum" (matched filter)-type estimator, using the solution of a weighted LS optimisation. It was applied to MIMO radar in [31]. Unlike the spatiotemporal approaches discussed in Section 4.4.1, IAA supports the estimation of Doppler and delay (in addition to DOA and β) and so can be applicable to the original signal model in Eq. (4.16). However, the derivation of IAA is based on a different signal model, which views the received signal as a linear combination of contributions from *all* points in a discrete DOA-Doppler-delay parameter grid (but where only K coefficients, $\beta(\theta_q, l_p, \mathcal{F}_h)$, are non-zero):

$$\mathbb{X} = \sum_{p=1}^{P} \sum_{q=1}^{Q} \sum_{h=1}^{H} \beta(\theta_q, l_p, \mathcal{F}_h) \underline{S}_{Rx}(\theta_q) \underline{S}_{Tx}^{H}(\theta_q) \mathbb{M}(\mathbb{J}_L^T)^{l_p} \text{diag}\{\underline{\mathcal{F}}_h\} + \mathbb{N}$$

$$(4.67)$$

where p is the delay index; $P - 1$ denotes the maximum delay between the echoes from various range bins and the first received echo (i.e. from the closest range bin to the radar); q is the azimuth index; Q is the number of potential targets from different directions in the same range bin; h is the Doppler index; and H denotes the number of bins in the Doppler interval of interest. Moreover, $\mathbb{J}_L^T \in \mathcal{R}^{(L \times L)}$ acts to right-shift (i.e. delay) the columns of $\mathbb{M}$, which is zero-padded as follows:

$$\mathbb{M} = [\underline{a}[1], \underline{a}[2], \dots, \underline{a}[\mathcal{N}_s], \quad \mathbb{O}_{N_{Tx} \times \mathcal{N}_s}] \qquad (4.68)$$

where the $(N_{Tx} \times \mathcal{N}_s)$ matrix of zeros, $\mathbb{O}_{N_{Tx} \times \mathcal{N}_s}$, has been appended under the assumption that the maximum relative path delay is less than $\mathcal{N}_s T_c$, leading to $L = 2\mathcal{N}_s$ total snapshots.

It is important to point out that (4.67) should be connected to the original model given in Eq. (4.16), by observing that $\beta(\theta_q, l_p, \mathcal{F}_h)$ is only non-zero for those points in the parameter grid corresponding to true target parameters (which are assumed to be aligned to the grid).

So, in IAA the initial "delay and sum" estimate of β is obtained by evaluating the following cost function across the entire DOA-Doppler-delay parameter grid:

$$\hat{\beta}_{IAA}^{[1]}(\theta, l, \mathcal{F}) = \frac{\underline{h}_{IAA}^{H}(\theta, l, \mathcal{F}) \text{vec}\{\mathbb{X}\}}{\|\underline{h}_{IAA}(\theta, l, \mathcal{F})\|^2} \qquad (4.69)$$

where $\hat{\beta}_{IAA}^{[n]}$ denotes the n-th iteration of IAA and $\underline{h}_{IAA}(\theta, l, \mathcal{F})$ is defined as

$$\underline{h}_{IAA}(\theta, l, \mathcal{F}) \triangleq \text{vec}\{\underline{S}_{Rx}(\theta) \underline{S}_{Tx}^{H}(\theta) \mathbb{M}(\mathbb{J}_L^T)^{l} \text{diag}\{\underline{\mathcal{F}}\}\}. \qquad (4.70)$$

In order to iteratively refine the β estimates, the first step is to use $\hat{\beta}_{IAA}^{[n]}$ to refine the following estimated covariance matrix (based on the structure of $\text{vec}\{\mathbb{X}\}$):

$$\hat{\mathbb{R}}_{IAA}^{[n]} \triangleq \sum_{p=1}^{P} \sum_{q=1}^{Q} \sum_{h=1}^{H} |\hat{\beta}_{IAA}^{[n]}(\theta_q, l_p, \mathcal{F}_h)|^2 \underline{h}_{IAA}(\theta_q, l_p, \mathcal{F}_h) \underline{h}_{IAA}^{H}(\theta_q, l_p, \mathcal{F}_h).$$

$$(4.71)$$

Then, the *interference* covariance matrix is estimated as follows:

$$\hat{\mathbb{Z}}^{[n]}(\theta,l,\mathcal{F}) \triangleq \hat{\mathbb{R}}^{[n]}_{IAA} - |\hat{\beta}^{[n]}_{IAA}(\theta,l,\mathcal{F})|^2 \underline{h}_{IAA}(\theta,l,\mathcal{F})\underline{h}^{H}_{IAA}(\theta,l,\mathcal{F}).$$

(4.72)

Finally, the IAA method solves the following weighted LS optimisation problem:

$$\min_{\beta} \left((\text{vec}\left\{\mathbb{X}\right\} - \beta\underline{h}_{IAA}(\theta,l,\mathcal{F}))^H \mathbb{Z}^{-1}(\text{vec}\{\mathbb{X}\} - \beta\underline{h}_{IAA}(\theta,l,\mathcal{F})) \right)$$

(4.73)

which is expressed as a function of the true *theoretical* interference covariance matrix $\mathbb{Z}$ and yields the theoretical solution:

$$\beta_{IAA}(\theta,l,\mathcal{F}) = \frac{\underline{h}^{H}_{IAA}(\theta,l,\mathcal{F})\mathbb{R}^{-1}_{IAA}\,\text{vec}\{\mathbb{X}\}}{\underline{h}^{H}_{IAA}(\theta,l,\mathcal{F})\,\mathbb{R}^{-1}_{IAA}\underline{h}_{IAA}(\theta,l,\mathcal{F})}.$$

(4.74)

However, in practice, $\mathbb{R}_{IAA}$ in Eq. (4.74), is not exactly known and so we must use the previously-obtained estimate, $\hat{\mathbb{R}}^{[n-1]}_{IAA}$, which is itself a function of the previously-obtained β estimates (see Eq. (4.71)). Of course, this recursive relationship provides the basis for the iterative approach, which can be written as:

$$\hat{\beta}^{[n]}_{IAA}(\theta,l,\mathcal{F}) = \frac{\underline{h}^{H}_{IAA}(\theta,l,\mathcal{F})(\hat{\mathbb{R}}^{[n-1]}_{IAA})^{-1}\,\text{vec}\{\mathbb{X}\}}{\underline{h}^{H}_{IAA}(\theta,l,\mathcal{F})(\hat{\mathbb{R}}^{[n-1]}_{IAA})^{-1}\underline{h}_{IAA}(\theta,l,\mathcal{F})}$$

(4.75)

with initial condition:

$$\hat{\mathbb{R}}^{[0]}_{IAA} = \mathbb{I}_{N_{Rx}L}$$

(4.76)

as required by Eq. (4.69).

A summary of the IAA method is given in Table 4.3. This is a method that has been shown to achieve extremely powerful parameter estimation performance, but at the expense of very high computational cost, particularly in the three-parameter (i.e. θ, l, $\mathcal{F}$) case (see Section 4.4.3).

4.4.3 *Simulation studies*

In order to compare the spatiotemporal arrayed MIMO approach introduced in Section 4.4.1 with the IAA presented in Section 4.4.2, two different simulation scenarios will be considered in this section: one with stationary targets and one with moving targets. For all simulations, a carrier frequency

Table 4.3   IAA for MIMO radar multi-target parameter estimation.

---

initialise:

$$\hat{\mathbb{R}}^{[0]}_{IAA} = \mathbb{I}_{N_{Rx}L}$$

repeat:

$$\beta^{[n]}_{IAA}(\theta,l,\mathcal{F}) = \frac{\underline{h}^{H}_{IAA}(\theta,l,\mathcal{F})(\hat{\mathbb{R}}^{[n-1]}_{IAA})^{-1}\,\mathrm{vec}\{\mathbb{X}\}}{\underline{h}^{H}_{IAA}(\theta,l,\mathcal{F})(\hat{\mathbb{R}}^{(n-1)}_{IAA})^{-1}\underline{h}_{IAA}(\theta,l,\mathcal{F})}$$

$$\hat{\mathbb{R}}^{[n]}_{IAA} \triangleq \sum_{p=1}^{P}\sum_{q=1}^{Q}\sum_{h=1}^{H} |\beta^{[n]}_{IAA}(\theta_q,l_p,\mathcal{F}_h)|^2 \underline{h}_{IAA}(\theta_q,l_p,\mathcal{F}_h)\underline{h}^{H}_{IAA}(\theta_q,l_p,\mathcal{F}_h)$$

until: (desired number of iterations completed)

---

of $F_c = 2\,\mathrm{GHz}$ and chip period of $T_c = 0.8138\mu s$ are used. Due to the orthogonality of the elements of the transmitted waveforms $\underline{m}(t)$, it follows from Eq. (4.5) that the signal to noise ratio of the k-th target return at each receive antenna is given by:

$$\mathrm{SNR}_k = \frac{|\beta_k|^2}{\sigma_n^2}. \tag{4.77}$$

Therefore, for notational simplicity, we define the "reference" SNR as:

$$\mathrm{SNR}_0 \triangleq \frac{\mathrm{Tr}\left\{\mathcal{E}\left\{\underline{m}(t)\underline{m}^H(t)\right\}\right\}}{\sigma_n^2} \tag{4.78}$$

such that from Eq. (4.77), $\mathrm{SNR}_k = |\beta_k|^2\mathrm{SNR}_0$. As described in Section 4.4.1.2, when spatial and/or Doppler smoothing are applied, relevant subvector dimensions are chosen using the approximate optimal value suggested in [27] (except when targets are known in advance to be stationary, then no Doppler dimension is required and the parameter d in Fig. 4.11 is equal to 1, i.e. $d = 1$).

4.4.3.1   *Simulated environment 1: Stationary targets*

The stationary-target scenario used here was taken from [31], wherein the IAA was applied to MIMO radar. Precise parameter values were not provided in [31], Fig. 1(a), but the values used here are approximately the same. In this scenario, an $N_{Rx} = 5$ element receiver ULA and $N_{Tx} = 5$ element transmitter ULA (stretched by a factor of 5 along the x-axis) are employed. The target environment comprises $K = 29$ stationary targets

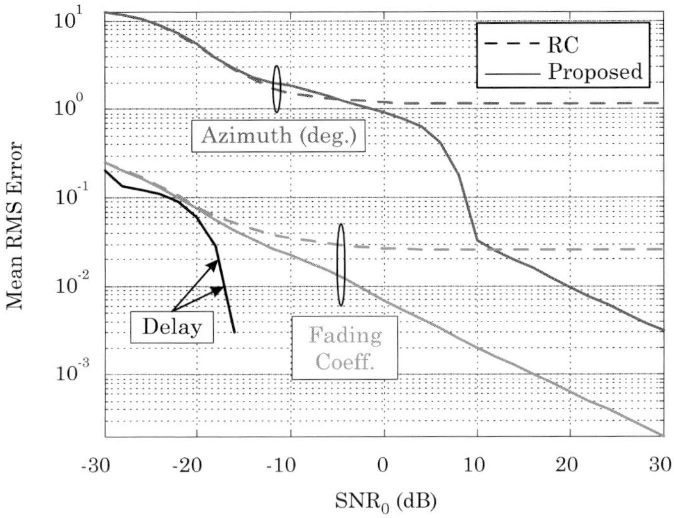

Fig. 4.12   Estimation error comparison of the spatiotemporal arrayed MIMO (proposed) approach and a similar range compression (RC) based approach (1000 trials).

with azimuth values between $60°$ and $120°$ and relative delays between $3T_c$ and $23T_c$. Fading coefficient magnitudes take values between approximately 0.01 and 1.

Figure 4.12 shows the RMSE of the spatiotemporal arrayed MIMO (proposed) approach presented in Section 4.4.1, across all $K = 29$ targets, for -30 dB$\leq SNR_0 \leq 30$ dB. Also plotted is the performance of the same approach, but with Eq. (4.58) replaced by traditional range compression. In both cases, $L = 6350$ snapshots ($\approx 5ms$ observation interval) and $\mathcal{N}_c = 127$. Delays are estimated in the same way for both approaches and Fig. 4.12 indeed shows identical delay estimation performance. At low SNRs (where noise effects dominate), overall estimation performance is similar. However, at higher SNRs, the performance degradation due to "leakage" from other delays becomes significant and the advantage of using the proposed approach is evident.

Although the IAA method is too computationally burdensome for Doppler-delay-DOA estimation performance evaluation, it is possible to investigate under the stationary targets assumption.

The IAA method is particularly well suited to scenarios where very small numbers of snapshots are available (but SNR is sufficiently high), while the proposed method requires a larger number of snapshots (but can operate well at lower SNRs). Therefore, to assess the algorithms' intrinsic

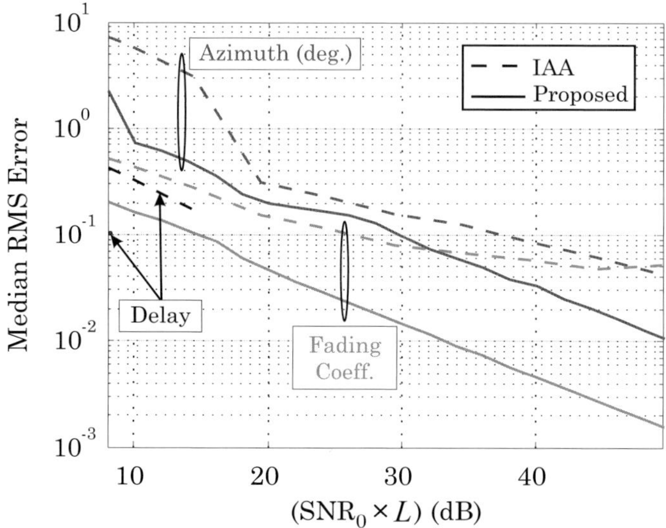

Fig. 4.13   Estimation error comparison of the proposed method and the IAA (500 trials).

capabilities, the two methods are evaluated using different numbers of snapshots, but compared under a constant $(\text{SNR} \times L)$ constraint. As noted in [31], IAA is found to fail to resolve all targets under this scenario (even at SNRs as high as 30 dB). To prevent spurious spectral peaks from skewing results, Fig. 4.13 therefore shows the median RMSE from the $K = 29$ targets. For IAA, Gold sequences of length 63 chips were transmitted, leading to $L = 87$ received snapshots.

It can be seen that the two algorithms have similar DOA estimation performance for $(\text{SNR} \times L)$ between 20 dB and 30 dB. However, the proposed approach shows superior estimation performance across all parameters, particularly for higher $(\text{SNR} \times L)$.

4.4.3.2   *Simulated environment 2: Moving targets*

To demonstrate the full parameter estimation capabilities of the proposed method (in the presence of moving targets), the array configuration of Fig. 4.14 will be used.

The target environment comprises $K = 27$ moving and stationary targets, located across the full 360° azimuth, with radial velocities between -40 ms^{-1} and 60 ms^{-1} (true parameters are marked in Fig. 4.16). Fading coefficient magnitudes were selected independently from a uniform distribution

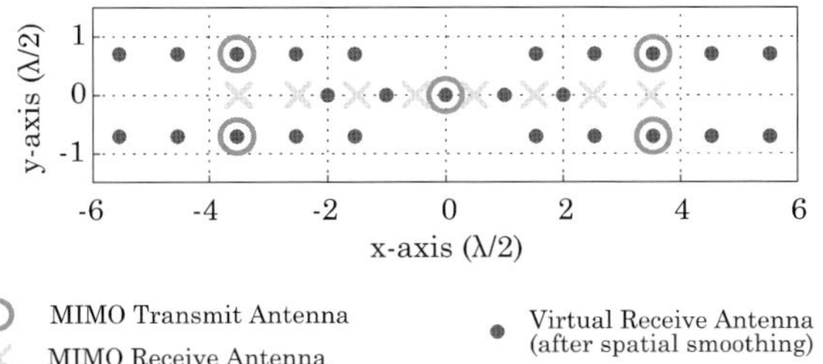

Fig. 4.14   Arrayed MIMO configuration for Simulated Environment 2. Transmit array is an $N_{Tx} = 5$ element X-shaped array and receive array is an $N_{Rx} = 8$ element uniform linear array.

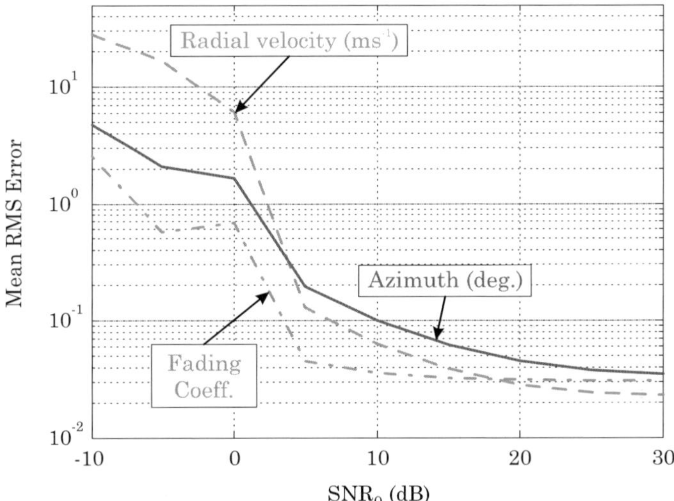

Fig. 4.15   Estimation error evaluation of the proposed method (600 trials). No delay estimation errors occured for $\mathrm{SNR}_0 \geq -10$ dB.

on the interval $[\sqrt{0.1}, 1]$ (such that $\frac{1}{10}\mathrm{SNR}_0 \leq \mathrm{SNR}_k \leq \mathrm{SNR}_0$) and fixed for all trials (true parameters marked in Fig. 4.18). All 27 targets span just two relative delays ($8T_c$ and $9T_c$) and numerous instances of each type of "coherent sources" problems are present. Specifically: 14 targets share identical Doppler and delay with some other target(s); 13 share identical DOA and delay; and eight share identical DOA and Doppler.

In Fig. 4.15, accurate parameter estimation is evident for SNR_0 greater than approximately 5 dB. However, performance is seen to plateau at very high SNRs due to Doppler effects within a symbol period (assumed in Eq. (4.50) to be negligible) taking effect. However, with estimation errors

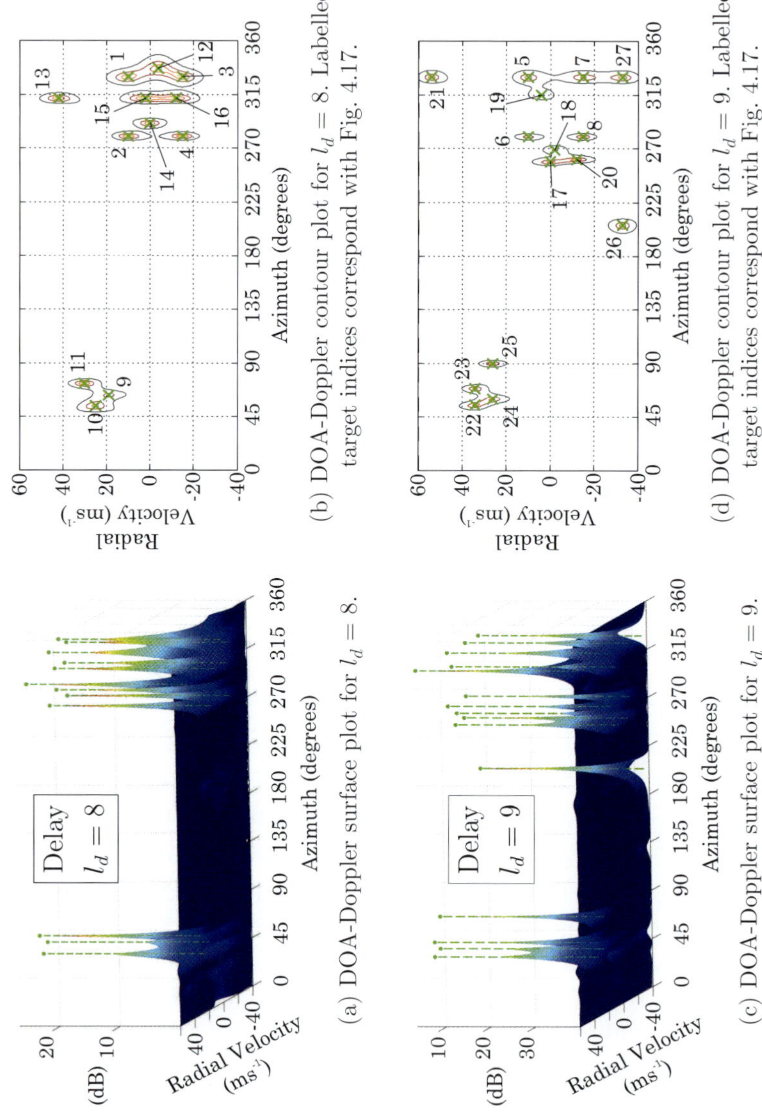

(a) DOA-Doppler surface plot for $l_d = 8$.

(b) DOA-Doppler contour plot for $l_d = 8$. Labelled target indices correspond with Fig. 4.17.

(c) DOA-Doppler surface plot for $l_d = 9$.

(d) DOA-Doppler contour plot for $l_d = 9$. Labelled target indices correspond with Fig. 4.17.

Fig. 4.16   Joint DOA-Doppler estimation for delays $l_d = 8$ and $l_d = 9$ (with true parameters marked in green). $SNR_0 = 10$ dB.

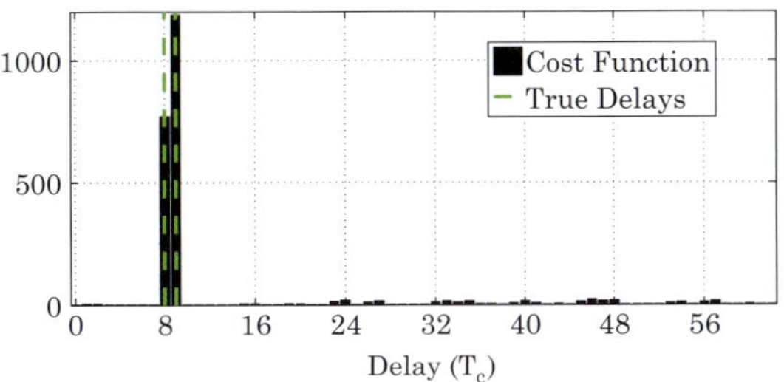

Fig. 4.17   Distinct delays correctly estimated at $8T_c$ and $9T_c$ (SNR$_0$ = 10 dB).

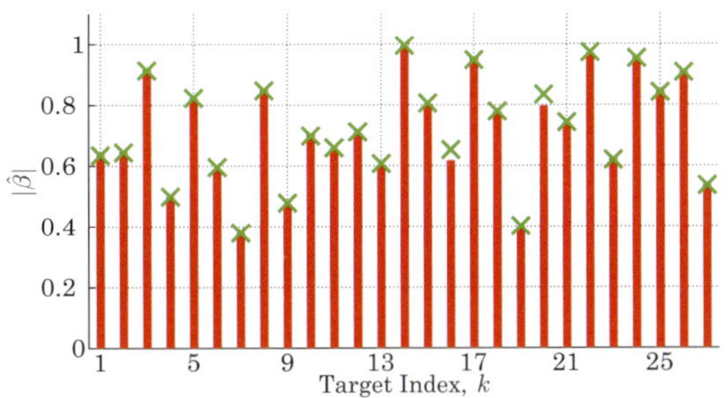

(a) Fading coefficient *magnitudes* for all targets.

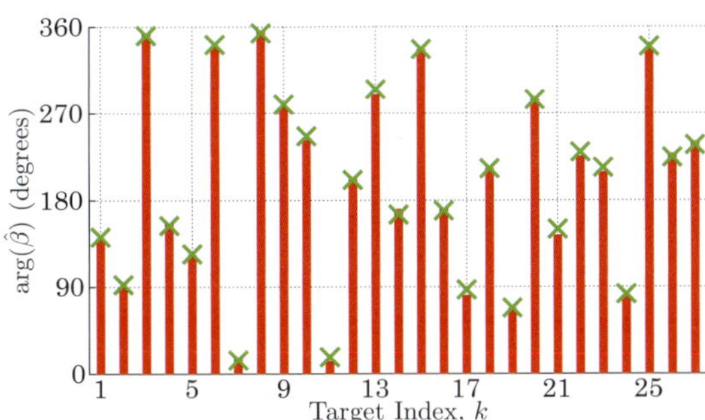

(b) Fading coefficient *phases* for all targets.

Fig. 4.18   Complex path fading coefficient estimates (with true parameters marked with a green "X"). SNR$_0$ = 10 dB.

still measuring only in the hundredths of degrees azimuth/ms^{-1} radial velocity (and with absolute RMS β error lower bounded by $\approx$0.0287), such performance may be tolerable in a number of applications.

Illustrative results from a single trial (with SNR$_0$ = 10 dB) are presented in Figs. 4.16–4.18.

4.4.4   *Complexity analysis*

In this subsection, the number of complex multiplications will be used as a measure of computational complexity of the spatiotemporal arrayed MIMO approach. Eigendecomposition or inversion of an $(N \times N)$ matrix are both assumed to have complexity $\mathcal{O}(N^3)$. Assuming the number of azimuth and Doppler search points (denoted by K_θ and $K_{\mathcal{F}}$, respectively) are sufficiently large, the $\mathcal{O}(\mathcal{N}_c(N_{Rx}L + K_\tau^2))$ delay and fading coefficient estimation stages are dominated by the $\mathcal{O}(K_\tau(\frac{1}{\mathcal{N}_c}N_{Rx}L)^3 + KK_\theta K_{\mathcal{F}}\frac{1}{\mathcal{N}_c}N_{Rx}L)$ DOA-Doppler search cost.

By comparison, the IAA method with $\mathcal{N}_{it}$ iterations is $\mathcal{O}(\mathcal{N}_{it}((N_{Rx}L)^3 + K_\tau K_\theta K_{\mathcal{F}}(N_{Rx}L)^2))$. Since $\mathcal{N}_c > K_\tau$ and $N_{Rx}L \gg K$, it is clear that the complexity of IAA is greater by a factor greatly exceeding $\mathcal{N}_c^2 \mathcal{N}_{it}$.

4.5   Conclusions

In this chapter, the received signal model of a MIMO radar operating in a multi-target scenario has been presented, for both static and moving targets. Based on this model two cases were investigated.

In the first case it was assumed that the target echoes arrive with equal delays or these delays are ignored. In this case these delays are not taken into account in the modelling while the targets are assumed static. Based on this, a number of multi-target parameter estimation techniques, such as LS, Capon, APES, CAPES, CAML and Kai's "iterative" and the so called "multidimensional optimal" method, were presented and supported by comparative computer simulation studies. The main conclusion is that first five methods suffer from the problem of mutual interference effects among the echoes, especially in the case of closely-spaced multiple targets. In Kai's "iterative" and "multidimensional optimal" methods [19] this problem is addressed by using a different modelling in handling interference from other targets, in conjunction with novel interference suppression approaches for parameter estimation. Overall, the "multidimensional optimal" method was found to outperform all the other methods while LS showed the worst performance.

In the second case the target echoes' delays are taken into account in the signal model while the targets were assumed moving. In this case a novel subspace-based method was presented that blends

- the spatiotemporal structure of the 3D datacube that is formed after the discretisation process of the received signal, with

- the concept of the equivalent "virtual" SIMO representation of a MIMO radar system,

in order to provide enhanced fundamental detection and resolution capabilities, compared to any approach which only exploits the receiver array geometry. In particular, this method is based on a single long vector $\underline{x}_{\mathrm{aug}}$ and replaces an exhaustive three-parameter search, which is prohibitively complex to compute in practice, with an equivalent two-stage estimation procedure. In the first stage only the delays are estimated and then, in the second stage, a joint Doppler-DOA estimation process is provided for all targets. Furthermore, having estimated all other target model parameters, it was shown that estimation of path fading coefficients then follows in a straightforward manner. Computer simulations were carried out to evaluate the performance of this spatiotemporal MIMO radar approach and was shown to outperform IAA which is one of the most powerful but complex existing algorithms.

4.A   Appendix: Equivalent two-stage estimation

In this appendix, it will be shown that the proposed in Section 4.4.1 two-stage estimation procedure is fundamentally equivalent to an exhaustive three-parameter search based on $\underline{h}_{\mathrm{aug}}(\theta, \mathcal{F}, l)$. The structure of $\underline{x}_{\mathrm{aug}}$ is recalled from Eq. (4.50) as follows:

$$
\begin{aligned}
\underline{x}_{\mathrm{aug}} &= \sum_{k=1}^{K} \beta_k \underline{h}_{\mathrm{aug}}(\theta_k, \mathcal{F}_k, l_k) + \underline{n}_{\mathrm{aug}} \\
&= \sum_{k=1}^{K} \beta_k \left(\underline{h}(\theta_k, \mathcal{F}_k) \otimes \mathbb{J}^{l_k} \underline{c} \right) + \underline{n}_{\mathrm{aug}}.
\end{aligned}
\tag{4.79}
$$

First of all, it is useful to recognise that there is intrinsic redundancy in the $2 N_{Rx}\mathcal{N}_s\mathcal{N}_c$-dimensional observation space because $(\mathbb{I}_{N_{Rx}\mathcal{N}_s} \otimes \mathbb{C})$ spans only $N_{Rx}\mathcal{N}_s\mathcal{N}_c$ dimensions. Therefore, the remaining $N_{Rx}\mathcal{N}_s\mathcal{N}_c$ dimensions are known to contain only noise, which may be completely eliminated *a priori*, using $(\mathbb{I}_{N_{Rx}\mathcal{N}_s} \otimes \mathbb{P}_{\mathbb{C}})\underline{x}_{\mathrm{aug}}$. Clearly, these dimensions are then totally redundant so, equivalently, we may directly *extract* the relevant $N_{Rx}\mathcal{N}_s\mathcal{N}_c$-dimensional space using:

$$
\begin{aligned}
\overbrace{\underline{x}_{half}}^{(\mathcal{N}_c \times 1)} &\triangleq (\mathbb{I}_{N_{Rx}\mathcal{N}_s} \otimes \mathbb{E}_{\mathbb{C}}^{T}) \overbrace{\underline{x}_{\mathrm{aug}}}^{(2\mathcal{N}_c \times 1)} \\
&= \sum_{k=1}^{K} \beta_k (\underline{h}(\theta_k, \mathcal{F}_k) \otimes \mathbb{J}^{l_k} \underline{c}_{half}) + \underline{n}_{half}
\end{aligned}
\tag{4.80}
$$

where $\mathbb{E}_{\mathbb{C}}$ is a $(2\mathcal{N}_c \times \mathcal{N}_c)$ orthonormal basis that spans the same subspace as the columns $\mathbb{C}$. Clearly, $\mathbb{C}_{half} \triangleq \mathbb{E}_{\mathbb{C}}^T \mathbb{C}$ therefore describes an identical manifold shape as $\mathbb{C}$, but now with no redundancy. Similarly, noise properties are unchanged. Thus, Eqs. (4.79) and (4.80) are fundamentally equivalent with respect to parameter estimation capabilities.

In fact, due to the columns of $\mathbb{C}$ being almost orthogonal, it is possible to extract and process an $N_{Rx}\mathcal{N}_s$-dimensional space associated with just one desired delay, denoted l_d, at each time. More specifically, for complete suppression of leakage terms, we must only consider spaces orthogonal to $(\mathbb{I}_{N_{Rx}\mathcal{N}_s} \otimes \mathbb{C}_{l_d})$ (where $\mathbb{C}_{l_d}$ is defined as $\mathbb{C}$ with the column $\mathbb{J}^{l_d}\underline{c}$ removed). Then, to eliminate intrinsic subspace redundancy (as per Eq. (4.80)), we must only consider spaces lying within the span of $(\mathbb{I}_{N_{Rx}\mathcal{N}_s} \otimes \mathbb{C})$.

To *isolate* this subspace, we could therefore use $(\mathbb{I}_{N_{Rx}\mathcal{N}_s} \otimes \mathbb{P}_{\mathbb{C}_{l_d}}^{\perp} \mathbb{P}_{\mathbb{C}})$ but, of course, the subspace may be explicitly *extracted*. An intuitive way to achieve this is to left-multiply $\underline{x}_{\text{aug}}$ by $(\mathbb{I}_{N_{Rx}\mathcal{N}_s} \otimes \mathbb{P}_{\mathbb{C}_{l_d}}^{\perp} \mathbb{J}^{l_d}\underline{c})^T$, which may be interpreted as a superresolution leakage-cancelling beamformer:

$$\underline{\breve{y}}_{l_d} \triangleq \left(\mathbb{I}_{N_{Rx}\mathcal{N}_s} \otimes \mathbb{P}_{\mathbb{C}_{l_d}}^{\perp} \mathbb{J}^{l_d}\underline{c}\right)^T \underline{x}_{\text{aug}}$$
$$= A_{l_d} \sum_{\{l_k = l_d\}} \beta_k \sqrt{2\mathcal{N}_c}\,\underline{h}(\theta_k, \mathcal{F}_k) + \underline{\breve{n}}_{l_d} \tag{4.81}$$

where the small attenuation, A_{l_d}, is assumed negligible:

$$A_{l_d} \triangleq \frac{1}{\sqrt{2\mathcal{N}_c}} \left\|\mathbb{P}_{\widehat{\mathbb{C}}_{l_d}}^{\perp} \mathbb{J}^{l_d}\underline{c}\right\| \approx 1. \tag{4.82}$$

Furthermore, note that $\mathbb{C}$ has been replaced by $\widehat{\mathbb{C}}$, which comprises only the K_τ columns associated with the $K_\tau \leq \mathcal{N}_c$ distinct delays existing in the signal environment (estimated first). Using $\widehat{\mathbb{C}}$ therefore minimises any attenuation due to A_{l_d} and also ensures that only K_τ subsequent $(\theta, \mathcal{F})$ searches must be computed. Furthermore, estimating delays separately causes no fundamental degradation in delay estimation performance, since inclusion of $\underline{h}(\theta, \mathcal{F})$ in the manifold vector cannot increase the angle between different columns of $\mathbb{C}$, since these are already (approximately) orthogonal.

Since $A_{l_d} \approx 1$, computing $\underline{\breve{y}}_{l_d}$ in Eq. (4.81) therefore amounts to perfect isolation of signal terms associated with delay l_d. The manifold of $\sqrt{2\mathcal{N}_c}\underline{h}(\theta, \mathcal{F})$ has an identical shape to that of $\underline{h}_{\text{aug}}(\theta, \mathcal{F}, l_d)$. Therefore, it is clear to see that Eqs. (4.79) and (4.81) are fundamentally equivalent with respect to parameter estimation performance.

References

[1] D. W. Bliss and K. W. Forsythe, "Multiple-input multiple-output (MIMO) radar and imaging: degrees of freedom and resolution," in *2004 Conference Record of the Thirty-Seventh Asimolar Conference on Signals, Systems and Computers*, vol. 1, Pacific Grove, CA, Nov. 2003, pp. 54–59.

[2] D. J. Rabideau and P. Parker, "Ubiquitous MIMO multifunction digital array radar," in *2004 Conference Record of the Thirty-Seventh Asilomar Conference on Signals, Systems and Computers*, vol. 1, Pacific Grove, Nov. 2003, pp. 1057–1064.

[3] A. M. Haimovich, R. S. Blum, and L. J. Cimini, "MIMO radar with widely separated antennas," *IEEE Signal Processing Magazine*, vol. 25, no. 1, pp. 116–129, Jan. 2008.

[4] N. Levanon, *Radar Principles*, 1st ed., ser. A Wiley-Interscience publication. Hoboken, US: John Wiley & Sons, 1988.

[5] H. L. V. Trees, *Detection, Estimation, and Modulation Theory, vol. III*. Hoboken, US: John Wiley & Sons, 1968.

[6] J. Li and P. Stoica, "MIMO radar with colocated antennas," *IEEE Signal Processing Magazine*, vol. 24, no. 5, pp. 106–114, Sep. 2007.

[7] H. Krim and M. Viberg, "Two decades of array signal processing research: the parametric approach," *IEEE Signal Processing Magazine*, pp. 67–94, Jul. 1996.

[8] L. Xu, J. Li, and P. Stoica, "Radar imaging via adaptive MIMO techniques," in *14th European Signal Processing Conference*, Florence, Italy, Sep. 2006.

[9] ——, "Adaptive techniques for MIMO radar," in *Proc. 4th IEEE Workshop on Sensor Array and Multi-Channel Signal Processing*, Waltham, MA, Jul. 2006, pp. 258–262.

[10] J. Li, P. Stoica, L. Xu, and W. Roberts, "On parameter identifiability of MIMO radar," *IEEE Signal Processing Letters*, vol. 14, no. 12, pp. 968–971, Dec. 2007.

[11] L. Xu, J. Li, and P. Stoica, "Target detection and parameter estimation for MIMO radar systems," *IEEE Transactions on Aerospace and Electronic Systems*, vol. 44, no. 3, pp. 927–939, Jul. 2008.

[12] J. Chen, G. Hong, and S. Weimin, "Angle estimation using ESPRIT without pairing in MIMO radar," *Electronics Letters*, vol. 44, no. 24, pp. 1422–1423, Nov. 2008.

[13] A. Hassanien and S. A. Vorobyov, "Direction finding for MIMO radar with colocated antennas using transmit beamspace preprocessing," in *3rd IEEE International Workshop on Computational Advances in Multi-Sensor Adaptive Processing*, Dec. 2009, pp. 181–184.

[14] N. Liu, L.-R. Zhang, J. Zhang, and D. Shen, "Direction finding of MIMO radar through ESPRIT and Kalman filter," *Electronics Letters*, vol. 45, no. 17, pp. 908–910, Aug. 2009.

[15] J. Li and P. Stoica, "An adaptive filtering approach to spectral estimation and SAR imaging," *IEEE Transactions on Signal Processing*, vol. 44, no. 6, pp. 1469–1484, Jun. 1996.

[16] P. Stoica, H. Li, and J. Li, "A new derivation of the APES filter," *IEEE Signal Processing Letters*, vol. 6, no. 8, pp. 205–206, Aug. 1999.

[17] A. Jakobsson and P. Stoica, "Combining Capon and APES for estimation of spectral lines," *Circuits, Systems and Signal Processing*, vol. 19, no. 2, pp. 159–169, 2000.

[18] L. Xu, P. Stoica, and J. Li, "A diagonal growth curve model and some signal-processing applications," *IEEE Transactions on Signal Processing*, vol. 54, no. 9, pp. 3363–3371, Sep. 2006.

[19] K. Luo and A. Manikas, "Superresolution multitarget parameter estimation in MIMO radar," *IEEE Transactions on Geoscience and Remote Sensing*, vol. 51, no. 6, pp. 3683–3693, Jun. 2013.

[20] R. O. Schmidt, "Multiple emitter location and signal parameter estimation," *IEEE Transactions on Antennas and Propagation*, vol. 34, pp. 276–280, Mar. 196.

[21] A. Manikas, *Differential Geometry in Array Processing*.   Imperial College Press, 2004.

[22] I. Bekkerman and J. Tabrikian, "Target detection and localization using MIMO radars and sonars," *IEEE Transactions on Signal Processing*, vol. 54, no. 10, pp. 3873–3883, Oct. 2006.

[23] B. Friedlander, *Adaptive signal design for MIMO radars*, J. Li and P. Stoica, Eds.   Wiley-IEEE Press, Oct. 2008.

[24] M. Wax and T. Kailath, "Detection of signals by information theoretic criteria," *IEEE Transactions on Acoustics, Speech and Signal Processing*, vol. 33, no. 2, pp. 387–392, Apr. 1985.

[25] G. Efstathopoulos and A. Manikas, "Extended array manifolds: functions of array manifolds," *IEEE Transactions on Signal Processing*, vol. 59, no. 7, pp. 3272–3287, Jul. 2011.

[26] H. Commin and A. Manikas, "The figure of merit 'C' for comparing super-resolution direction-finding algorithms," in *Sensor Signal Processing for Defence (SSPD)*, Sep. 2010, pp. 1–5.

[27] A. Gershman and V. Ermolaev, "Optimal subarray size for spatial smoothing," *IEEE Signal Processing Letters*, vol. 2, no. 2, pp. 28–30, Feb. 1995.

[28] T. J. Shan, M. Wax, and T. Kailath, "On spatial smoothing for direction of arrival estimation of coherent signals," *IEEE Transactions on Acoustics, Speech and Signal Processing*, vol. 33, no. 4, pp. 806–811, Aug. 1985.

[29] H. Commin and A. Manikas, "Virtual SIMO radar modelling in arrayed MIMO radar," in *Sensor Signal Processing for Defence (SSPD)*, Sep. 2012, pp. 1–6.

[30] T. Yardibi, J. Li, P. Stoica, M. Xue, and A. Baggeroer, "Source localization and sensing: a nonparametric iterative adaptive approach based on weighted least squares," *IEEE Transactions on Aerospace and Electronic Systems*, vol. 46, no. 1, pp. 425–443, Jan. 2010.

[31] W. Roberts, P. Stoica, J. Li, T. Yardibi, and F. Sadjadi, "Iterative adaptive approaches to MIMO radar imaging," *IEEE Journal of Selected Topics in Signal Processing*, vol. 4, no. 1, pp. 5–20, Feb. 2010.

Chapter 5

Beamforming for Wake Wave Detection and Estimation
— An Overview —

Karen Mak and Athanassios Manikas

Communications and Array Processing,
Department of Electrical and Electronic Engineering,
Imperial College London

In this chapter different types of ship wake waves are described, as well as environment and synthetic aperture radar (SAR) parameters that allow the observation of certain wakes in SAR imagery. A number of common wake wave detection algorithms for 2D SAR imagery are studied. Then examples of SAR systems using two or more SAR beamformers[1] for ocean applications are presented. In particular, the use of two SAR beamformers for the formation of interferograms is given.

5.1   Introduction

In general a SAR transmits electromagnetic chirp signals which are reflected by scatterers present in the area of interest being illuminated. These backscattered signals are then received. Different types of scatterers will reflect the transmitted chirp signals differently and therefore will produce different radar returns. Depending on these radar returns, scatterers will appear with varying intensities in the final SAR image formed after

---

[1]It is assumed that a beam is created using a planar array, where beamforming is performed by applying weights to the elements of the array such that the desired beam is created.

159

processing of the raw SAR signal. If there is significant radar return, and therefore a high radar cross section per unit area, the corresponding scatterers will appear bright in the SAR images due to enhanced backscattering. However, if there is not significant radar return, and therefore a low radar cross section per unit area, the corresponding scatterers will appear dark in the SAR images due to reduced backscattering [1].

SAR is not limited to the use of a single beamformer to form a single beam for transmitting and receiving, where in this chapter, a beamformer is defined as a planar array of N elements which are weighted to form the required beam(s) for transmitting and receiving. Two or more SAR beamformers are used in the area of interferometric SAR (InSAR) where interferograms are created from the phase difference between two or more SAR images. These interferograms can then be used to derive digital elevation maps (DEMs) allowing height estimation of the imaged area of choice.

In terms of ocean applications, different configurations of SARs have been used to form a SAR interferometer for wave height [2] as well as ocean surface current determination, including the velocity of the moving ocean surface [3]. SAR technology has also been used in the more specific area of ship wake wave detection and estimation, where main applications include surveillance and maritime security and safety.

Ship wake waves are visible in SAR imagery due to the changes they create on the ocean surface and therefore the transmitted signals. Different components of ship wake waves have different characteristics in SAR imagery, making them identifiable. Although in some cases, detection of the ship itself would suffice, there are many advantages in detecting the ship's wake waves rather than the physical ship. One advantage is that wake waves can last for many hours and can stretch for several kilometres. Also, in some cases the wake waves are more prominent in SAR images compared to the physical ship. Another advantage is related to the way SAR collects data by moving in the cross-range direction, which often results in the ship to be displaced, or Doppler-shifted, in the image. This, therefore, causes the location of the ship in the cross-range direction of the image to be incorrect. Although this displacement also affects the ocean surface, and therefore the wake wave returns, the amount of Doppler shift is not as significant as it is in the case of the ship, where the amount of Doppler shift depends on the ship's direction and speed. Therefore the wake waves often provide a better estimate of the true location of the ship. However, ship wake waves are not present if the ship is small or slow moving. Despite this, if the wake waves are present and can be imaged with good resolution, many parameters about the ship can be estimated. Examples include the speed, location and the direction at which the ship is pointing, i.e. its heading.

This chapter is structured as follows. In Section 5.2, ship wake wave types will be presented and discussed before environmental conditions and SAR parameters for the imaging of certain wake waves are given in Section 5.3. Then in Section 5.4, the detection of ship wake waves will be investigated. In particular, the pre-processing, transform and post-processing stages will be investigated before the estimation of parameters from ship wake waves is given in Section 5.5. In Section 5.6, SAR systems for ocean applications will be described and finally the chapter is concluded in Section 5.7.

5.2 Types of ship wake waves

There exist different types of ship wake waves which can be classified into three categories, depending on how they are formed. These are [4]:

- ship-generated surface waves,
- turbulent wakes and
- ship-generated internal waves.

5.2.1 *Ship-generated surface wakes*

Ship-generated surface wakes can be further categorised into

- bright narrow V-wakes and
- the components of the Kelvin wake.

Narrow V-wake:
As a SAR transmits chirps at an incidence angle θ_i to an assumed calm ocean surface, specular reflection rarely occurs. Specular reflection only occurs when the incident and reflected transmitted electromagnetic waves form an angle with respect to the perpendicular of the surface of interest and when these two angles are equal. Therefore it is assumed that the received radar returns are due to Bragg scattering.

Bragg scattering occurs when the wavelengths of the ocean waves match an integer multiple of one half of the SAR wavelength, resulting in the contribution of each wave adding constructively so that significant radar return is received by the SAR [1]. The wavelength of the ocean waves moving towards or away from the SAR for Bragg scattering is defined [5] as follows

$$\lambda = \frac{\lambda_r}{2\sin\theta_i} \tag{5.1}$$

where
$$\begin{cases} \lambda & = & \text{wavelength of the ocean waves} \\ \lambda_r & = & \text{wavelength of the radar transmitted signal} \\ \theta_i & = & \text{incident angle.} \end{cases}$$

Due to the significant radar return, and therefore high radar cross section per unit area, these Bragg waves appear bright in SAR images, where the Bragg waves are formed by the wind or ship.

Narrow V-wakes can be modelled based on Bragg scattering. They have a characteristic bright V shape [6], as shown in Fig. 5.1, with an angle $2\alpha_v$, where the half-angle α_v is given below [4]

$$\alpha_v = \tan^{-1}\left(\frac{C_g}{V_s}\cos\theta\right) \tag{5.2}$$

where
$$\begin{cases} C_g & = & \text{the group velocity of the Bragg waves} \\ V_s & = & \text{speed of the ship} \\ \theta & = & \text{the radar azimuth direction with respect to the} \\ & & \text{ship track.} \end{cases}$$

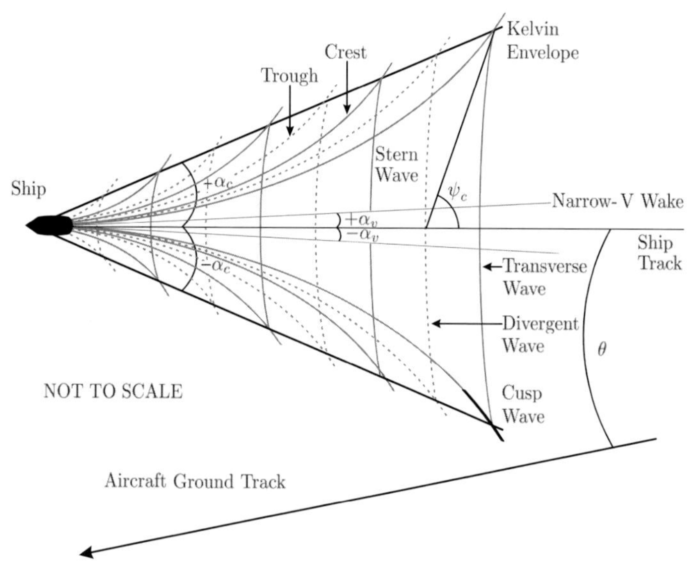

Fig. 5.1   Ship-generated surface wakes.

This half-angle α_v, will decrease in the SAR imagery if the SAR collects the data by moving in the same direction as the ship, and will increase if the SAR moves in the opposite direction to the ship [4], where the amount of decrease or increase depends on the velocities of the SAR and the ship. However, narrow V-wakes can sometimes be obscured by other wake wave types, for example the turbulent wake, which will be described in Section 5.2.2. They can also be obscured by other components along the ship track, such as surface foam generated in the near-field area, where the near-field area is an imaginary area extending around the ship.

Kelvin wakes:
Another category of ship-generated surface wakes are the components of the Kelvin wake, as illustrated in Fig. 5.1, which travels at the same speed as the ship. The Kelvin wake pattern is formed due to the water flow around the ship of interest [1]. The components of a Kelvin wake include the stern and cusp waves, which are all confined to a V-shaped pair of lines called the Kelvin envelope. In an ideal case the Kelvin envelope has a characteristic angle of about $2\alpha_c$, where the half-angle α_c is given as [4]

$$\alpha_c = \pm \sin^{-1}\left(\frac{1}{3}\right) = \pm 19.5°. \tag{5.3}$$

It is at this ideal characteristic angle that divergent waves, which interfere with the transverse wave to form the Kelvin arms on either side of the ship of interest, have their largest amplitude [7]. These waves are the cusp waves [4].

Although in Fig. 5.1 these waves are illustrated as individual wave fronts, in SAR imagery they are often observed as a single bright line, due to the SAR's inability to resolve the individual short wavelength wave fronts. When the waves have a wavelength less than λ_c, which is defined as [4]

$$\lambda_c = \frac{4\pi V_s^2}{3g} \tag{5.4}$$

$$\text{where} \begin{cases} \lambda_c &= \quad \text{wave wavelength} \\ V_s &= \quad \text{speed of the ship} \\ g &= \quad \text{acceleration gravity} \end{cases}$$

then the wave fronts form an angle with respect to the ship track of ψ_c, which is given as follows

$$\psi_c = \tan^{-1}\sqrt{2} = 54.7°. \tag{5.5}$$

If the waves have a wavelength larger than λ_c, the wave fronts form an angle with respect to the ship track that is greater than ψ_c. The longest of these waves are the *stern* waves, which are wave fronts perpendicular to the ship track and have a wavelength λ_s given by the following expression [4]

$$\lambda_s = \frac{2\pi V_s^2}{g} \tag{5.6}$$

$$\text{where} \begin{cases} V_s &= \text{speed of the ship} \\ g &= \text{acceleration gravity.} \end{cases}$$

It can be seen from Eq. (5.6) that the speed of the ship can also be calculated from λ_s.

5.2.2 *Turbulent wakes*

Turbulent wakes are observed in SAR imagery as a central, dark and narrow line behind the ship of interest along the ship track. In general, features that appear dark in SAR images suggest reduced roughness. Therefore turbulent wakes appear dark in SAR imagery as they reduce the ocean surface roughness and so appear smooth when the surrounding ocean is sufficiently rough in comparison. As well as the dark line along the ship track, a bright line on one or both sides of the turbulent wake is sometimes observed, as shown in Fig. 5.2. These bright lines are from the narrow V-wake.

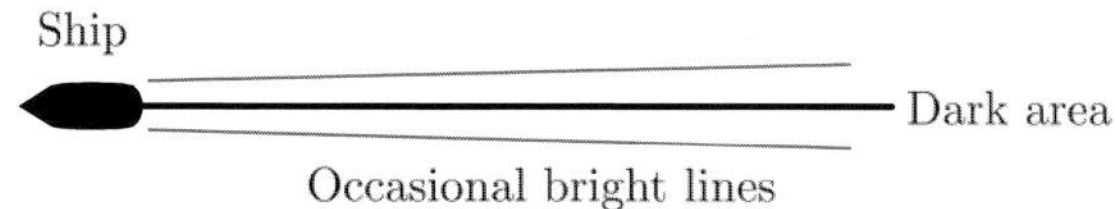

Fig. 5.2   The turbulent wake.

5.2.3 *Ship-generated internal wake waves*

Depending on the conditions when the SAR data is collected, ship-generated internal wake waves may be present in the imagery. For example, they can sometimes be seen in the SAR imagery of coastal areas. These wake waves, which are located between layers of water that have a different density or temperature, are shown in Fig. 5.3, and have a half-angle of α_s given by

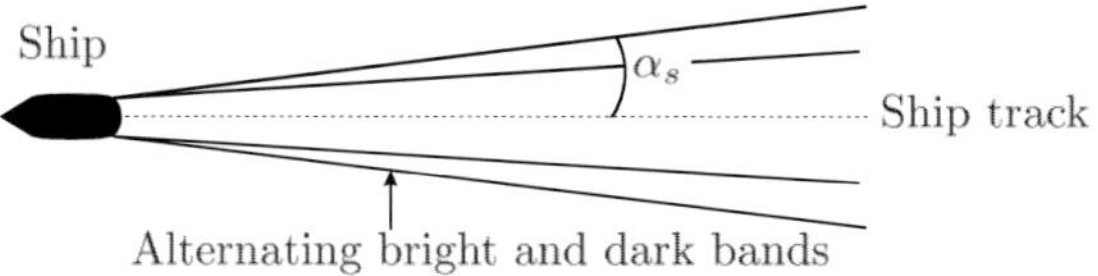

Fig. 5.3   Internal waves created by the ship.

the following [4] equation:

$$\alpha_s = \tan^{-1}\left(\frac{C_{iw}}{V_s}\right) \tag{5.7}$$

$$\text{where} \begin{cases} \alpha_s &= \text{ship wake half-angle} \\ C_{iw} &= \text{internal wave phase speed} \\ V_s &= \text{ship velocity.} \end{cases}$$

5.3   Environmental conditions and SAR parameters for wake wave imaging

As shown in Section 5.2, there are different types of wake waves. However, in SAR wake wave images, not all wake wave types can be observed. The type(s) of wakes observed depends on the size of the ship as well as the sea state. Sea state includes sea surface roughness, which is related to the wind speed and strength.

In terms of environmental conditions, narrow V-wakes are usually observed in low wind conditions, around $3\,\text{m/s}$, whereas Kelvin wakes can still be observed at moderate wind conditions, from about $3\,\text{m/s}$ to $10\,\text{m/s}$. This is also the case for the internal waves created by the ship [4]. The turbulent wakes are only observed when they appear smoother than the surrounding ocean surface and therefore appear darker than the surrounding ocean surface in SAR imagery. This occurs at moderate wind speeds between $3\,\text{m/s}$ and $10\,\text{m/s}$.

In terms of SAR parameters, both narrow-V wakes and Kelvin wakes can be observed at a radar frequency in the L-band (between 1 and $2\,\text{GHz}$). The Kelvin wake, in particular the bright arms forming the characteristic $39°$ angle [8], can be observed in the X-band (between 8 and $12\,\text{GHz}$). Turbulent wakes and internal waves created by the ship can also be observed in both the L-band and the X-band.

Polarisation can also affect how well the wake waves are imaged [9]. In VV polarised SAR images, where vertically (V) polarised electromagnetic waves are transmitted and only vertically (V) polarised reflections are received, the ship wakes can be clearly seen, whereas HH polarisation, where horizontally (H) polarised electromagnetic waves are transmitted and only horizontally (H) polarised reflections are received, produces images where the physical ship can be clearly observed with rarely seen wake waves. VH polarisation, where vertically (V) polarised electromagnetic waves are transmitted and only horizontally (H) polarised reflections are received, provides a better contrast between the ship and the ocean surface, showing the ship as a small, bright cluster of pixels and the ocean surface as a dark background.

Ship wake waves can also often be observed when the ship travels in approximately the same direction as the SAR. This is especially true in the case narrow V-wakes [5].

5.4　Detection approaches for wake waves

There are many techniques in the existing literature for the detection of ship wakes in SAR imagery [8][10]. Most of these techniques include one or more of the following three stages:

- the pre-processing stage,
- the transform stage and
- the post-processing and detection stage.

In the transform stage it is desirable to have a higher signal-to-noise ratio (SNR) in the transform space compared to the original SAR image space, as well as a concentration of the signal of interest around a point for easier detection. As it can be assumed that the wake waves are approximately linear features in the SAR images, transforms such as the Radon transform and Hough transform are often used since these transform linear features into points or spikes, for detection. Examples of the types of techniques used in each stage of pre-processing, transform and post-processing for detection are given in Fig. 5.4, where the techniques in bold boxes will be discussed in more detail.

5.4.1　*Pre-processing stage*

One aim of the pre-processing stage is to reduce the amount of speckle in the image and therefore increase the SNR. There are many techniques that

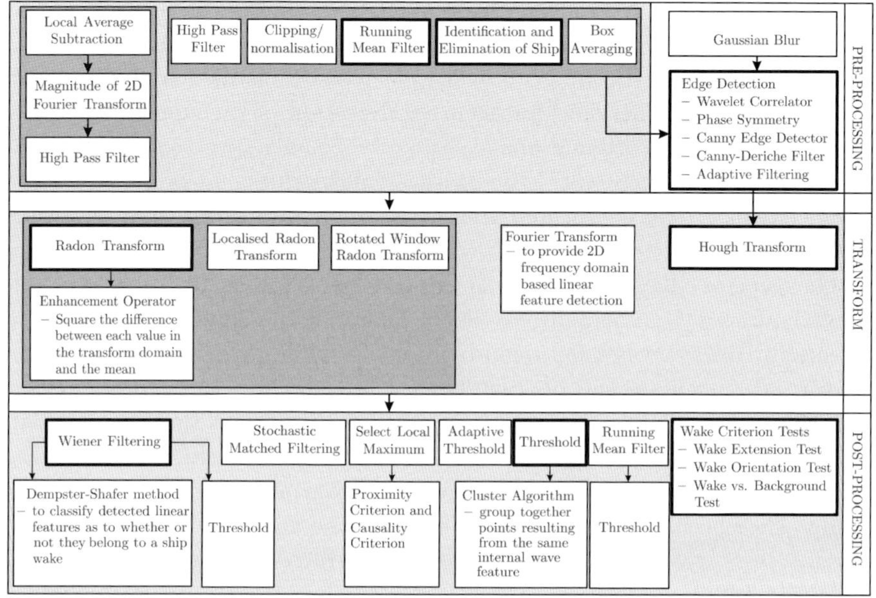

Fig. 5.4   Ship wake wave detection — stages and representative techniques.

can be applied, which in some cases also increase the visibility of the wake waves in the image.

One such technique is the running mean filter. First a window with dimensions of $n \times n$ pixels is defined and centred on a pixel in the image. The mean of the pixel amplitudes in the window is then subtracted from the amplitude of the centre pixel. The window is then moved through the image and the subtraction repeated for all pixels [11]. This has the effect of increasing the image SNR.

Another aim of the pre-processing stage is to remove the bright pixels corresponding to the ship returns. This is achieved by detecting the bright pixels and then replacing the pixel amplitude values with the mean pixel amplitude value of the entire image.

5.4.2   Transform stage

After the pre-processing stage, the image is represented in a different plane. Two common transforms are the Radon transform and the related Hough transform, where both can be used due to the assumption that wake waves are linear features [12]. However, in the Hough transform case, an edge detector needs to be additionally applied in the pre-processing stage to detect the edges corresponding to the wake waves. Sudden changes due to noise fluctuations in the image intensity can also be mistaken for a local edge by edge detectors, therefore for accurate detection of the edges with a low false alarm rate, the original image cannot be heavily speckled. Edge detection algorithms also require some form of despeckling in order to ensure enhancement of the edges, and thus enhance the appearance of the wake waves.

One type of edge detector is the Canny edge detector. In [13] it is implemented after a Gaussian filter, which smooths the image prior to edge detection. However, as mentioned before, the sudden changes in the image intensity due to noise can be mistaken for a local edge, therefore one solution is to increase the amount of smoothing performed by the Gaussian filter. However, in this case, only the strong edges in the image will be detected by the Canny edge detector. Another type of edge detector is the Canny–Deriche filter, which is derived from the Canny edge detector and has a better performance [14]. However, it still behaves poorly when the image is highly corrupted by speckle. Symmetry can also be used in order to identify edges of the wake waves. This is the idea of phase symmetry [15], where phase information is used to measure the symmetry due to patterns in phase that are seen at points of symmetry and asymmetry. Another technique that can be used for edge detection is the Wavelet transform, as described for SAR images in [16] and [17].

After edge detection has been performed, the edge elements are transformed into another plane using the Hough transform [18], which is related to the Radon transform and will be described later. Like the Radon transform, the Hough transform also transforms a line in the original (x, y) plane into a single point in the transform plane. This is achieved by creating a binary edge map, where a pixel in the (x, y) plane is given a value of 0 if it is not located on an edge and a value of 1 if it is located on an edge. Then all the possible lines that pass through each pixel are considered. A process known as "voting" is implemented so that the number of pixels that are located on a line with equation

$$\rho = x \cos \theta + y \sin \theta \tag{5.8}$$

is related to the number of "votes" in the corresponding (ρ, θ) bin in the transform plane. Here ρ is the distance from the origin of the image to the line of interest and θ is the angle between the x-axis and the normal of the line of interest. Therefore if there are a large number of pixels on a particular line, there will be a large number of "votes" in the line's corresponding bin in the transform plane [8].

If the Radon transform is used, an edge detector is not required. In general a straight line in the (x, y) plane can be written as Eq. (5.8), and the Radon transform maps the (x, y) coordinates of the line into (ρ, θ) coordinates in the Radon plane. Mathematically the Radon transform [11], applied to the image $g(x, y)$ to be transformed, is given as

$$R(\rho, \theta) = \iint_D g(x, y)\delta(\rho - x \cos \theta - y \sin \theta)dydx \qquad (5.9)$$

where
$$\begin{cases}
R(\rho, \theta) &= \text{normal distance from the origin to the line} \\
D &= \text{image plane} \\
\delta(\cdot) &= \text{Delta function} \\
\rho &= \text{normal distance from the origin to the line} \\
\theta &= \text{angle between the } x\text{-axis and the normal of the} \\
& \quad \text{line.}
\end{cases}$$

The properties of the Radon transform include:

(1) A straight line in the (x, y) plane is represented as a point (ρ, θ) in the Radon plane, due to the integration of the image intensity along all the lines in the image. Therefore detection of the bright or dark linear features of wake waves becomes detection of the positive or negative spikes in the Radon plane.

(2) A point in the (x, y) plane is represented as a sinusoid in the Radon plane. Therefore in the use of the Radon transform there are detection problems due to speckle or natural or man-made non-linear features in the image before transformation, as these can swamp the peaks corresponding to the linear wake wave features. This is one of the reasons for the detection and removal of the bright returns of the physical ship in the pre-processing stage, as it is represented by bright sinusoids in the Radon plane. These may affect the accuracy of the detection of the peaks corresponding to the wake waves, especially if the ship's bright pixels are brighter than the components of its corresponding wake.

(3) Lines in the (x, y) plane that cross the point (x_0, y_0) correspond to points along the $\rho = x_0 \cos \theta + y_0 \sin \theta$ curve in the Radon plane.

(4) The integration performed by the Radon transform has the effect of averaging the noise intensity fluctuations resulting in a higher SNR in the Radon plane compared to in the original plane. Due to this

averaging effect the Radon transform is more immune to speckle noise compared to the related Hough transform.

The continuality of the line does not affect the Radon transform, however, there may be difficulty in the detection of peaks in the Radon plane corresponding to linear features whose dimensions are a lot smaller than the whole image. Even if these linear features appear a lot brighter or darker than the surrounding background, longer, less bright or dark lines may be represented as peaks with the same magnitude as the short linear features in the Radon plane [19]. In addition to this, the transform does not give information about the length or the positions of the end-points of the linear features.

Sometimes the Radon transform is normalised by calculating the square root of the length of the integration associated with the elements in the Radon plane and dividing the elements by it. This has the effect of creating a constant probability of false alarm. Moreover, the performance of the Radon transform also depends on the size of the image on which the transform is being performed. In general if the image has dimensions of the same order as the wake wave of interest, the performance is increased for the detection of that particular wake wave. This leads to the idea of the localised Radon transform, where the Radon transform is computed on overlapping sub-images [12]. In general, the greater the difference in the dimensions of the sub-image and the wake of interest, the lower the SNR. However, very small sub-images have the effect of removing the noise fluctuations in the image, which reduces the averaging factor [20].

5.4.3 *Post-processing*

After the image has been transformed, processing and analysis are applied for detection. In the transformed space the lines corresponding to the wake waves are transformed to a positive or negative peak, therefore the detection problem is to detect these peaks. One method is to first apply a running mean filter, which has the effect of enhancing the peaks, and then to apply a threshold of $m + k\sigma$ [21], where m is the mean of the transformed image, k is a constant (defined as between 3 and 4 in [8] and greater than 3 if the noise is Gaussian in [21]) and σ is the standard deviation of the transformed image. Any bin in the transformed image with a modulus that is greater than this threshold is then considered to be a positive or negative peak corresponding to a bright or dark line in the original image.

However, a single peak in the transformed image does not always occupy a single bin, and therefore cannot be modelled as a single dirac. Instead the shape and magnitude of a peak in the transformed image can be matched to a predetermined ideal peak modelled as a Gaussian function. In [11] this Gaussian is defined as having a magnitude five times greater than the standard deviation of the noise in the transform domain. Peaks in the transform plane with a magnitude within a certain range centred on the ideal model peak amplitude are then processed using two Wiener filters: one for the positive peaks and one for the negative peaks. This has the effect of increasing their sharpness. Then a threshold can be applied to detect these enhanced peaks.

Another method that can be used is stochastic matched filtering [22], where the aim is to expand a noise corrupted signal data matrix $\mathbb{Y}$ formed from the collected SAR data at a single beamformer into a sum of uncorrelated random variable weighted basis functions. The basis functions are chosen such that the SNR of the signal is increased after processing. By using the maximum likelihood criterion, an inequality is formed from which a decision D_0 corresponding to Hypothesis H_0, where no signal of interest is present in the observation, and D_1 corresponding to the Hypothesis H_1, where the signal of interest is present in the observation, can be made. Therefore, by using this threshold, detection of the signals of interest can be achieved.

Criterion tests can also be applied in order to decide whether a wake wave is present or not. In [23] three criterion tests are used: these are the wake extension test, the wake orientation test and the wake versus background test. These are applied to two transformed images, where one is from the transform of the left half of the SAR image and the other from the right half. For the wake extension test, a wake candidate from each transformed image is selected and then compared to see if they have the same length from the origin to the wake and the same angle measured between the wake normal and the x-axis. If these lengths and angles are equal, the two wake candidates form a single extended wake and are rejected. For the wake orientation test, the orientation of the candidate wake is compared to the ship's orientation, from which the decision of whether the candidate wake is a wake wave can be made. The type of wake wave can also be determined. For the wake versus background test, the normalised radar cross section of the candidate wake is compared with that of the background ocean surface. Along with the wake orientation test, this allows the determination of the type of wake.

However, there are problems related to the detection of ship wake waves for the estimation of parameters. One problem is the speckle noise present in the SAR imagery, which may hide the wake waves in the imagery. This is particularly the case when the wake waves are from small, slow moving ships. Algorithms can be used in order to reduce the amount of speckle; however, they may also reduce the intensity of the wake wave features and the speckle may not be completely removed. It is also assumed that the ship wake waves are linear features and therefore linear detectors such as the Radon transform are used. However this is not always true. For example, there is a curvature in the wake wave when the ship changes direction or speed or due to the ship's yaw and sway. This curvature can create problems with the use of the Radon transform. However, the localised Radon transform is the solution to this problem. Another problem is false alarms, which may be due to non-wake linear features in the SAR imagery. Non-linear features can also cause problems in the detection process. These features could either be natural or man-made. One particular example are oil spills, which after being transformed may swamp the peaks corresponding to the linear features resulting in difficulties in peak detection.

5.5   Estimation of parameters from ship wake waves

There are a number of parameters that can be estimated from ship wake waves. These include:

- the speed of the ship,
- the moving direction of the ship,
- the heading of the ship, defined as the direction the ship points towards,
- the course of the ship, defined as the angle between the ship's direction and a reference, for example true north, and
- the beam of the ship, defined as the width of the ship.

5.5.1   *Parameter estimation from Kelvin envelope*

The Kelvin envelope, in particular the local wavelength, can be used in order to estimate the speed of its corresponding ship. The local wavelength can be derived by calculating where the turbulent and divergent waves of the Kelvin wake cross the Kelvin envelope. This occurs at half the characteristic angle of the Kelvin envelope, α_c. The difference between the values at consecutive wave crests allows the derivation of the local wavelength λ_{α_c} on the Kelvin arms given as follows

$$\lambda_{\alpha_c} = \frac{4\pi V_s^2}{\sqrt{3}g} \tag{5.10}$$

$$\text{where} \begin{cases} V_s & = & \text{speed of the ship} \\ g & = & \text{acceleration gravity.} \end{cases}$$

Rearranging gives the speed of the ship

$$V_s = \sqrt{\frac{\sqrt{3}g\lambda_{\alpha_c}}{4\pi}} \tag{5.11}$$

$$\text{where} \begin{cases} \lambda_{\alpha_c} & = & \text{local wavelength on the Kelvin arms} \\ g & = & \text{acceleration gravity.} \end{cases}$$

5.5.2 *Parameter estimation from stern waves*

The ship's speed can also be estimated using the wavelengths of the stern waves [4] given in Eq. (5.6) in Section 5.2.1 rearranged as shown below

$$V_s = \sqrt{\frac{g\lambda_s}{2\pi}} \tag{5.12}$$

where λ_s is the wavelength of the stern waves and g represents the acceleration gravity.

Comparing Eqs. (5.11) and (5.12) the difference is the factors ($\frac{\sqrt{3}}{4}$ in Eq. (5.11) and $\frac{1}{2}$ in Eq. (5.12)); however, $\frac{\sqrt{3}}{4} \approx \frac{1}{2}$.

The displacement or Doppler-shift between the ship and its wake wave in the SAR image can also be used in the estimation of the ship's speed. The displacement can be described [4] using the following equation

$$\Delta X = \frac{R}{V} V_r \tag{5.13}$$

$$\text{where} \begin{cases} R & = & \text{distance between the ship and the SAR (slant} \\ & & \text{range)} \\ V & = & \text{velocity of the SAR} \\ V_r & = & \text{velocity of the ship in the range direction.} \end{cases}$$

Therefore the velocity of the ship in the range direction can be determined from the displacement from its wake wave.

5.5.3   *Parameter estimation from turbulent wake*

The turbulent wake CAN also allow estimation of the ship's parameters including its propulsion system [24]. In particular the beam of the ship, B, can be estimated, where the beam is defined as the width of the widest part of the ship. From values of the width W of the turbulent wake at different distances aft of the ship, a relationship between the two can be generated in the form

$$W(x_L L) = w_B B \tag{5.14}$$

where
$$\begin{cases} x_L & = & \text{weight of the length of the ship giving the distance} \\ & & \text{aft of the ship} \\ L & = & \text{length of the ship} \\ w_B & = & \text{weight of the beam of the ship giving the width of} \\ & & \text{the turbulent wake at a specific distance } x_L L \text{ aft} \\ & & \text{of the ship} \\ B & = & \text{beam of the ship.} \end{cases}$$

From [25] it is suggested that at a distance of four ship lengths aft of the ship the turbulent wake width is approximately four ship beams. By writing the relationship between the turbulent wake width and the ship's beam in the form

$$W(x) = (A x B^{\alpha-1})^{\frac{1}{\alpha}} \tag{5.15}$$

where
$$\begin{cases} A & = & \text{constant of proportionally} \\ x & = & \text{distance aft of the ship} \\ \alpha & = & \text{a constant between 4 and 5} \end{cases}$$

an expression for the constant of proportionality A can be derived and the width [26] of the turbulent wake can be given as

$$W(x) = \frac{w_B}{\left(\frac{x_L L}{B}\right)^{\frac{1}{\alpha}}} B^{\frac{(\alpha-1)}{\alpha}} x^{\frac{1}{\alpha}}. \tag{5.16}$$

If $W(x)$ is known, the parameter α and the beam of the ship can be estimated [26].

Using the quasi- or pseudo-maximum likelihood estimator, the beam of the ship can be estimated from the turbulent wake by first obtaining [26]

$$\widetilde{W} = \log W(x) \tag{5.17a}$$

$$= \log\left(\frac{w_B}{\left(\frac{x_L L}{B}\right)^{\frac{1}{\alpha}}}\right) + \log\left(B^{\left(\frac{\alpha-1}{\alpha}\right)}\right) + \log\left(x^{\frac{1}{\alpha}}\right) \tag{5.17b}$$

$$= \frac{1}{\alpha} \log x + \left(\frac{\alpha - 1}{\alpha} \right) \log B + \log \left(\frac{w_B}{\left(\frac{x_L L}{B} \right)^{\frac{1}{\alpha}}} \right) \tag{5.17c}$$

$$= a\xi + b \tag{5.17d}$$

$$\text{where} \begin{cases} a &= \frac{1}{\alpha} \\ \xi &= \log x \\ b &= \left(\frac{\alpha-1}{\alpha} \right) \log B + \log \left(\frac{w_B}{\left(\frac{x_L L}{B} \right)^{\frac{1}{\alpha}}} \right). \end{cases}$$

Then by applying the least squares method, a and b can be evaluated allowing an estimation of α and B using [26].

$$\widehat{\alpha} = \frac{1}{a} \tag{5.18}$$

$$\widehat{B} = \exp \left(\frac{b}{1 - \left(\frac{1}{\widehat{\alpha}} \right)} \right) \left(\frac{L}{B} \right)^{\frac{1}{(\widehat{\alpha}-1)}} \left(\frac{x_L}{w_B^{\widehat{\alpha}}} \right)^{\frac{1}{(\widehat{\alpha}-1)}} \tag{5.19}$$

$$\text{where} \begin{cases} \widehat{\alpha} &= \text{estimate of the parameter } \alpha \\ \widehat{B} &= \text{estimate of } B, \text{ the beam of the ship.} \end{cases}$$

Therefore the beam of the ship can be estimated from the turbulent wake.

Other parameters derived from the ship wake waves can also be used for ship parameter estimation. One example is the orientation of the wake. This can be used for the estimation of the ship's course, where the course is the angle [10] between a reference, for example true north, and the path of the ship.

5.6 SAR for ocean applications

In this section, examples of SAR systems for ocean applications will be described. In particular systems that use more than one beamformer will be examined. In order to understand these systems for ocean applications, an introduction to InSAR will be given.

5.6.1 *Interferometric SAR*

InSAR involves the use of two different observations of an area of interest. These two observations can either come from:

(1) the use of two beamformers simultaneously, i.e. a single-pass SAR interferometer, or

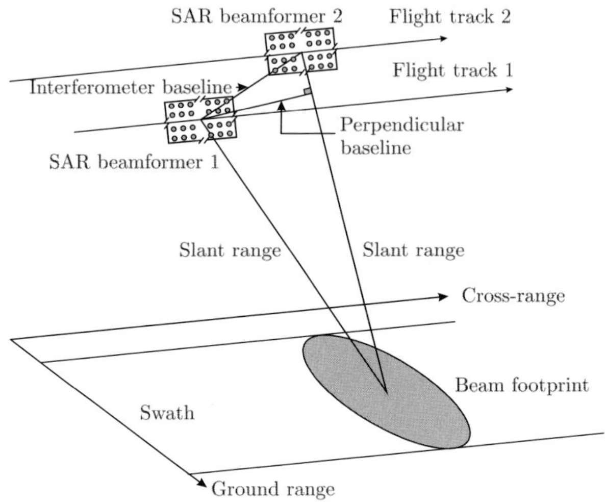

Fig. 5.5   Geometry of an InSAR in the case where two SARs are utilised simultaneously.

(2) the use of one beamformer that images the same area of interest at different times, i.e. a repeat pass SAR interferometer.

Examples of time intervals when the second option is used include one day for ERS-1, 35 days for ERS-2 and a multiple of 35 days for ENVISAT [27].

The geometry when two SAR beamformers are used simultaneously is illustrated in Fig. 5.5, where the interferometric baseline is defined as the distance between the two beamformers located in the plane perpendicular to the orbit's plane, and the perpendicular baseline is defined as the interferometer baseline's projection perpendicular to the slant range. The slant range is the distance between the single SAR beamformer and a particular scatterer on the Earth's surface and the swath is the total area imaged by the SAR system.

A SAR interferogram is produced after processing of the two SAR images with a pixel-by-pixel multiplication of one of the images by the complex conjugate of the second. The interferogram amplitude is therefore the multiplication of the amplitude of the first and second images, and the interferometric phase is the phase difference between the two images. The interferometric phase is used to produce the SAR interferogram, where the topography is shown by fringes representing contour lines. The altitude between the fringes is such that there is a 2π change in the interferogram phase after the effects of the Earth's curvature on the phase change has

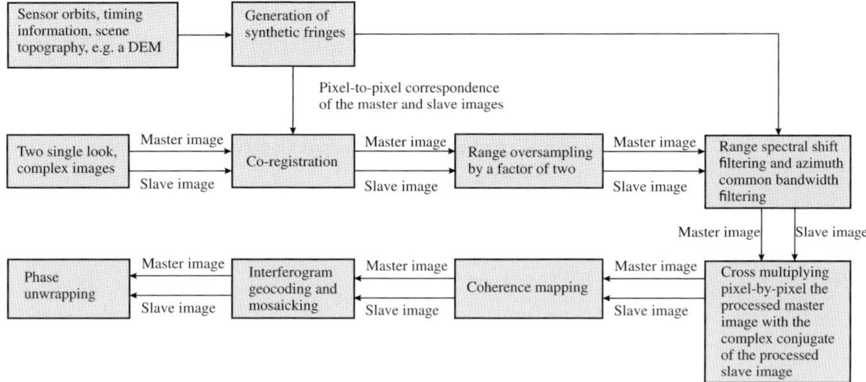

Fig. 5.6   Flow diagram of the formation of a SAR interferogram.

been removed i.e. after interferogram flattening has been performed. This altitude difference is called the altitude of ambiguity. A conceptual flow diagram of the formation of a SAR interferogram from two single look, complex images [27] is shown in Fig. 5.6 and its main blocks are discussed below.

Coregistration: Due to the use of two different observations, there are differences between the two obtained images. These differences could be due to differences in sensor altitude, orbit skew and baseline-induced deformations, to name but a few. Coregistration takes these into account. Here one of the complex images is chosen as a reference and is called the master image. The other image, called the slave image, is then pixel-by-pixel aligned to the master image. This ensures that both images now have the same reference and so each scatterer return from the ground at a certain range and azimuth location corresponds to the same pixel in both the master and slave images.

Generation of synthetic fringes: Generation of the synthetic fringes provides the pixel-to-pixel correspondence between the master and slave images in the coregistration step, as well as information of the spectral shift filtering step. In order to generate the synthetic fringes, information about the sensor orbits, timing information and scene topography are required in order to estimate the interferometric phase. This information could come from a digital elevation model (DEM). The interferogram vector phase can be estimated using the following equation

$$\underline{\psi} = \frac{4\pi}{\lambda}[\underline{R}_M - \underline{R}_S] \tag{5.20}$$

$$\text{where} \begin{cases} \lambda & = & \text{wavelength of transmitted electromagnetic signal} \\ \underline{R}_M & = & (M \times 1) \text{ vector of the slant ranges measured from} \\ & & \text{the master SAR to all } M \text{ targets on the ground} \\ \underline{R}_S & = & (M \times 1) \text{ vector of the slant ranges measured from} \\ & & \text{the slave SAR to all } M \text{ targets on the ground.} \end{cases}$$

Range oversampling: This is performed on both images after coregistration. As mentioned before, a SAR interferogram is obtained from the pixel-by-pixel multiplication of the two observations after processing. In order to avoid uncorrelated contributions when this multiplication is computed, range oversampling by a factor of two is implemented on both images. This step allows the generation of high-quality SAR interferograms.

Range spectral shift and azimuth common band filtering: The aim of these is to remove the uncorrelated spectral contributions in both images while keeping the mostly correlated contributions. Therefore the phase noise terms are removed.

Interferogram computation: Here the processed master image is pixel-by-pixel multiplied with the complex conjugate of the processed slave image. This produces an interferogram which has the same ground range and cross-range reference as the master SAR image, where the interferometric phase is derived from the difference between the phase of the master and slave images.

Generation of coherence maps: This can be performed before or after the phase unwrapping stage and the information provided by the maps can be used to analyse the imaged area, as areas with low coherence appear dark whereas areas with high coherence appear bright. Masking of the incoherent areas can also be performed [28].

Interferogram geocoding and mosaicking: Often smaller interferograms are used in order to form a longer interferogram. However, these smaller interferograms may overlap and be skewed with respect to each other and therefore cannot be simply joined together. Therefore mosaicking needs to be applied [27]. Geocoding is then applied to the mosaicked interferogram by resampling and mapping it onto the surface of the earth [27].

Phase unwrapping: As mentioned before, the altitude between each fringe of a SAR interferogram is the altitude of ambiguity. Due to the periodicity of the transmitted signals, slant ranges with a difference of an integer multiple of the wavelength of the transmitted signal will give the same phase change between the received chirp signal and the corresponding transmitted chirp signal. Therefore the interferogram so far gives a measurement of altitude from the interferogram phase, but with any integer number of altitudes of ambiguity removed. The aim of phase unwrapping is to add to the interferometric fringes the correct integer number of altitudes of ambiguity [29], thus giving the unambiguous phase of ψ, as given in Eq. (5.21), to each pixel:

$$\psi = \phi + 2\pi n \qquad (5.21)$$

$$\text{where} \begin{cases} \phi & = & \text{the wrapped phase, i.e. the phase values with any} \\ & & \text{integer number of altitudes of ambiguity removed} \\ n & = & \text{the integer number of } 2\pi \text{ cycles to be added to the} \\ & & \text{wrapped phase.} \end{cases}$$

As SAR interferometry allows the determination of heights (and also changes in heights), DEMs can be derived from SAR interferometry, where DEMs use the phase information provided by SAR interferograms and converts them to elevation. The accuracy of the formed DEMs depends on the baseline, i.e. the distance between the two SARs, as well as on the imaging environment. By combining ascending and descending DEMs, where ascending DEMs go from south to north and descending DEMs go from north to south, more precise DEMs can be achieved [30].

5.6.2 *SAR interferometry configurations for ocean applications*

SAR interferometry can be used for ocean applications, as the surface, motion and height of the waves all create phase differences between the transmitted and received signals [2]. Depending on the geometry of the SAR systems used to create a SAR interferometer, either along-track or across-track SAR interferometry can be achieved. Figure 5.7 shows an along-track InSAR, WHERE it can be seen that the two beamformers will illuminate the same area but with a time delay determined by the distance between them and their horizontal velocities. There will also be a phase difference due to the Doppler velocity of the scatterers. Along-track InSAR has been used for ocean surface current measurements, including the velocity of the moving ocean surface [3].

In across-track interferometry two or more SARs are configured such that the separation between them is perpendicular to the flight direction i.e. the cross-range direction as shown in Fig. 5.8. Therefore in the case of

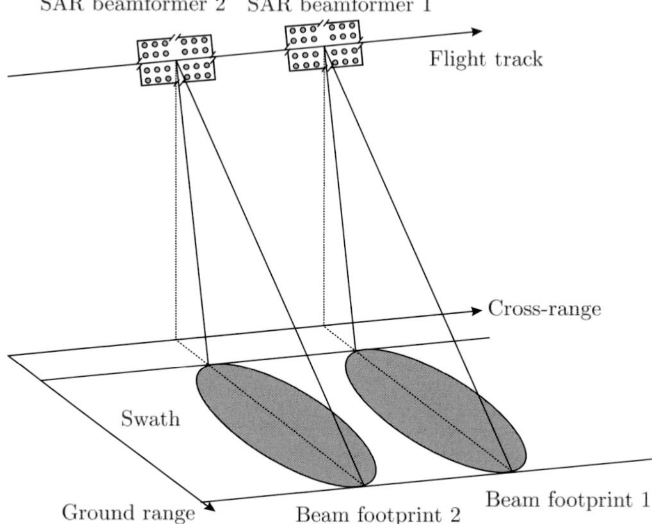

Fig. 5.7   Along-track InSAR.

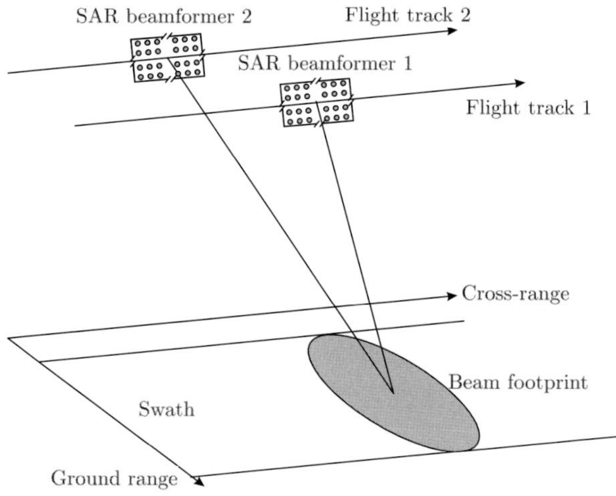

Fig. 5.8   Across-track InSAR.

across-track InSAR, the differences in phase between the observations of the two beamformers are due to topographic elevations and so across-track InSAR, has been used for sea surface topography determination and for the formation of DEMs [2].

There have been uses of tandem configurations of more than one SAR beamformer in order to achieve across-track and along-track interferometry, where one example is the use of TerraSAR-X and TanDEM-X by the European Space Agency (ESA) to form a high-resolution, single-pass SAR interferometer. In the particular case of TerraSAR-X and TanDEM-X satellites, not only is across-track and along-track interferometry achievable, but a bistatic SAR is also achieved [31].

An extension to the along-track SAR interferometer is the dual-beam interferometer where two dual-beam beamformers in an along-track InSAR configuration are used. Each beamformer produces a fore and aft beam, where the fore beam has a squint angle of $+20°$ and the aft beam has a squint angle of $-20°$ [32]. Therefore two interferograms can be created with this configuration, with one formed from the two fore beams and one formed from the two aft beams. A dual-beam interferometer is illustrated in Fig. 5.9, where it can be seen that as it moves in the cross-range direction, each illuminated area in the swath will be imaged from two different directions, due to the orientation of the fore and aft beams.

As mentioned before, in along-track InSAR the two beamformers illuminate the same area but with a time delay determined by the distance between them and their horizontal velocity. The differences in the phase

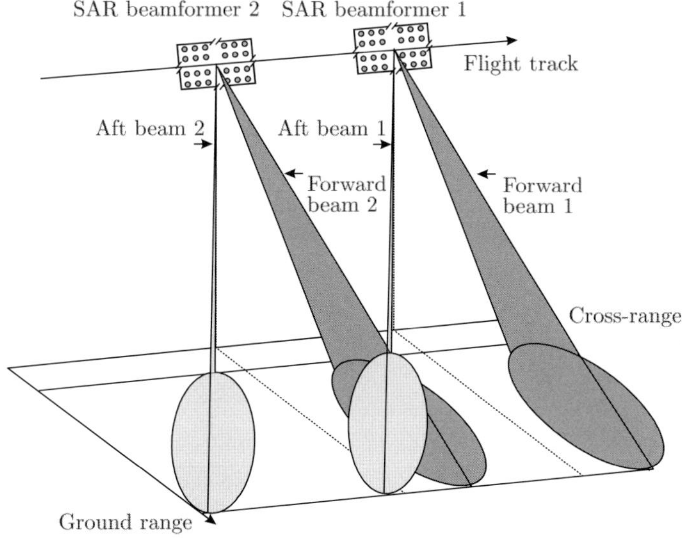

Fig. 5.9   The dual-beam interferometer.

in the observations between the beamformers are then due to the Doppler velocity of the surface scatterers. It is from this Doppler velocity measurement that estimation of the surface current can be made [33]. First an interferogram is formed from the two fore beams and the same is repeated for the two aft beams. Then for each interferogram the phases are related to radial velocities. Using the radial velocities from the two interferograms and the fore and aft beam's squint angles, the surface velocity components can be determined, while making the assumption that they are confined to the horizontal plane [32]. However, as the phases in the interferograms contain both contributions from the surface current and the propagating surface waves, the velocities also include these contributions. Therefore in order to get the surface current estimate, the contributions from the propagating surface waves need to be taken into account.

As mentioned before, across-track InSAR is often used for sea surface topography determination as the phase differences between the two beamformers in the InSAR are due to the topographic elevations. Altimeters are also often used for ocean wave height determination. However, altimeters tend to have small footprints, and therefore only a fraction of the area of interest is imaged during each pass [34]. An extension to the across-track InSAR and the altimeter is the wide swath ocean altimeter [35], which allows a larger area of the ocean to be imaged at one time compared to that of conventional altimeters for ocean topographic mapping [35, 36]. Figure 5.10 shows the InSAR component of the wide swath ocean altimeter from which it can be seen that the two beamformers alternatively illuminate the swaths on the left and right, with the swaths located closely on either side of the nadir.

An extension to the wide swath ocean altimeter is the Wavemill concept [37], which not only allows ocean topography determination due to the across-track interferometer, but also combines the ocean surface current measurement ability of along-track interferometry. This hybrid is achieved by separating the two beamformers in the wide swath ocean altimeter in both the across- and along-track directions [34]. The configuration of the Wavemill is shown in Fig. 5.11 (adapted from [37]), where each beam is squinted by 25°, as shown for beamformer 2.

As the Wavemill progresses in the flight direction, both an across-track and along-track interferometer is created, due to the locations of the beamformers parallel to and perpendicular to the flight direction. Therefore both ocean surface current measurements and topography measurements can be obtained.

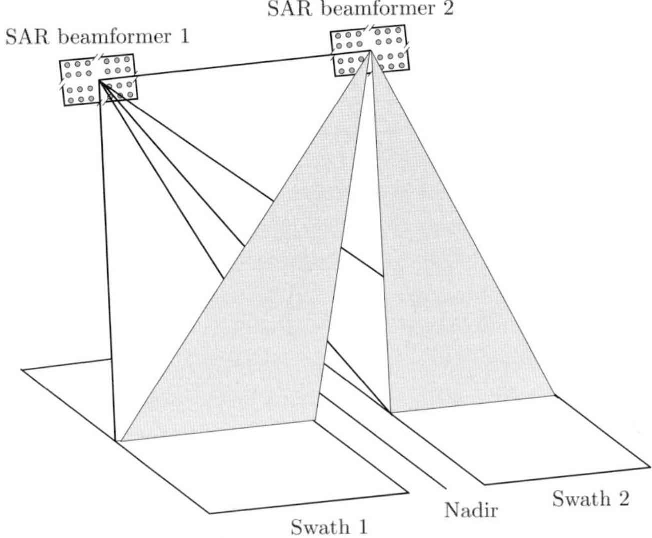

Fig. 5.10   The wide swath ocean altimeter.

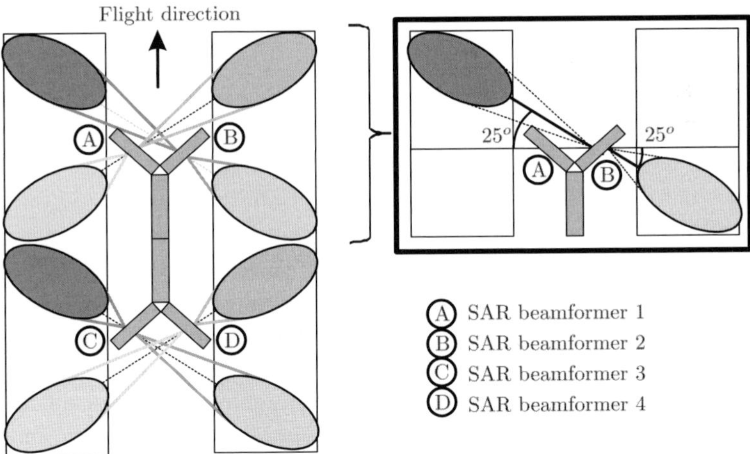

Fig. 5.11   Geometry of the Wavemill, where each beam has a 25° squint.

Another configuration where both across-track and along-track interferometric data can be produced is the interferometric cartwheel [38], where an example using three satellites is illustrated in Fig. 5.12, with $t_1 < t_2 < t_3 < t_4$. The use of the interferometric cartwheel allows the

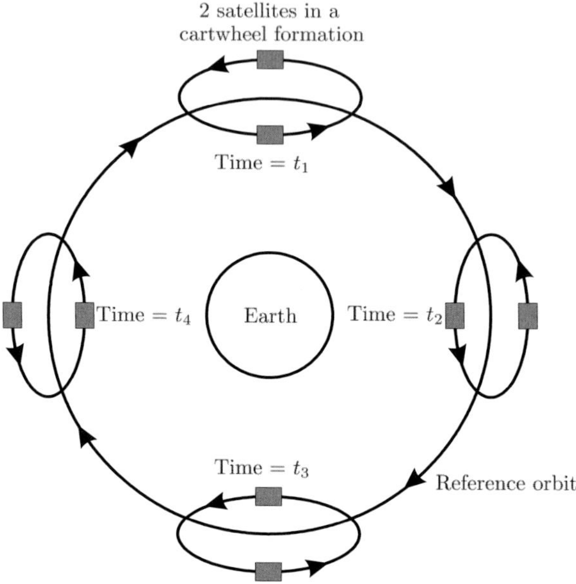

Fig. 5.12   An interferometric cartwheel with two satellites.

baselines to be varied, giving different sensitivities of the cartwheel. However these baselines also vary during the orbit, which complicates interferometric processing. It also has the advantage of allowing single-pass interferometry decreasing scene de-correlation compared to a multi-pass interferometer [38]. The use of several SARs simultaneously allows multiple imaging of an area of interest, after which the combination of the images can increase the range and cross-range resolutions [39].

In terms of ocean applications, current measurements can be made. The separation between two satellites in the whole configuration in the cross-range direction is sufficient such that phase differences due to the movement of the ocean surface can be used to obtain measurements [40].

5.7   Summary and conclusions

In this chapter, different types of ship wake waves have been described along with environmental and SAR parameters that allow the observation of certain wakes in SAR imagery. A number of techniques used in the pre-processing, transform, post-processing and detection stages for wake wave detection using 2D SAR images have been described, along with the estimation of ship parameters from the wake waves. Examples of SAR systems for ocean applications, including ocean surface current and topography measurements, were then given, including the generation of interferograms from SAR data.

References

[1] A. M. Reed and J. H. Milgram, "Ship wakes and their radar images," *Annual Review of Fluid Mechanics*, vol. 34, no. 1, pp. 469–502, 2002.

[2] J. Schulz-Stellenfleth and S. Lehner, "Ocean wave imaging using an airborne single pass across-track interferometric SAR," *IEEE Transactions on Geoscience and Remote Sensing*, vol. 39, no. 1, pp. 38–45, Jan. 2001.

[3] J. A. Johannessen, B. Chapron, F. Collard, V. Kudryavtsev, A. Mouche, D. Akimov, and K.-F. Dagestad, "Direct ocean surface velocity measurements from space: Improved quantitative interpretation of Envisat ASAR observations," *Geophysical Research Letters*, vol. 35, no. 22, Nov. 2008.

[4] J. D. Lyden, R. R. Hammond, D. R. Lyzenga, and R. A. Shuchman, "Synthetic aperture radar imaging of surface ship wakes," *Journal of Geophysical Research: Oceans*, vol. 93, no. C10, pp. 12 293–12 303, Oct. 1988.

[5] M. Balser, C. Harkless, W. McLaren, and S. Schurmann, "Bragg-wave scattering and the narrow-vee wake," *IEEE Transactions on Geoscience and Remote Sensing*, vol. 36, no. 2, pp. 576–588, Mar. 1998.

[6] J. Tunaley, "The narrow-v wake," LRDC Technical Report, Tech. Rep., Oct. 2009.

[7] I. Hennings, R. Romeiser, W. Alpers, and A. Viola, "Radar imaging of Kelvin arms of ship wakes," *International Journal of Remote Sensing*, vol. 20, no. 13, pp. 2519–2543, 1999.

[8] A. Arnold-bos, A. Khenchaf, and A. Martin, in *An Evaluation of Current Ship Wake Detection Algorithms in SAR Images*, Jan. 2006.

[9] R. Touzi, F. Charbonneau, R. Hawkins, K. Murnaghan, and X. Kavoun, "Ship-sea contrast optimization when using polarimetric SARs," in *IEEE International Geoscience and Remote Sensing Symposium*, vol. 1, Jul. 2001, pp. 426–428.

[10] D. J. Crisp, "The state-of-the-art in ship detection in synthetic aperture radar imagery," Defense Science and Technology Organisation, Department of Defense, Australian Government, Tech. Rep., 2004.

[11] M. Rey, J. Tunaley, J. T. Folinsbee, P. Jahans, J. Dixon, and M. Vant, "Application of Radon transform techniques to wake detection in SEASAT - SAR images," *IEEE Transactions on Geoscience and Remote Sensing*, vol. 28, no. 4, pp. 553–560, Jul. 1990.

[12] A. Copeland, G. Ravichandran, and M. Trivedi, "Localized Radon transform-based detection of ship wakes in SAR images," *IEEE Transactions on Geoscience and Remote Sensing*, vol. 33, no. 1, pp. 35–45, Jan. 1995.

[13] M. Ali and D. Clausi, "Using the Canny edge detector for feature extraction and enhancement of remote sensing images," in *IEEE International Geoscience and Remote Sensing Symposium*, vol. 5, Jul. 2001, pp. 2298–2300.

[14] E. Bourennane, P. Gouton, M. Paindavoine, and F. Truchetet, "Generalization of Canny–Deriche filter for detection of noisy exponential edge," *Signal Processing*, vol. 82, no. 10, pp. 1317–1328, Oct. 2002.

[15] Z. Xiao, Z. Hou, C. Miao, and J. Wang, "Using phase information for symmetry detection," *Pattern Recognition Letters*, vol. 26, no. 13, pp. 1985–1994, Oct. 2005.

[16] J. M. Kuo and K.-S. Chen, "The application of wavelets correlator for ship wake detection in SAR images," *IEEE Transactions on Geoscience and Remote Sensing*, vol. 41, no. 6, pp. 1506–1511, Jun. 2003.

[17] M. Tello Alonso, C. Lopez-Martinez, J. Mallorqui, and P. Salembier, "Edge enhancement algorithm based on the wavelet transform for automatic edge detection in SAR images," *IEEE Transactions on Geoscience and Remote Sensing*, vol. 49, no. 1, pp. 222–235, Jan. 2011.

[18] J. Skingley and A. Rye, "The Hough transform applied to SAR images for thin line detection," *Pattern Recognition Letters*, vol. 6, no. 1, pp. 61– 67, Jun. 1987.

[19] L. M. Murphy, "Linear feature detection and enhancement in noisy images via the Radon transform," *Pattern Recognition Letters*, vol. 4, no. 4, pp. 279–284, Sep. 1986.

[20] A. Scherbakov, R. Hanssen, G. Vosselman, and R. Feron, "Ship wake detection using Radon transforms of filtered SAR imagery," in *Microwave Sensing and Synthetic Aperture Radar*, G. Franceschetti, C. J. Oliver, F. S. Rubertone, and S. Tajbakhsh, Eds., vol. 2958, Dec. 1996, pp. 96–106.

[21] A. Arnold-Bos, A. Martin, and A. Khenchaf, "Obtaining a ship's speed and direction from its Kelvin wake spectrum using stochastic matched filtering," in *IEEE International Geoscience and Remote Sensing Symposium*, Jul. 2007, pp. 1106–1109.

[22] F. Chaillan and P. Courmontagne, "On the use of the stochastic matched filter for ship wake detection in SAR images," in *OCEANS 2006*, Sept 2006, pp. 1–6.

[23] I.-I. Lin, L. K. Kwoh, Y.-C. Lin, and V. Khoo, "Ship and ship wake detection in the ERS SAR imagery using computer-based algorithm," in *IEEE International conference on Geoscience and Remote Sensing*, vol. 1, Aug. 1997, pp. 151–153.

[24] J. Tunaley, "Theory of the turbulent far-wake," LRDC Technical Report, Tech. Rep., Jan. 2011.

[25] J. H. Milgram, R. A. Skop, R. D. Peltzer, and O. M. Griffin, "Modeling short sea wave energy distributions in the far wakes of ships," *Journal of Geophysical Research: Oceans*, vol. 98, no. C4, pp. 7115–7124, Apr. 1993.

[26] G. Zilman, A. Zapolski, and M. Marom, "The speed and beam of a ship from its wake's SAR images," *IEEE Transactions on Geoscience and Remote Sensing*, vol. 42, no. 10, pp. 2335–2343, Oct. 2004.

[27] A. Ferretti, A. Monti-Guarnieri, C. Prati, F. Rocca, and D. Massonet, *InSAR Principles-Guidelines for SAR Interferometry Processing and Interpretation.* ESA Publications, ESTEC, Feb. 2007, vol. TM-19.

[28] R. Abdelfattah and J.-M. Nicolas, "Topographic SAR interferometry formulation for high-precision DEM generation," *IEEE Transactions on Geoscience and Remote Sensing,* vol. 40, no. 11, pp. 2415–2426, Nov. 2002.

[29] S. Shiping, "DEM generation using ERS-1/2 interferometric SAR data," in *IEEE International Geoscience and Remote Sensing Symposium,* vol. 2, Jul. 2000, pp. 788–790.

[30] M. Coltelli, G. Fornaro, G. Franceschetti, R. Lanari, G. Puglisi, E. Sansosti, and M. Tesauro, "ERS-1/ERS-2 tandem data for digital elevation model generation," in *IEEE International Geoscience and Remote Sensing Symposium,* vol. 2, Jul. 1998, pp. 1088–1090.

[31] A. Moreira, G. Krieger, I. Hajnsek, D. Hounam, M. Werner, S. Riegger, and E. Settelmeyer, "TanDEM-X: a TerraSAR-X add-on satellite for single-pass SAR interferometry," in *IEEE International Geoscience and Remote Sensing Symposium,* vol. 2, Sep. 2004, pp. 1000–1003.

[32] J. Toporkov, D. Perkovic, G. Farquharson, M. Sletten, and S. Frasier, "Sea surface velocity vector retrieval using dual-beam interferometry: first demonstration," *IEEE Transactions on Geoscience and Remote Sensing,* vol. 43, no. 11, pp. 2494–2502, Nov. 2005.

[33] S. Frasier and A. Camps, "Dual-beam interferometry for ocean surface current vector mapping," *IEEE Transactions on Geoscience and Remote Sensing,* vol. 39, no. 2, pp. 401–414, Feb. 2001.

[34] C. Buck, "An extension to the wide swath ocean altimeter concept," in *IEEE International Geoscience and Remote Sensing Symposium,* vol. 8, Jul. 2005, pp. 5436–5439.

[35] B. Pollard, E. Rodriguez, L. Veilleux, T. Akins, P. Brown, A. Kitiyakara, M. Zawadski, S. Datthanasombat, and A. Prata, "The wide swath ocean altimeter: radar interferometry for global ocean mapping with centimetric accuracy," in *IEEE Aerospace Conference Proceedings,* vol. 2, Mar. 2002, pp. 1007–1020.

[36] E. Rodriguez and B. Pollard, "Centimetric sea surface height accuracy using the wide-swath ocean altimeter," in *IEEE International Geoscience and Remote Sensing Symposium,* vol. 5, Jul. 2003, pp. 3011–3013.

[37] J. Marquez, B. Richards, and C. Buck, "Wavemill: A novel instrument for ocean circulation monitoring," in *European Conference on Synthetic Aperture Radar,* Jun. 2010, pp. 1–3.

[38] H. Runge, R. Bamler, J. Mittermayer, F. Jochim, D. Massonnet, and E. Thouvenot, "The interferometric cartwheel for Envisat," in *3rd IAA Symposium on Small Satellites for Earth Observation,* Apr. 2001.

[39] D. Massonnet, "Capabilities and limitations of the interferometric cartwheel," *IEEE Transactions on Geoscience and Remote Sensing,* vol. 39, no. 3, pp. 506–520, Mar. 2001.

[40] R. Romeiser, "On the suitability of a TerraSAR-L interferometric cartwheel for ocean current measurements," in *IEEE International Geoscience and Remote Sensing Symposium,* vol. 5, Sep. 2004, pp. 3345–3348.

Chapter 6

Towed Arrays: Channel Estimation, Tracking and Beamforming

Vidhya Sridhar, Marc Willerton and Athanassios Manikas

Communications and Array Processing,
Department of Electrical and Electronic Engineering,
Imperial College London

Ocean towed arrays find applications in a variety of areas such as defence, oil and gas exploration and geological and marine life studies. The elements/sensors of the towed arrays are usually hydrophones that receive acoustic energy which characterise the feature or environment being monitored. This acoustic energy may be generated by the feature under study itself (such as a marine animal) or be a reflected signal emitted by an acoustic source (such as a sparker or boomer). The major challenge in towed array processing is the receiver positional uncertainties resulting from the array's flexible structure in combination with the ship's turning maneuvers or water currents. These uncertainties can cause a significant reduction in the performance of the receiver array of hydrophones. Initially, in this chapter, we briefly explore and classify prevalent towed array signal processing techniques developed to address these uncertainties and efficiently perform tracking, channel estimation and reception/beamforming. Then, two specific techniques, namely subspace pilot calibration and H^∞-based robustification, are presented, evaluated and compared using both "synthetic" and real data.

6.1 Introductory concepts and classification

A towed array comprises an array of hydrophones that are attached with equidistant spacing on a cable structure, which may be several miles long

189

and is towed by a submarine or ship. The acoustic signals received at these hydrophones are collected and routed to the towing vessel where they are analysed. Owing to this structure, the sensor/hydrophone positions are inherently unstable due to the motion of the ship coupled with ocean currents and other myriad influences.

The history of towed arrays can be traced back to World War I [1]. Considerable interest was shown in the science of listening to ships and the detection of submarines. This was based on the fact that surface warships and submarines produce noises due to inevitable activities such as propulsion of the ship and the operation of their internal machinery. These noises form a unique acoustic signature and was used to identify features such as the class of vessel, hull type, power setting of the engine driving the propeller, etc. Hence, this acoustic signature became an important weapon in underwater stealth combat. During this period, the conventional SOund Navigation And Ranging (SONAR) system, consisting of hydrophones located on the ship or submarine itself, was developed. This was not quite effective since it was blind in a certain sector due to the hull obstructing coverage and suffered from severe *self-noise* [2]. Towed arrays were introduced to overcome the problem of *self-noise* by getting away from their own ship noise. From these beginnings, technology for towed arrays has come a long way with the advent of optical fibres for cables and highly precise signal processing algorithms.

Apart from defence applications, towed arrays are also used for oil and gas exploration in marine waters. They are also employed for many applications in oceanography such as the study of the spatial characteristics of the ocean floor and marine life.

Another new area of research is that of a Spatially Referenced Towed Array (SPARTA). A SPARTA system consists of two platforms working together — a small fixed position line array (subarray) of passive equispaced hydrophones and a conventional towed array (subarray). Using arrays mounted on platforms with relative motion makes it possible to achieve larger apertures and hence better reception.

Though towed arrays were developed as a solution to the *self-noise* problem, their structure brings certain issues of their own. The motion of the towing vessel and external factors such as water drag results in uncertainties in the array sensor positions. This results in the failure of typical array signal processing algorithms since they are built on the premise of a constant and known array geometry. A number of towed array signal processing

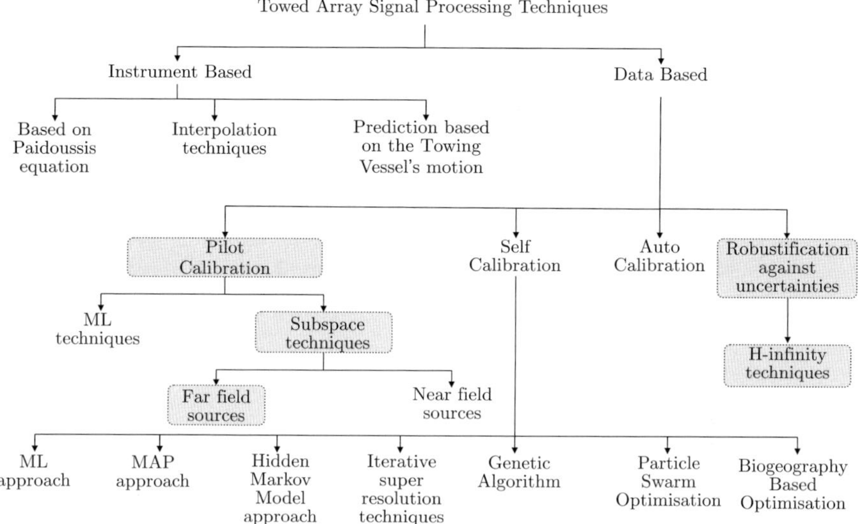

Fig. 6.1 Classification of towed array signal processing techniques. The highlighted techniques are studied in this chapter.

techniques have been developed to address this issue. Figure 6.1 illustrates the two main families of techniques for towed array signal processing:

- instrument-based and
- data-based.

In Sections 6.2 and 6.3, prevalent instrument-based and data-based towed array signal processing techniques are presented. This is followed, in Sections 6.4 and 6.5, by the framework of the underlying signal model employed for the underwater environment and the simulation framework employed to generate synthetic data. The subspace pilot calibration technique and robustification technique based on the H^∞ model are introduced in Sections 6.6 and 6.7 respectively. In Section 6.8, experimental results based on both synthetic and real data are presented and discussed. Finally, in Section 6.9, the conclusions of this chapter are presented.

6.2 Family of instrument-based calibration techniques

Instrument-based calibration techniques, as the name suggests, rely on positioning sensors, such as depth sensors or compasses, placed across the towed

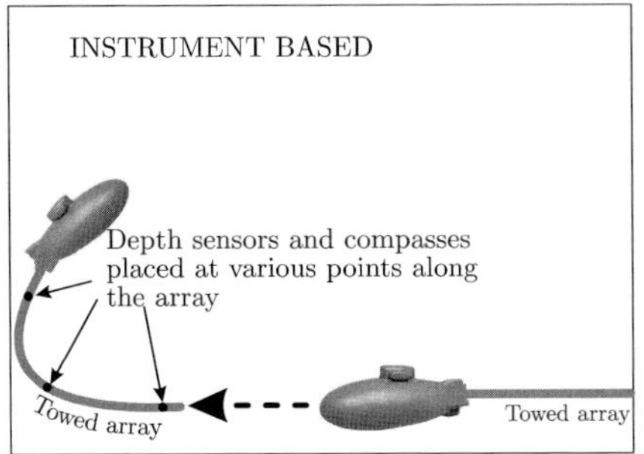

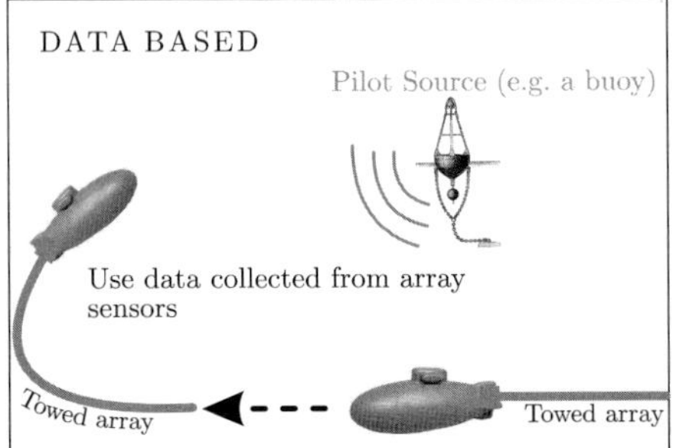

Fig. 6.2   Illustrative representation of instrument-based and data-based techniques.

array as illustrated in Fig. 6.2. These sensors are used to determine the exact receiver array positions in order to perform signal reception (beamforming).

However, due to mechanical considerations, only a few such instrument-based sensors can be placed across the array to estimate their positions. There are different techniques in the literature which use this data to estimate the remaining receiver positions.

A class of instrument-based techniques employ the *Paidoussis equation*, which describes the dynamic behaviour of a thin flexible array towed through the water [3]. This equation can be discretised to form a state space model with the transverse displacements of the array forming the states. Along with the measurements from the positioning sensors and this model, a recursive filter such as the Kalman filter can be used to estimate the transverse displacements and hence the array shape. In [4], the discretised Paidoussis equation is used along with the tow-point compass sensor readings to solve for the tow-point induced motion. Then, the tail compass sensor data is used to finely adjust the overall array shape.

Another class of techniques [5] uses an interpolation scheme to obtain the remaining array positions. One such interpolation technique is the twisted quartic spline approximation to a space curve [5]. There have also been other techniques based on investigations to mathematically express the shape of the towed array in terms of the towing vessel's path or from an accelerometer at the leading edge of the towed array [6].

6.3   Family of data-based calibration techniques

Data-based calibration techniques rely on the data at the receiver array from known, or unknown, source(s) rather than employing special instrument-based sensors (see Fig. 6.2) and can be further classified as follows:

- pilot calibration,
- self-calibration,
- auto-calibration or
- techniques providing robustification against uncertainties.

6.3.1 *Pilot calibration*

Pilot calibration approaches utilise the signals from one or more sources located at known directions. Some techniques in this class use the *maximum likelihood (ML) algorithm* to estimate the sensor positions as in [7]. In [8], an iterative damped Gauss–Newton-type algorithm based on an ML estimator is employed. An improvement on the ML is proposed in [9], where the requirement of a minimal number of calibration sources is relaxed. In [10], using a Fourier series parameterisation of the array manifold, an ML solution employing a large number of sources is proposed. One class of techniques uses far field sources and exploits the *signal subspace* information from the receiver covariance matrix to estimate the array positions [11]. With multiple *far field* sources, global calibration can also be performed, i.e. apart from location uncertainites, other uncertainties such as gain, phase, frequency and timing can be tackled [12]. In [13], in addition to gain, phase and location uncertainties, mutual coupling uncertainties are also handled by employing three time-disjoint calibrating sources. Calibrating sources can also be present in the *near–far field* presenting more practical scenarios [14].

6.3.2 *Self-calibration*

Self-calibration approaches perform the estimation of array uncertainties simultaneously with the standard operation of the array. For instance, in a self-calibrated direction finding (DF) system, the directions of the unknown targets are estimated in parallel with the array uncertainties from the same set of received signals. This implies that the number of unknowns in self-calibration approaches is higher than the available equations (formed from a number of independent measurements) and thus the problem cannot be analytically solved. However, the answers can be provided by solving a constrained, or unconstrained, optimisation problem using a suitably designed cost function.

The first class of such techniques uses an *iterative ML* approach to simultaneously estimate the direction of arrival (DOA) and sensor locations [15]. In [16], the self-calibration is based on the total least-squares (TLS) and a constrained least-squares approach. The algorithm is iterated until the DOA estimates and the sensor location estimates converge.

Another important class of self-calibration techniques is based on the use of *maximum a posteriori (MAP) techniques*, as proposed in [17], where an *a priori* distribution of the perturbation parameters is assumed to be available and a MAP estimator is designed on this basis. To simplify the optimisation problem, the idea of noise subspace fitting is introduced which helps to decouple the estimation of the two sets of unknown parameters [17, 18].

Alternative techniques use the *hidden Markov model (HMM) approach* where the distortion of the array from a linear geometry is modelled as a hidden Markov chain [19]. A measurement sequence is formed from the Fourier coefficients of the various sensor outputs and in conjunction with the HMM model the likelihood of the array shape is estimated using probability theory.

Although, superresolution techniques, such as MUltiple SIgnal Classification (MUSIC) suffer from the presence of array uncertainties, they form the basis of another class of iterative self-calibration techniques. In [20], in the presence of a single moving source, as a first step, the nominal array geometry is used to first estimate the source DOA using MUSIC. In the second step, this estimated DOA and a known/reference sensor position are used along with projection techniques to obtain a large set of points on which the array positions are predicted to lie. The nominal geometry is projected onto the predicted space to obtain an improved estimate of the array sensor coordinates. These steps are iterated until stability is achieved in the array shape. A similar approach is followed in [21] with the exception of a different optimisation function in the second step to estimate the array positions. The second step in [21] selects the set of array positions that maximises the orthogonality between the signal and noise subspaces. There are other techniques, such as in [22], which use *genetic algorithms* for solving a non-linear constrained optimisation problem.

A more recent class of self-calibration techniques is based on swarm intelligence procedures. Swarm intelligence approaches refer to a kind of problem-solution strategies that emerge from the exchange of information between various simple information-processing units. These approaches are used to solve very difficult non-deterministic polynomial-type (NP) class problems in a variety of fields. *Particle swarm optimisation* (PSO) is one such technique where each "particle" is a possible solution and is associated with a "position" and a "velocity". All the particles in the swarm are iterated over by evaluating a "fitness function" and assigned new positions and velocities until a common goal, or sufficient fitness value, is achieved. In [23] and [24], the ML self-calibration problem is solved using the aforesaid PSO technique, each employing different fitness functions.

Finally, adapted from the study of geographical distribution of biological species, evolutionary algorithms such as *biogeography-based optimisation* (BBO) [25] could also be explored. In [26], for a multiple-input multiple-output (MIMO) radar parameter estimation problem, the BBO algorithm is used to solve a nonlinear optimisation problem based on a cost function which incorporates all the targets' directions.

6.3.3 *Auto-calibration*

While in the pilot and self-calibration techniques, the array elements are "receivers" and there are some nominal values available (i.e. the uncertainties are assumed to be small), in the auto-calibration techniques the array elements are transceivers and unknown. Thus, auto-calibration utilises no external sources for calibration and instead uses its array elements themselves. One way to do this is to allow the antenna array elements to transmit in turns, with the remaining elements operating as the receiver array. For more information see Chapter 7 of this book.

6.3.4 *Robustification against uncertainties*

Robustification against uncertainties refers to techniques which provide a robust framework to "shield" the estimation and reception procedures from uncertainties rather than estimating the system uncertainties quantitatively. One such class of techniques attempts to use the H^∞-based state space model to incorporate the uncertainties within the array signal processing framework [27, 28]. From this framework, estimators that minimise the maximum energy gain from the system disturbances (location uncertainties in this case) are designed such that the desired signal is obtained as output. Later in this chapter, the aforesaid H^∞ state space technique is utilised to process signals obtained at a towed array in motion.

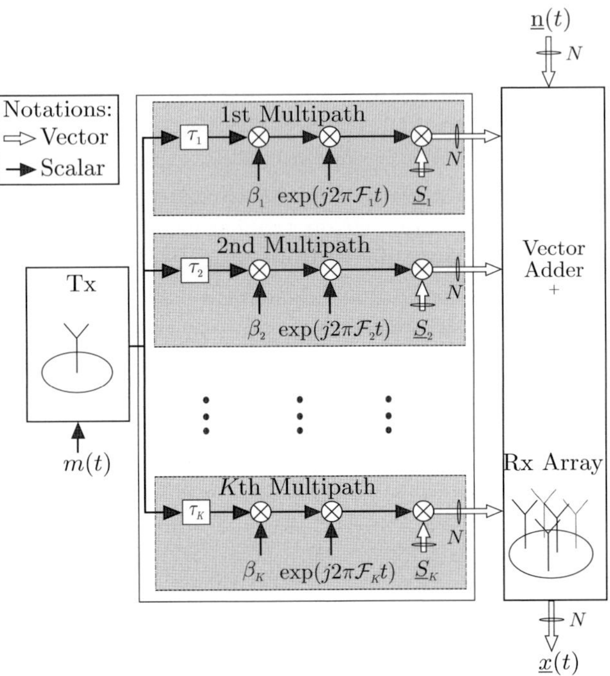

Fig. 6.3   SIMO underwater acoustic channel model for a towed array system.

6.4   Towed array signal model

Consider an array of N hydrophones operating in the presence of K paths (also known as eigenrays) from a single underwater acoustic source. This acoustic channel can be modelled as a single-input multiple-output (SIMO) acoustic channel as represented by Fig. 6.3, where the received acoustic signal vector $\underline{x}(t)$ can be modelled as

$$\underline{x}(t) = \sum_{i=1}^{K} \beta_i \exp(j2\pi\mathcal{F}_i t)\underline{S}_i \odot \underline{m}(t, \underline{\tau}_i) + \underline{n}(t) \tag{6.1}$$

with

$$\underline{m}(t, \underline{\tau}_i) = [m(t - \tau_{i1}), m(t - \tau_{i2}), \ldots, m(t - \tau_{iN})]^T, \tag{6.2}$$

$$\underline{\tau}_i = [\tau_{i1}, \tau_{i2}, \ldots, \tau_{iN}]^T. \tag{6.3}$$

In the above, it is assumed that the i-th eigenray arrives at the k-th sensor of the array with a propagation delay of τ_{ik}. Hence, the vector $\underline{\tau}_i$ in

Eq. (6.3) denotes the path propagation delay across all sensors and $\underline{m}(t, \underline{\tau}_i)$ in Eq. (6.2) the message received at all sensors with varying delays. Furthermore, the array manifold vector is parameterised by θ_i, ϕ_i, ρ_i and F_c, that is $\underline{S}_i \triangleq \underline{S}(\theta_i, \phi_i, \rho_i, F_c)$, where the parameters θ_i, ϕ_i, ρ_i represent the azimuth angle, elevation angle and range of the i-th eigenray respectively, while F_c represents the carrier frequency. Note that the range refers to the distance between the last ocean surface or bottom (seabed) bounce along the path and the array reference point. In addition, the i-th eigenray is characterised by the complex path gain β_i and Doppler frequency $\mathcal{F}_i$ arising due to the motion of the receiver array which is given by

$$\mathcal{F}_i \triangleq -\frac{v_i F_c}{c}. \tag{6.4}$$

In Eq. (6.4), v_i is the radial velocity of the array with respect to the i-th multipath/eigenray and c is the speed of sound underwater. Furthermore, in Eq. (6.1), $\underline{n}(t) \in \mathcal{C}^{N \times 1}$ models the additive white Gaussian noise of zero mean with covariance matrix

$$\mathbb{R}_{\mathrm{nn}} = \sigma_{\mathrm{n}}^2 \mathbb{I}_N, \tag{6.5}$$

where σ_{n}^2 denotes the noise power.

In general,

$$m(t - \tau_{i1}) \neq m(t - \tau_{i2}) \neq \cdots \neq m(t - \tau_{iN}). \tag{6.6}$$

However, in the towed array, the acoustic signal varies slowly as it moves across the array, which implies that

$$m(t - \tau_{i1}) \approx m(t - \tau_{i2}) \approx \cdots \approx m(t - \tau_{iN}), \tag{6.7}$$

i.e.

$$\underline{m}(t, \underline{\tau}_i) = m(t - \tau_i)\underline{1}_N. \tag{6.8}$$

Thus Eq. (6.1) is simplified to

$$\boxed{\underline{x}(t) = \sum_{i=1}^{K} \beta_i \exp(j2\pi\mathcal{F}_i t)\underline{S}_i m(t - \tau_i) + \underline{n}(t).} \tag{6.9}$$

The array manifold vector $\underline{S}_i$ of the i-th path is based on either a spherical wave propagation model or a simplified plane wave propagation model. This depends on the location of the last surface or bottom bounce along the path being at *near field* or *far field* respectively of the array and can be written as follows:

$$\underline{S}_i = \begin{cases} \exp(-j[\underline{r}_x, \underline{r}_y, \underline{r}_z]\underline{k}(\theta_i, \phi_i)) & \text{— Far Field} \\ \rho_i^\alpha \cdot \underline{\rho}^{-\alpha} \odot \exp\left(-j\frac{2\pi F_c}{c}(\rho_i \cdot \underline{1}_N - \underline{\rho})\right) & \text{— Near Field} \end{cases}, \quad (6.10)$$

with

$$\underline{\rho} = \sqrt{\rho_i^2 \cdot \underline{1}_N + \underline{r}_x^2 + \underline{r}_y^2 + \underline{r}_z^2 - \frac{\rho_i c}{\pi F_c}[\underline{r}_x, \underline{r}_y, \underline{r}_z]\underline{k}(\theta_i, \phi_i)} \quad (6.11)$$

where α represents the path loss exponent. Note that $\rho_i \to \infty$ in the far field. Here, $[\underline{r}_x, \underline{r}_y, \underline{r}_z] = [\underline{r}_1, \underline{r}_2, \ldots, \underline{r}_N]^T \in \mathcal{R}^{N \times 3}$ is a matrix containing the Cartesian coordinates of the sensors in metres and $\underline{k}(\theta_i, \phi_i)$ is the wavenumber vector and is defined as follows

$$\underline{k}(\theta_i, \phi_i) = \frac{2\pi F_c}{c}[\cos(\theta_i)\cos(\phi_i), \sin(\theta_i)\cos(\phi_i), \sin(\phi_i)]^T. \quad (6.12)$$

6.5　Synthetic data generation and BellHop framework

In order to evaluate the towed array signal processing techniques, synthetic data adhering to the model described in Section 6.4 needs to be generated. Towards this, the BellHop framework that models acoustic propagation in underwater ocean environments by means of beam tracing to predict acoustic pressure fields [29] was employed. Using this framework, several types of beams can be implemented with geometrical and physics-based spreading laws using features such as varying *altimetry, bathymetry* and *sound profile,* directional sources, geoacoustic properties for bounding media, ocean top and bottom reflection coefficients.

In the BellHop framework, the underwater environment, or terrain, is assumed to be known in terms of the sound speed profile and information about the ocean bottom. Furthermore, by utilising the source and receiver locations, BellHop provides the resultant *eigenrays* (i.e. *multipaths*). Each of these eigenrays are characterised by the following parameters:

- time delay,
- path gain,
- departure angle from source,
- arrival angle at receiver,
- number of surface and bottom bounces along the ray path.

In addition, Doppler effects are incorporated if there is a relative motion between the source and the towed array. The generated eigenrays form the *impulse response function* of the channel given by

$$\underline{h}(t) = \sum_{i=1}^{K} \beta_i \exp(j2\pi\mathcal{F}_i t)\underline{S}_i \delta(t - \tau_i) \tag{6.13}$$

where $\delta(t)$ is the delta function. For most practical testing scenarios, the impulse response can be assumed to be time-invariant over the observation interval. To obtain the receiver time series signal, the source time series is convolved with the impulse response function i.e. $\underline{x}(t) = \underline{h}(t) * \underline{m}(t)$ which provides Eq. (6.9).

A sample set of synthetic towed array hydrophone data generated using the BellHop framework was made available[1] for evaluation. The datasets provided employed a far field source and a simulated towed array of $N = 64$ hydrophones in the North Atlantic environment. The source produced continuous wave (CW) signals at 1200, 1300 and 1400 Hz of equal amplitude. This is equivalent to a message signal of 100 Hz being amplitude modulated with the carrier frequency of 1300 Hz. The channel impulse response was assumed to be time invariant across the whole aperture of the array and throughout the dataset. Figure 6.4 shows the curved geometry of the 64-element towed array on the x-y plane at the start of its motion where the position of the i-th sensor $\underline{r}_i = [r_{x_i}, r_{y_i}, r_{z_i}]^T$ for this geometry at the start may be derived as follows

$$\underline{r}_i = \begin{bmatrix} \cos\left(\frac{30°}{N}\right), & \sin\left(\frac{30°}{N}\right), & 0 \\ -\sin\left(\frac{30°}{N}\right), & \cos\left(\frac{30°}{N}\right), & 0 \\ 0, & 0, & 0 \end{bmatrix} \left(\underline{r}_{i-1} - \begin{bmatrix} 60.16 \\ 0 \\ 0 \end{bmatrix} \right) + \begin{bmatrix} 60.16 \\ 0 \\ 0 \end{bmatrix} \tag{6.14}$$

where $N = 64$ and $\underline{r}_1 = [0, 0, 0]^T$. The rotation angle $\frac{30°}{N}$ is derived from the fact that the array aperture subtends $30°$ of the arc on the x-axis and the value 60.16 metres is the radius of the circle on which the arc of the towed array lies. As shown in the figure, a source is located at a range of 10 km and an azimuth angle of $30°$.

To check that the provided data $\underline{x}[n]$ (discretised version of $\underline{x}(t)$ modelled by Eq. (6.9)) contains the above-described environment, the data

---

[1] These datasets were provided by the Defence Science and Technology Laboratory (DSTL) and were made available from the UDRC.

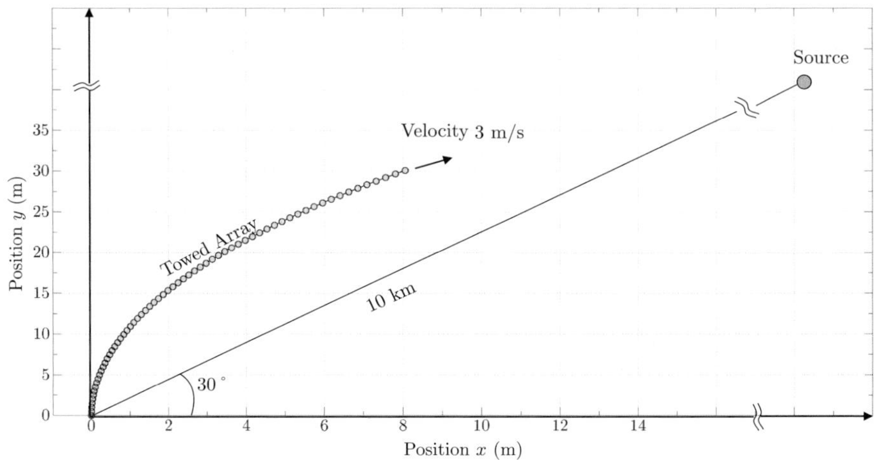

Fig. 6.4   Array geometry and source position at start of dataset.

was analysed using MUSIC-like cost functions. Firstly, the following two-dimensional cost function was evaluated for the carrier frequency of 1300 Hz, assuming the far field scenario (i.e. $\rho_i \to \infty$)

$$\xi(\theta, \phi) = -10 \times \log_{10} \left(\underline{S}^H (\theta, \phi) \, \mathbb{E}_n \mathbb{E}_n^H \, \underline{S} (\theta, \phi) \right), \qquad (6.15)$$

where $\mathbb{E}_n$ is the matrix with columns the eigenvectors corresponding to the "noise" eigenvalues of the covariance matrix

$$\mathbb{R}_{xx} = \frac{1}{L} \sum_{l=1}^{L} \underline{x}(t_l) \, \underline{x}^H (t_l) \qquad (6.16)$$

for $L = 500$ snapshots (samples).

Figure 6.5 shows the result of Eq. (6.15) for the first observation interval. The results show the presence of multipaths at an azimuth angle of 30° and elevation angles closely spaced between −30° and 30°. Figure 6.6 shows the result of Eq. (6.15) for an azimuth angle of 30° and indicates the elevations of the strongest paths.

Furthermore, in the far field, at an azimuth angle of 30° and elevation angle of 4° (corresponding to one of the strongest paths), the expression

$$\xi(f) = -10 \times \log_{10} \left(\underline{S}^H (f) \, \mathbb{E}_n \mathbb{E}_n^H \, \underline{S} (f) \right), \qquad (6.17)$$

was evaluated and the results are shown in Fig. 6.7 indicating the presence of the three frequencies at 1200, 1300 and 1400 Hz.

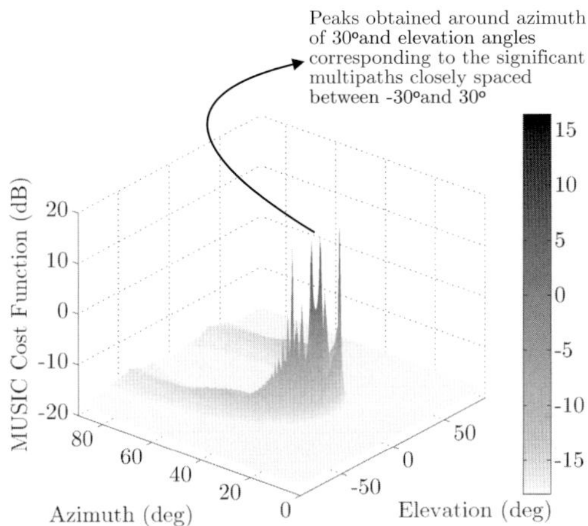

Fig. 6.5 MUSIC cost function versus azimuth and elevation angles for the first observation interval (first set of 500 snapshots).

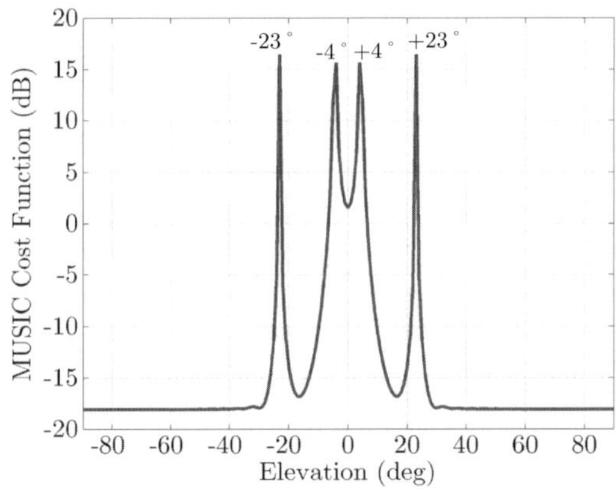

Fig. 6.6 MUSIC cost function versus elevation angle at a fixed azimuth of 30 degrees for the first observation interval (first set of 500 snapshots).

Finally, to study the effect of range on the MUSIC cost function, for the carrier frequency of 1300 Hz and at an elevation angle of the path at $4°$, the following expression was evaluated

$$\xi(\theta, \rho) = -10 \times \log_{10} \left(\underline{S}^H (\theta, \rho) \, \mathbb{E}_n \mathbb{E}_n^H \, \underline{S} (\theta, \rho) \right). \qquad (6.18)$$

This evaluation yields a maximum value at a range of 10 km and azimuth angle of $30°$ as shown in Fig. 6.8. The presented results indicate that the

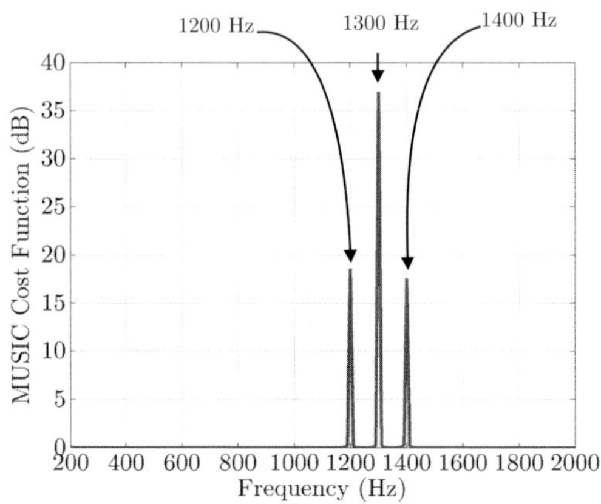

Fig. 6.7 MUSIC cost function versus frequency for the first observation interval of 500 snapshots.

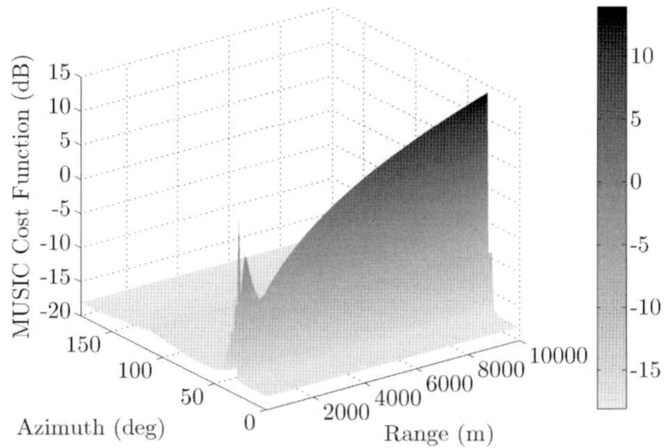

Fig. 6.8 MUSIC cost function versus azimuth angle and range at the first observation interval of 500 snapshots.

synthetic towed array data provided by DSTL, using the BellHop framework, corresponds to the underwater acoustic scenario described at the beginning of the section and is in line with the model presented in Section 6.4.

It is important to point out that if the above analysis is carried out for the 50th observation interval (50th set of 500 snapshots), then, the results of Fig. 6.9 are obtained instead of the result of Fig. 6.5 indicating the failure of the estimation process. This is due to the uncertainties in the location

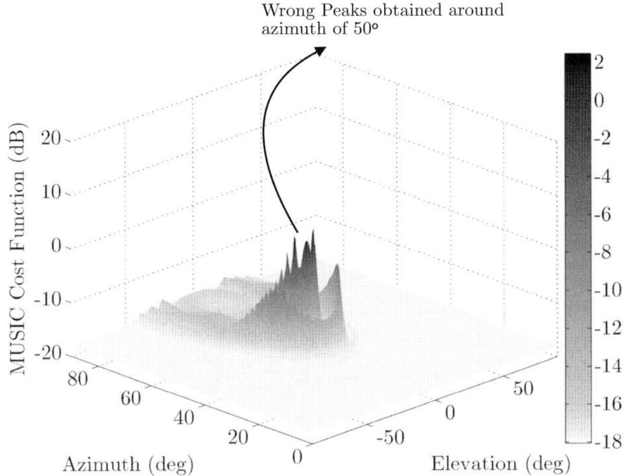

Fig. 6.9 MUSIC cost function versus azimuth and elevation angles at the 50th observation interval of 500 snapshots.

positions $[\widetilde{\underline{r}}_x, \widetilde{\underline{r}}_y, \widetilde{\underline{r}}_z]$ arising from the motion of the towed array. A subspace pilot calibration technique and an H^∞-based robustification technique for handling these uncertainties are presented in the next two sections.

6.6 Subspace pilot calibration techniques

The subspace pilot calibration techniques may use a single acoustic pilot source located in the far field of the towed array to estimate the sensor positions over a small observation interval. As the name suggests, this technique is based on finding the subspace spanned by the manifold vector of the pilot source [12]. This can be done by using the eigendecomposition of THE receiver covariance matrix [11, 12]. Without any loss of generality, the position of the first sensor is assumed to be $(0, 0, 0)$, i.e. the Cartesian origin (array reference point). Based on the model given by Eq. (6.9), from the eigendecomposition of the covariance matrix of the data received at the towed array from an acoustic pilot source arriving through one path, it can be shown that the principal eigenvector $\underline{E}_{\mathrm{s}}$, corresponding to the principal eigenvalue, belongs to the linear subspace spanned by the manifold vector $\underline{S}$,

$$\underline{E}_{\mathrm{s}} \in \mathcal{L}\{\underline{S}\}, \tag{6.19}$$

that is, $\underline{S}$ and $\underline{E}_{\mathrm{s}}$ are collinear. Since the first array element is assumed to be the "reference point", by normalising the elements of $\underline{E}_{\mathrm{s}}$ by its first element c_{ref}, the relationship between $\underline{S}$ and $\underline{E}_{\mathrm{s}}$ can be written as follows

$$\angle \underline{S} = \angle \left(\frac{1}{c_{\mathrm{ref}}} \underline{E}_{\mathrm{s}} \right). \tag{6.20}$$

Also, in a towed array, the distance between successive array elements is fixed and equal to

$$\Delta = \left\| \underline{r}_l - \underline{r}_{l-1} \right\|, \forall l \in [2, N], \tag{6.21}$$

where $[\underline{r}_x, \underline{r}_y, \underline{r}_z] = [\underline{r}_1, \ldots, \underline{r}_N]^T$ which implies

$$(\mathbb{T}\underline{r}_x)^2 + (\mathbb{T}\underline{r}_y)^2 = \Delta^2 \cdot \underline{1}_{N-1}, \tag{6.22}$$

where $(\cdot)^2$ indicates the element-wise square and $\mathbb{T}$ is a $(N-1) \times N$ matrix given by

$$\mathbb{T} = \begin{bmatrix} 1, & -1, & 0, & \ldots, & 0, & 0 \\ 0, & 1, & -1, & \ldots, & 0, & 0 \\ \vdots & \vdots & \ddots & \ddots & \vdots & \vdots \\ 0, & 0, & \ldots, & 1, & -1, & 0 \\ 0, & 0, & \ldots, & 0, & 1, & -1 \end{bmatrix}. \tag{6.23}$$

Hence, from the structure of the manifold vector for a far field source as described in Eq. (6.10), in conjunction with Eq. (6.20) and the constraint of constant spacing between consecutive sensors, the sensor locations can be estimated.

Based on the above, the steps involved in the sensor position estimation are as follows:

(1) For an observation interval of L snapshots, obtain the receiver covariance matrix.
(2) Obtain the principal eigenvector $\underline{E}_{\mathrm{s}}$ of the covariance matrix. This is collinear to the manifold vector. Evaluate $\underline{\psi}$, the element-wise phase of $\underline{E}_{\mathrm{s}}$ i.e.

$$\underline{\psi} = \angle \underline{E}_{\mathrm{s}} + 2\pi \underline{K} \tag{6.24}$$

where $\underline{K}$ is the array of integers to correct phase wrapping. This array of integers can be estimated using the nominal locations of the sensors $[\widehat{\underline{r}}_x, \widehat{\underline{r}}_y, \widehat{\underline{r}}_z]$ as follows

$$\underline{K} = \left\lfloor \frac{- [\widehat{\underline{r}}_x, \widehat{\underline{r}}_y, \widehat{\underline{r}}_z] \, k\,(\theta, \phi)}{2\pi} \right\rfloor \tag{6.25}$$

assuming the error in sensor positions is less than a wavelength.

(3) From Eq. (6.10) and Eq. (6.20), the following relation is obtained when $\underline{r}_z = \underline{0}_N$

$$\mathbb{T}\underline{r}_x \cos\theta \cos\phi + \mathbb{T}\underline{r}_y \sin\theta \cos\phi = -\frac{c}{2\pi F_c}\mathbb{T}\underline{\psi}, \qquad (6.26)$$

Note that (θ, ϕ) denotes the known direction of the pilot or calibrating source. Equation (6.26) provides $N-1$ equations for $2N-2$ unknowns ($\underline{r}_x$ and $\underline{r}_y$, where $\underline{r}_1 = [0,0]$ and $\underline{r}_z = \underline{0}_N$). It can be proven that the Eq. (6.26) can be solved in conjunction with Eq. (6.22) to yield the following result

$$\left[\underline{r}_x, \underline{r}_y\right] = \frac{1}{\cos\phi}\mathbb{L}\begin{bmatrix} 0, & 0 \\ \underline{A}, & \underline{B} \end{bmatrix}\begin{bmatrix} \cos\theta, & \sin\theta \\ -\sin\theta, & \cos\theta \end{bmatrix} \qquad (6.27)$$

where $\underline{A} = \frac{c}{2\pi F_c}\mathbb{T}\underline{\psi}$ and $\underline{B} = \sqrt{\Delta^2 \cos^2\phi \cdot \underline{1}_{N-1} - (\frac{c}{2\pi F_c}\mathbb{T}\underline{\psi})^2}$ (note that $\sqrt{\cdot}$ of a vector is element-wise) and $\mathbb{L}$ is the $N \times N$ lower triangular matrix given by

$$\mathbb{L} = \begin{bmatrix} 1, & 0, & \ldots, & 0, & 0 \\ 1, & 1, & \ldots, & 0, & 0 \\ \vdots & \vdots & \ddots & \vdots & \vdots \\ 1, & 1, & \ldots, & 1, & 0 \\ 1, & 1, & \ldots, & 1, & 1 \end{bmatrix}. \qquad (6.28)$$

6.7 Robustification techniques: The H^∞ state space model

In practical situations, a pilot source may not be present in order to estimate the array sensor uncertainties and, furthermore, no statistical information may be available. In such scenarios, robustification techniques can be employed to simultaneously carry out typical array signal processing tasks and handle uncertainties in the absence of a viable pilot source. H^∞ estimation techniques are known to be robust to uncertainties in model design and lack of statistical information about the environment. In this chapter, a state space model is presented for the received signal at the towed array and placed in the framework of an H^∞ approach [28]. Following this, an H^∞-*based* a posteriori estimator is found to "filter" out the uncertainties and yield the calibrated signal.

For K paths from an underwater acoustic source, the received signal vector $\underline{x}(t)$ given by Eq. (6.9) can be rewritten in a compact format as

$$\underline{x}(t) = (\mathbb{G} \odot \mathbb{S})\,\underline{m}(t) + \underline{n}(t) \tag{6.29}$$

or equivalently,

$$\underline{x}(t) = \left(\mathbb{G} \odot (\widehat{\mathbb{S}} + \widetilde{\mathbb{S}})\right)\underline{m}(t) + \underline{n}(t) \tag{6.30}$$

where $\underline{m}(t)$ is the $K \times 1$ complex signal vector having $\beta_\ell \exp(j2\pi\mathcal{F}_\ell t)$ $m(t - \tau_\ell)$, $\ell \in [1, K]$ as its elements. Also, $\mathbb{G}$ is the $N \times K$ complex array response matrix, $\widehat{\mathbb{S}}$ is the $N \times K$ complex matrix with nominal manifold vectors as its columns, $\widetilde{\mathbb{S}}$ is the deviation of the manifold vectors due to position errors and $\underline{n}(t)$ is the additive white Gaussian noise vector. The deviation manifold vector $\widetilde{\mathbb{S}}$ is given by

$$\widetilde{\mathbb{S}} = \widehat{\mathbb{S}} \odot (\exp(-j[\widetilde{\underline{r}}_x, \widetilde{\underline{r}}_y, \widetilde{\underline{r}}_z][\underline{k}_1, \underline{k}_2, \dots, \underline{k}_K])) - \widehat{\mathbb{S}}, \tag{6.31}$$

where $[\widetilde{\underline{r}}_x, \widetilde{\underline{r}}_y, \widetilde{\underline{r}}_z]$ are the location uncertainties due to the towed array motion. The vector $\underline{k}_i$ denotes the wavenumber vector of the i-th incoming signal from direction (θ_i, ϕ_i). This can be expressed as a state space model as follows [28]

$$\underline{s}_{j+1} = \operatorname{diag}(\underline{a}_{j+1} \odot \underline{a}_j^*)\underline{s}_j + \mathbb{B}_j\underline{u}_j, \quad j \in [0, N-1] \tag{6.32}$$

$$x_j = \underline{g}_j^T \underline{s}_j + v_j, \quad j \in [1, N], \quad \underline{s}_0 = \underline{m}(t) \tag{6.33}$$

where $\underline{s}_j$ is a $K \times 1$ complex vector representing the state of the j-th sensor, $\underline{a}_j$ is the transpose of the j-th row of $\widehat{\mathbb{S}}$ with $\underline{a}_0 = \underline{1}_K$, $\underline{g}_j$ is the transpose of the j-th row of $\mathbb{G}$ and x_j is the data received on the j-th sensor, i.e. the j-th element of $\underline{x}(t)$. The $K \times 1$ complex vector $\underline{u}_j$ represents the process noise, i.e. the array uncertainties at the j-th sensor and noise effects. The scalar v_j represents the gain and phase errors. Also,

$$\mathbb{B}_j = \rho_j \mathbb{I}_K \tag{6.34}$$

where ρ_j is a positive constant imposing a bound on the process noise at every sensor. Here, the bound is assumed to be constant across all sensors i.e. $\rho_j = \rho; \forall j$.

Using this model, a set of recursive equations can be formed to "filter" out the uncertainties and noise on a snapshot by snapshot basis. In order to proceed with the estimation of the filtered signal, the following assumptions are made to provide the initial state:

- initial value of the sensor locations are available,
- initial source directions are available.

From the state space model given by Eqs. (6.32) and (6.33), an H^∞-*based* a posteriori estimator can be found that bounds the maximum energy gain from the disturbances to the estimation errors [30]. The steps to obtain reliable information about the desired signal from the uncalibrated array are as follows:

(1) Form $\widehat{\mathbb{S}}$ and $\widehat{\mathbb{G}}$ using the initial array locations $[\widehat{\underline{r}}_1, \widehat{\underline{r}}_2, \ldots, \widehat{\underline{r}}_N]$ and initial source directions $(\widehat{\theta}_1, \widehat{\phi}_1), (\widehat{\theta}_2, \widehat{\phi}_2), \ldots, (\widehat{\theta}_K, \widehat{\phi}_K)$.
(2) Take one snapshot $\underline{x}(t_l)$ of the received data across all sensors (where l is initialised to 1).
(3) Initialise $\hat{\underline{s}}_0$ as the initial guess of the message signal.

$$\hat{\underline{s}}_0 = (\widehat{\mathbb{G}} \odot \widehat{\mathbb{S}})^{\#} \underline{x}(t_l). \tag{6.35}$$

Also, initialise $\mathbb{P}_0$ as

$$\mathbb{P}_0 = \mu \mathbb{I}_K, \tag{6.36}$$

where μ is a positive constant. The parameter $\mathbb{P}_0$ represents the covariance of the initial estimation error. This is defined in terms of the system parameter μ which represents confidence in the initial message estimate. A higher μ is indicative of decreased confidence in the accuracy of the initial message estimate provided by $\hat{\underline{s}}_0$ (note that $\hat{\underline{s}}_j$ represents the estimate of $\underline{s}_j$).

(4) Compute the state of the remaining sensors, i.e. $\hat{\underline{s}}_1$ to $\hat{\underline{s}}_N$ using the following recursive equations:

$$\hat{\underline{s}}_{j+1} = \mathrm{diag}[\underline{a}_{j+1} \odot \underline{a}_j^*]\hat{\underline{s}}_j + \underline{K}_{j+1}(x_{j+1} - \underline{g}_{j+1}^T \mathrm{diag}[\underline{a}_{j+1} \odot \underline{a}_j^*]\hat{\underline{s}}_j), \tag{6.37}$$

with

$$\underline{K}_{j+1} = \mathbb{P}_{j+1}\underline{g}_{j+1}^*(1 + \underline{g}_{j+1}^T \mathbb{P}_{j+1}\underline{g}_{j+1}^*)^{-1}, \tag{6.38}$$

$$\mathbb{P}_{j+1} = \mathrm{diag}\left[\underline{a}_{j+1} \odot \underline{a}_j^*\right] \mathbb{P}_j \mathrm{diag}\left[\underline{a}_{j+1} \odot \underline{a}_j^*\right]^H + \mathbb{B}_j \mathbb{B}_j^H$$

$$- \mathrm{diag}\left[\underline{a}_{j+1} \odot \underline{a}_j^*\right] \mathbb{P}_j \left[\underline{g}_j^*, \mathbb{W}_j^H\right] \mathbb{R}_{e,j}^{-1} \begin{bmatrix} \underline{g}_j^T \\ \mathbb{W}_j \end{bmatrix}$$

$$\times \mathbb{P}_j \mathrm{diag}\left[\underline{a}_{j+1} \odot \underline{a}_j^*\right]^H,$$

$$\mathbb{R}_{e,j} = \begin{bmatrix} 1, & \underline{0}^T \\ \underline{0}, & -\epsilon^2 \mathbb{I} \end{bmatrix} + \begin{bmatrix} \underline{g}_j^T \\ \mathbb{W}_j \end{bmatrix} \mathbb{P}_j \left[\underline{g}_j^*, \mathbb{W}_j^H\right],$$

where ϵ is a positive constant. The system parameter ϵ represents the error tolerance level of the H^∞ filter. A higher ϵ implies that the filter's performance requirement is not stringent and vice versa. The weight matrix $\mathbb{W}_j$ is given by

$$\mathbb{W}_j = \mathrm{diag}(\underline{a}_j^*). \qquad (6.39)$$

Note that the above solution exists if and only if

$$\mathbb{P}_j^{-1} + \underline{g}_j^* \underline{g}_j^T - \epsilon^{-2} \mathbb{W}_j^H \mathbb{W}_j > 0, \quad j \in [0, N]. \qquad (6.40)$$

(5) Using the state of the sensors, the desired signal can be obtained as the average of the linear combinations of the states i.e.,

$$\widehat{\underline{m}}(t) = \frac{1}{N} \sum_{j=1}^{N} \widehat{\underline{z}}_j, \qquad (6.41)$$

where

$$\widehat{\underline{z}}_j = \mathbb{W}_j \widehat{\underline{s}}_j. \qquad (6.42)$$

Note that the "filtered" signal can be obtained instead by replacing $\mathbb{W}_j$ with $\underline{g}_j^T$ in all the above equations.

(6) Repeat steps 2 to 5 for every snapshot, i.e. $\forall l$ (i.e. $l = 2, \ldots, L$).

6.8   Experimental evaluation of techniques and discussion

In order to evaluate the two techniques described in Sections 6.6 and 6.7, the environment presented in Section 6.5 was employed. The pilot calibration technique treats the acoustic far field source at an azimuth angle of 30° as a pilot of known direction to estimate the position uncertainties of the towed array. On the other hand, the robustification technique considers everything unknown, and attempts to provide information about the far field source in a robust way, resisting the sensor position uncertainties.

6.8.1   *Experiments with subspace pilot calibration*

Consider the moving curved towed array of 64 sensors as described in Section 6.5, operating in the presence of a single acoustic source of known direction (30° azimuth) which transmits a CW signal of 1200 Hz. Furthermore, consider two levels of SNR (20 dB and −5 dB) and that the towed array has an angular velocity of 0.001 degree/snapshot. By partitioning the time to time frames (observation intervals) of $L = 100$ snapshots and applying repetitively the subspace pilot calibration technique presented in Section 6.6, the aim of this experiment is to track the hydrophone positions as the array is moving. The results are shown in Fig. 6.10. From this figure, it can be observed that the true array positions almost coincide with the estimated array positions. However, at a lower SNR of around -5 dB, there are larger estimation errors in the array positions. This is due to the fact that the estimation error of the subspace based estimator is inversely proportional to the level of SNR for a fixed L. Furthermore, it can be observed from Fig. 6.10 that the estimation error in one sensor propagates to the estimation in the next sensor leading to a reduction in accuracy across the array.

Finally, a number of Monte-Carlo experiments were carried out for various angular velocities per snapshot of the towed array and various observation intervals (number of snapshots L). The results in terms of root-mean-square error (RMSE) versus SNR $\times$ L are shown in Fig. 6.11 averaged over 100 experiments and for a fixed input SNR at 10. When the array is static and the observation interval is large, e.g. $L = 10^6$, the RMSE is almost zero (in the range of 10^{-8} metres) but for a small observation interval, e.g. $L = 100$, the RMSE rises to about 10^{-4} metres. Furthermore, if the angular velocity increases from 0 (no motion) to 10^{-4} degrees per snapshot (about $12\,\text{m/s}$), then Fig. 6.11 shows that for $L = 10^6$ the RMSE increases from almost zero (10^{-8}) to 10^{-5} metres. To combat this, one may choose a smaller observation interval for evaluating higher angular velocities. However, too short observation intervals may lead to a poorer approximation of the practical covariance matrix to the theoretical one.

6.8.2 Experiments with H^∞-based robustification technique

6.8.2.1 Experimental results using synthetic towed array data

As stated earlier, a pilot source may not be always present in order to estimate the uncertainties introduced by the motion of the towed array. In such scenarios, robustification techniques present a more practical solution to carry out typical array signal processing tasks to reasonable accuracy. In this section, a number of experiments are presented using the synthetic data to demonstrate the performance of the H^∞-based robustification technique described in Section 6.7. The aim is to show that this technique withstands

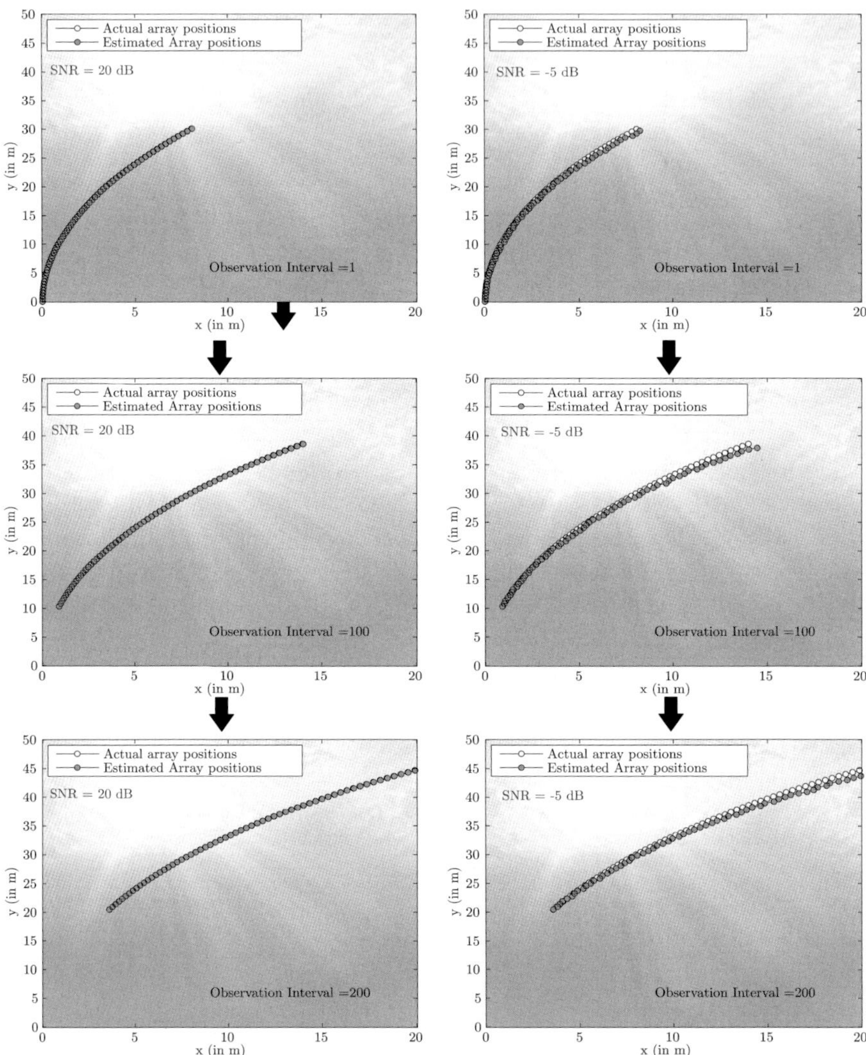

Fig. 6.10   Array tracking by subspace pilot calibration for a towed array angular velocity of 0.001 deg/snapshot at SNR of 20 dB (left) and -5 dB (right).

and overcomes the presence of unknown uncertainties in the positions of the 64-element moving towed array when it operates in the presence of the $30°$ unknown far field acoustic signal source for the environment presented in Section 6.5. Figure 6.12 shows the MUSIC cost function of the channel estimator obtained before and after the application of the H^∞ technique at the

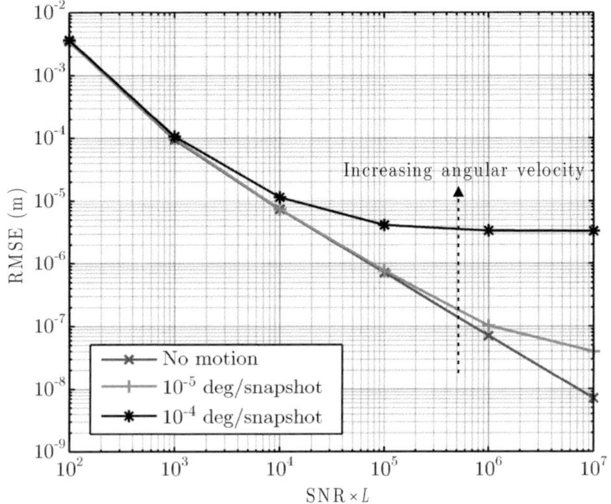

Fig. 6.11   RMSE performance of subspace pilot calibration technique. The input SNR is assumed equal to 10 and L taking values from 10 to 10^6.

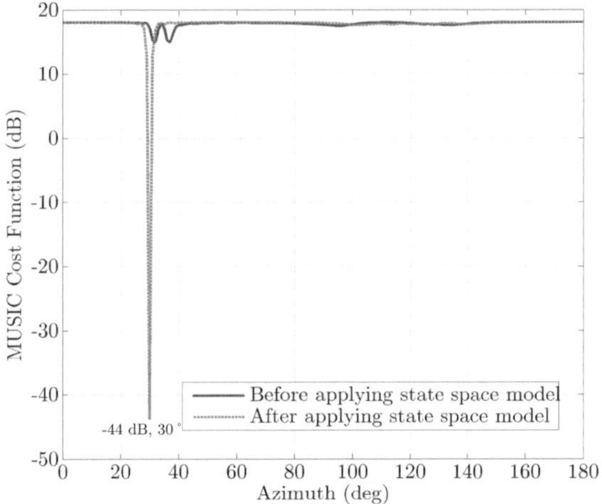

Fig. 6.12   MUSIC cost function outputs from the channel estimator with and without the application of the H^∞ technique for the 20th observation interval in an environment with a far field source at an azimuth of 30 degrees.

20th observation interval (time frame) of 100 snapshots. From this figure it can be seen that the MUSIC algorithm fails to estimate the direction of $30°$ during the 20th time frame due to the presence of array uncertainties due to the towed array's motion. However, the use of the H^∞ filter overcomes

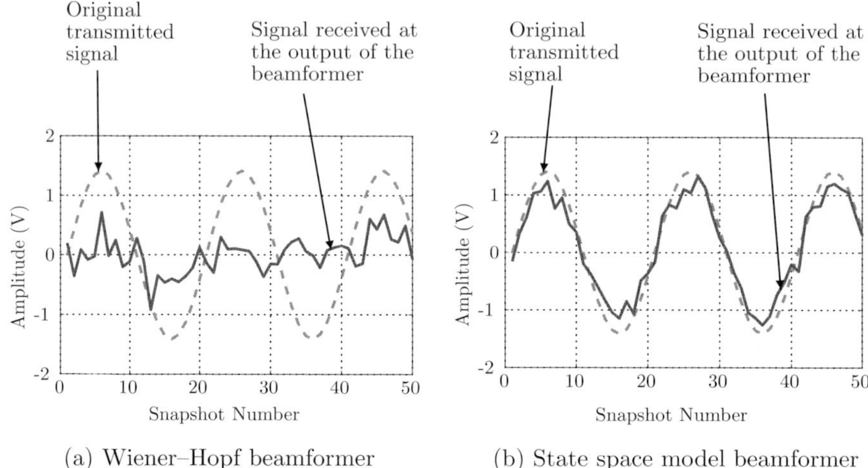

(a) Wiener–Hopf beamformer   (b) State space model beamformer

Fig. 6.13   Ouputs from the (a) Wiener–Hopf beamformer and (b) H^∞ state space model beamformer for the towed array.

the effects of the presence of uncertainties and helps the MUSIC algorithm to estimate correctly the direction of $30°$.

In a similar fashion Fig. 6.13 shows the output of a Wiener–Hopf-type towed array beamformer as well as the output of an H^∞ beamformer presented using the technique presented in Section 6.7. The signal transmitted by an underwater acoustic source is a 1200 Hz sinewave, shown as a dotted line in Fig. 6.13. The results clearly show that the state space-based H^∞ beamformer performs much better than conventional beamformers such as the Wiener–Hopf.

6.8.2.2   *Experimental results using real towed array data from sea trials*

As mentioned in Section 6.1, "listening to ships", i.e. tracking enemy vessels on sea using their ship noise as the source signal, was the origin of the applications of towed arrays. Towards this end, experiments were conducted at sea and data[2] collected from these experiments were also made available to the University Defence Research Centre (UDRC) for evaluation. The experiments were conducted using a passive towed array consisting of

---

[2]This data was provided by the DSTL (UK) based on sea trials conducted with a passive towed array in the southwestern approaches to the UK.

32 hydrophones with a nominal linear array geometry. This towed array, attached to a heavy tow cable, was deployed by a tow ship. Acoustic data received by the towed array in the nominal frequency band of 500–1000 Hz was collected and provided for post-processing. Two specific trials are examined here namely Trial 1 (called DW10B) and Trial 2 (called DW10B2).

In Trial 1, a 1000 tonne trawler moved past broadside of the towing ship. The towing ship remained static and hence the towed array remained linear.

In Trial 2, the 1000 tonne trawler continued to move past broadside. However, the towing ship underwent a five degree wiggle at start, i.e. a course change of five degrees to the left, 10 degrees to the right and five degrees to the left. Hence, the towed array was not linear anymore and experienced perturbations.

Figure 6.14 indicates the motion of the trawler in terms of its azimuth with respect to the tow ship in Trial 1 (a) and Trial 2 (b). The acoustic

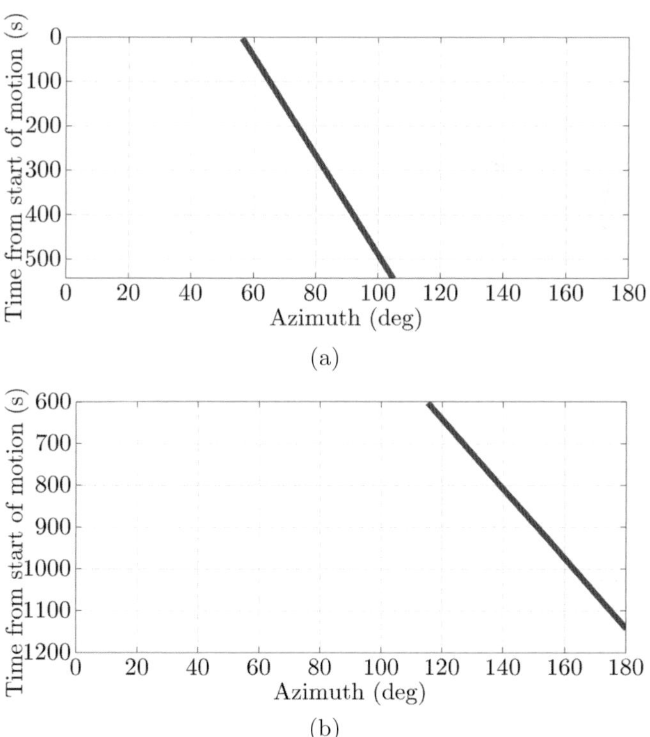

(a)

(b)

Fig. 6.14   Azimuth of the moving trawler with respect to the tow ship in (a) Trial 1 and (b) Trial 2.

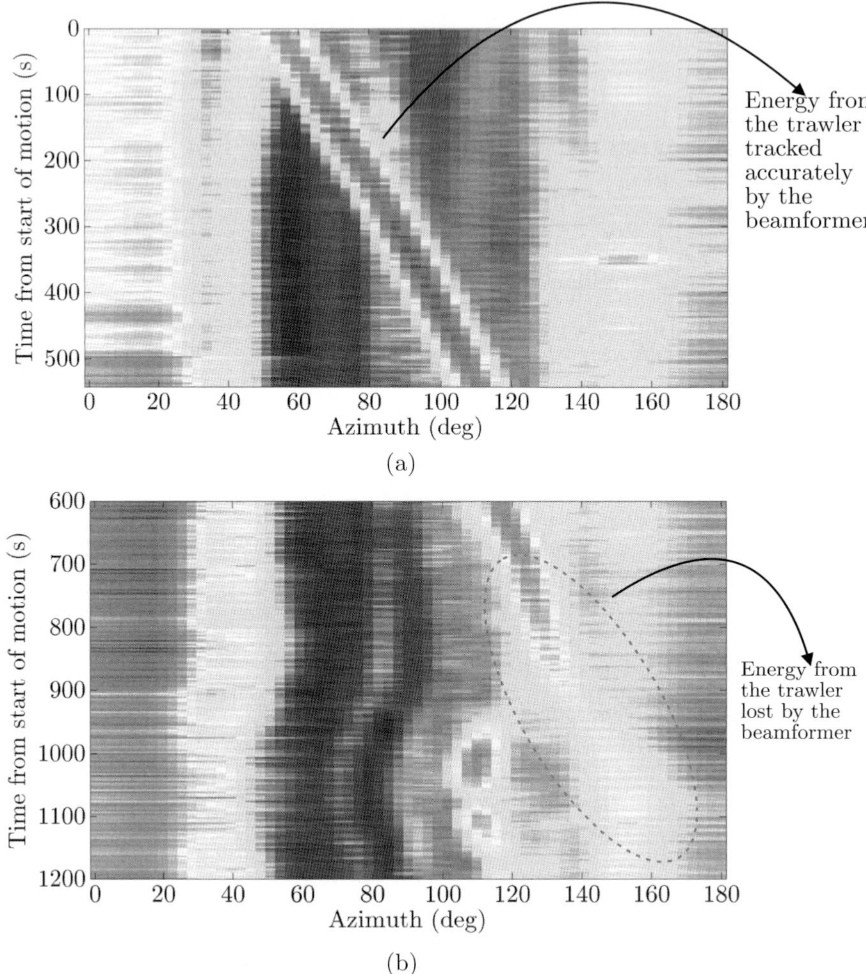

Fig. 6.15   With no post-processing, energy from the trawler being tracked by the steering vector beamformer in (a) Trial 1 and (b) Trial 2.

data from both these trials were analysed as shown in Figs. 6.15 and 6.16. These figures show the power at the output of a steering vector beamformer averaged across the ship noise frequency range of 500 to 1000 Hz at every azimuth.

In Fig. 6.15a (Trial 1), since the towing ship is static, the towed array remains linear. Hence, the beamformer utilises the known/true receiver array positions perfectly and the trawler's motion is tracked correctly.

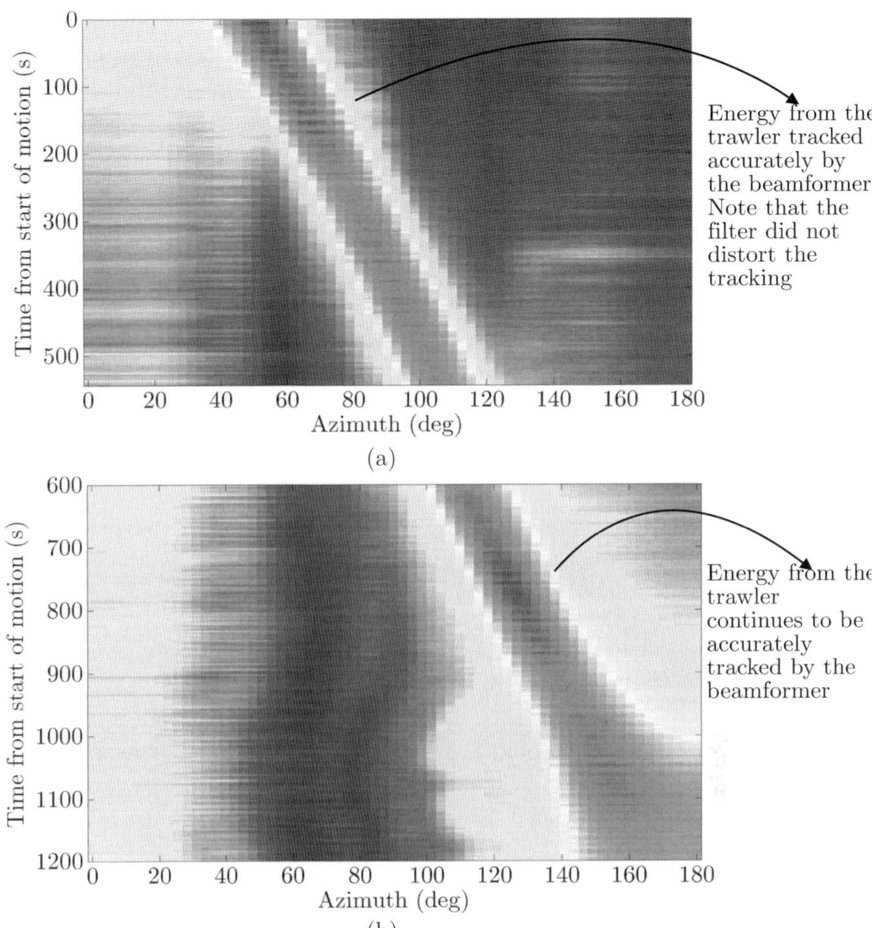

Fig. 6.16 After post-processing, energy from the trawler being tracked by the steering vector beamformer in (a) Trial 1 and (b) Trial 2.

However, in Fig. 6.15b (Trial 2), due to the wiggle of the towing ship, the towed array is not linear anymore. Hence, the uncertainties in the array positions lead to tracking failure by the beamformer as seen in Fig. 6.15b (Trial 2). On the other hand, in Fig. 6.16b (Trial 2), the beamformer at the receiver continues to track the trawler successfully due to the elimination of the position uncertainties by the H^∞ filter. Also, it can be observed that the filtering eliminates the self noise from the towing ship which is visible in the azimuth region between $0°$ and $25°$ in Figs. 6.15a and 6.15b. The employed filter assumes a knowledge of the nominal direction of the trawler within $\pm 10°$. These nominal values were assumed based on the assumption that the trawler continues to move in the same path. This can be relaxed if some knowledge of the message signal, i.e. the trawler noise in this scenario, is available. Also, the filter parameters, namely ρ, the upper bound of the measurement noise consisting of position errors and other external noise factors, μ representing knowledge of intial state, i.e. the source signal and ϵ representing the error tolerance level of the filter, can be tuned to produce more accurate results and tolerate further perturbations in the array geometry. Note that in this experiment, the chosen values of these parameters were $\rho = 0.001$, $\epsilon = 0.9$ and $\mu = 0.05$.

6.9   Conclusions

In this chapter, an overview of the prevalent towed array signal processing techniques was presented. In particular, two techniques, namely subspace pilot calibration and H^∞-based robustification were presented, evaluated

Table 6.1   Advantages and disadvantages of subspace-based and H^∞-based towed array signal processing techniques.

Technique	Advantages	Disadvantages
Subspace pilot calibration technique	• High accuracy as compared to other non-subspace type techniques. • Multiple calibrating sources augments number of uncertainties that can be handled.	• Requires an observation interval where the uncertainties remain constant. This limits the extent of variation. • With increasing uncertainties, number of calibrating sources for accurate estimation is often higher than the minimum.
H^∞-based robustification technique	• Eliminates the need for an interval where the uncertainties are required to be constant. Instead works on a snapshot by snapshot basis. • The model can be augmented to handle many types of uncertainties.	• Upper bounds on the parameter uncertainties need to be known and specified during model design.

and compared based on computer simulation studies carried out on both synthetic and real data. The results show that both these techniques were successfully able to handle the location uncertainties arising from the motion of the towed array with different levels of accuracy. Table 6.1 outlines a comparison of the advantages and disadvantages of each of these techniques which may aid in the selection of the technique as suited for the targetted towed array application.

References

[1] S. G. Lemon, "Towed-array history, 1917–2003," *IEEE Journal of Oceanic Engineering*, vol. 29, no. 2, pp. 365–373, Apr. 2004.

[2] M. Lasky, R. D. Doolittle, B. D. Simmons, and S. G. Lemon, "Recent progress in towed hydrophone array research," *IEEE Journal of Oceanic Engineering*, vol. 29, no. 2, pp. 374–387, Apr. 2004.

[3] D. A. Gray, B. D. O. Anderson, and R. R. Bitmead, "Towed array shape estimation using Kalman filters-theoretical models," *IEEE Journal of Oceanic Engineering*, vol. 18, no. 4, pp. 543–556, Oct. 1993.

[4] F. Lu, E. Milios, S. Stergiopoulos, and A. Dhanantwari, "New towed-array shape-estimation scheme for real-time sonar systems," *IEEE Journal of Oceanic Engineering*, vol. 28, no. 3, pp. 552–563, Jul. 2003.

[5] B. E. Howard and J. M. Syck, "Calculation of the shape of a towed underwater acoustic array," *IEEE Journal of Oceanic Engineering*, vol. 17, no. 2, pp. 193–203, Apr. 1992.

[6] A. P. Dowling, "The dynamics of towed flexible cylinders. Part 1. neutrally buoyant elements," *Journal of Fluid Mechanics*, vol. 187, pp. 507–532, Feb. 1988.

[7] B. C. Ng and C. M. S. See, "Sensor-array calibration using a maximum-likelihood approach," *IEEE Transactions on Antennas and Propagation*, vol. 44, no. 6, pp. 827–835, Jun. 1996.

[8] B. C. Ng and A. Nehorai, "Active array sensor location calibration," in *IEEE International Conference on Acoustics, Speech and Signal Processing*, Apr. 1993, pp. 21–24.

[9] M. Moebus, H. Degenhardt, and A. Zoubir, "Local array calibration using parametric modeling of position errors and a sparse calibration grid," in *IEEE 15th Workshop on Statistical Signal Processing*, Sep. 2009, pp. 453–456.

[10] M. A. Koelber and D. R. Fuhrmann, "Array calibration by Fourier series parameterization: scaled principal components method," in *IEEE International Conference on Acoustics, Speech and Signal Processing*, vol. 4, Apr. 1993, pp. 340–343.

[11] J. J. Smith, Y. H. Leung, and A. Cantoni, "The partitioned eigenvector method for towed array shape estimation," *IEEE Transactions on Signal Processing*, vol. 44, no. 9, pp. 2273–2283, Sep. 1996.

[12] N. Fistas and A. Manikas, "A new general global array calibration method," in *IEEE International Conference on Acoustics, Speech and Signal Processing*, vol. 4, Apr. 1994, pp. IV/73–IV/76.

[13] K. Stavropoulos and A. Manikas, "Array calibration in the presence of unknown sensor characteristics and mutual coupling," in *Proc. EUSIPCO*, vol. 3, Sep. 2000, pp. 1417–1420.

[14] M. Willerton and A. Manikas, "Array shape calibration using a single multi-carrier pilot," in *Sensor Signal Processing for Defence (SSPD 2011)*, Sep. 2011, pp. 1–6.

[15] A. J. Weiss and B. Friedlander, "Array shape calibration using sources in unknown locations — a maximum likelihood approach," *IEEE Transactions on Acoustics, Speech and Signal Processing*, vol. 37, no. 12, pp. 1958–1966, Dec. 1989.

[16] S. Hwang and D. B. Williams, "A constrained total least squares approach for sensor position calibration and direction finding," in *IEEE National Radar Conference*, Mar. 1994, pp. 155–159.

[17] M. Viberg and A. L. Swindlehurst, "A Bayesian approach to auto-calibration for parametric array signal processing," *IEEE Transactions on Signal Processing*, vol. 42, no. 12, pp. 3495–3507, Dec. 1994.

[18] B. Wahlberg, B. Ottersten, and M. Viberg, "Robust signal parameter estimation in the presence of array perturbations," in *IEEE International Conference on Acoustics, Speech and Signal Processing*, Apr. 1991, pp. 3277–3280.

[19] B. G. Quinn, R. F. Barrett, P. J. Kootsookos, and S. J. Searle, "The estimation of the shape of an array using a hidden Markov model," *IEEE Journal of Oceanic Engineering*, vol. 18, no. 4, pp. 557–564, Oct. 1993.

[20] G. Efstathopoulos and A. Manikas, "A blind array calibration algorithm using a moving source," in *5th IEEE Sensor Array and Multichannel Signal Processing Workshop*, Jul. 2008, pp. 455–458.

[21] A. J. Weiss and B. Friedlander, "Array shape calibration using eigenstructure methods," in *IEEE Twenty-Third Asimolar Conference in Signals, Systems and Computers*, Oct./Nov. 1989, pp. 925–929.

[22] J. Zhuo, C. Sun, and J. Feng, "Towed array shape self-calibration based on blind signal separation," in *Europe Oceans 2005*, vol. 1. IEEE, Jun. 2005, pp. 599–604.

[23] S. Wan, P.-J. Chung, and B. Mulgrew, "Array shape self-calibration using particle swarm optimization and decaying diagonal loading," in *Sensor Signal Processing for Defence (SSPD)*, Sep. 2010, pp. 1–5.

[24] B. Liao, G. Liao, and J. Wen, "Array calibration with sensor position errors using particle swarm optimization algorithm," in *IET International Radar Conference*, Apr. 2009, pp. 1–3.

[25] D. Simon, "Biogeography-based optimization," *IEEE Transactions on Evolutionary Computation*, vol. 12, no. 6, pp. 702–713, Dec. 2008.

[26] K. Luo and A. Manikas, "Superresolution multitarget parameter estimation in MIMO radar," *IEEE Transactions on Geoscience and Remote Sensing*, vol. 51, no. 6, pp. 3683–3693, Jun. 2013.

[27] T. Ratnarajah and A. Manikas, "An H^∞ approach to mitigate the effects of array uncertainties on the MUSIC algorithm," *IEEE Signal Processing Letters*, vol. 5, no. 7, pp. 185–188, Jul. 1998.

[28] ——, "A state space model for H^∞ type array signal processing," in *International Conference on Acoustics, Speech and Signal Processing*, vol. 5, Apr. 1997, pp. 3745–3748.

[29] O. C. Rodriguez, "General description of the BELLHOP ray tracing program," Universidade do Algarve, Tech. Rep., 2008.

[30] B. Hassibi, A. H. Sayed, and T. Kailath, "Linear estimation in Krein spaces—Part I: theory," *IEEE Transactions on Automatic Control*, vol. 41, no. 1, pp. 18–33, Jan. 1996.

Chapter 7

Array Uncertainties and Auto-calibration

Marc Willerton, Evangelos Venieris and Athanassios Manikas

Communications and Array Processing,
Department of Electrical and Electronic Engineering,
Imperial College London

In this chapter, an array auto-calibration approach is described which is capable of estimating geometrical (array shape), gain and phase uncertainties associated with an array of sensors. As opposed to other data-based array calibration techniques in the literature, no transmitting sources are required (either pilot sources or sources of opportunity). Instead, elements in the array operate as transceivers which are utilised to auto-calibrate the array. Hence, unlike in the case of pilot calibration approaches, all elements are at *unknown* locations. Under this scenario, the distance of the transmitting elements from the array is small compared to the aperture[1] of the Rx-array formed by the remaining elements. Hence, the standard plane wave propagation assumption used in array processing is no longer valid and a spherical wave propagation model should be considered. Monte-Carlo simulations are used to analyse the performance of the approach.

7.1 Introduction

Arrays of sensors are commonly employed for many applications across military, industry and research in both radio frequency (RF) and acoustic environments. Examples include electronic surveillance, mobile

---

[1]The "array aperture" is defined as the largest distance between any two sensors in the array.

communications, environmental protection, structural monitoring, wireless sensor networks, and localisation and tracking [1]. Traditionally, these arrays have small apertures (e.g. half-wavelength sensor spacing). However, large aperture arrays can also be used where elements could be spaced several hundreds of wavelengths apart [2]. The array manifold vector is one of the most important and fundamental concepts in array signal processing. This represents the response of the array under a plane [3] or spherical [2] wave propagation model and is a function of array parameters (e.g. geometry) and source parameters (e.g. source location and operating frequency). Typically it is assumed that array parameters are known. However, if parameters associated with the array of sensors in this model are imprecisely known (i.e. the array contains uncertainties) then this will lead to a rapid degradation in the detection, resolution and estimation capabilities of the array system [4, 5] and its overall performance.

In practical array systems, uncertainties often arise as a result of tolerances or imperfections introduced in the production process of the array. They also occur over time due to sensor ageing or environmental conditions. Broadly, they can be split into geometrical and electrical uncertainties. Geometrical uncertainties arise due to the location of the sensors in the array being imprecisely known. In contrast, electrical uncertainties arise as a result of unknown perturbations in the electronics of the array system. This could include having an imprecisely known gain or phase (i.e. complex gain) associated with the sensors in the array or mutual coupling effects arising due to the re-radiation of some of the signal energy received by the array elements. This chapter will consider gain, phase and sensor location uncertainties. Gain and phase uncertainties are typically considered to be direction-independent uncertainties (although nominal antenna gain may be directional). This implies that these uncertainties will degrade an array receiver to the same extent, independently of source direction. In contrast, geometrical uncertainties are direction-dependent uncertainties and so degrade the array as a function of the source direction. This chapter is concerned with the estimation of gain, phase and sensor location uncertainties to allow for an improvement in the performance of a given array system. This process is known as array calibration. Array calibration approaches in the literature are generally based on collecting data from the array in the presence of external emitters. The two main approaches are *pilot* calibration and *self*-calibration.

In *pilot* calibration, sources with known parameters (i.e. known location/direction) are used to estimate the array uncertainties analytically by exploiting the mathematical model of the array response and solving a set of linear equations. For example, in [6], [7] and [8], three or more far field pilot sources are used to estimate complex gain, mutual coupling effects and geometrical uncertainties. Furthermore, in [9] it is shown how a single moving pilot operating in the far field of the array with a known radial velocity can be used to estimate the array shape. In each case, a minimum of three distinct pilot locations are required to estimate 3D geometrical uncertainties and one extra is required if complex gain uncertainties must also be estimated. Only one pilot location is required for estimating complex gain uncertainties alone. In [10] and [11], Leshem *et al.* and Mengot *et al.* respectively propose different pilot-based calibration approaches for use in a multipath channel. In [12], an acoustic array calibration method based on time difference of arrival is proposed where only an approximate position of sources is required. This provides a pilot-based approach which may be more easily used in practical applications. However, it is well known that time difference of arrival methods suffer from bandwidth limitations and multipath effects. In all of these methods the sensor location uncertainties must be less than 50% of a wavelength (i.e. small). In addition, it is important to note that these approaches require pilot sources to be deployed at precisely known locations, which may be impractical and expensive in a number of scenarios.

In *self*-calibration, array uncertainties and source parameters (i.e. location/direction) are estimated simultaneously. Since there will now be many more unknowns than equations, a cost function must typically be optimised to solve this problem. In general, self-calibration cost functions are highly non-linear making traditional gradient based optimisation approaches unsuitable and others prone to divergence or have a large convergence time. In [13] and [14], Rockah and Schultheiss provide an initial analysis of the self-calibration problem by deriving Cramer–Rao bounds on the achievable calibration accuracies in the presence of geometrical array uncertainties. This is done in the case in which calibration sources are separated in the time or frequency domain and located in the near-far or far field of the array at unknown locations. Paulraj and Kailath in [15] proposed a self-calibration algorithm for a uniform linear array (ULA) based on exploiting the Toeplitz properties existing in the received covariance matrix to estimate complex gain uncertainties. This is achieved without needing to solve a cost function but is unsuitable in the presence of geometrical uncertainties or non-linear arrays which limits its use in practice. In [16] and [17], Weiss and Friedlander proposed objective cost functions to estimate geometrical array uncertainties based on the conditional maximum likelihood (CML) and the MUSIC cost function respectively. In both of these contributions, a cost function is constructed using a first-order Taylor approximation of the manifold vector and hence the approaches are only suitable under small sensor uncertainties. Furthermore, both approaches converge slowly. Each of the proposed cost functions provides many local maxima. Imposing a small uncertainty bound (i.e. assuming that an initial estimate of the unknowns is provided with a good degree of accuracy) enables the solution to fall within the global maxima instead of local ones. However, this clearly limits the usefulness of this type of algorithm. In [18], Flanagan and Bell demonstrate a self-calibration algorithm to allow the removal of the small perturbation approximation. This is done by combining a series of existing methods including Fistas and Manikas' global pilot calibration algorithm, Weiss and Friedlander's eigenstructure self-calibration algorithm and Viberg and Swindlehurst's MAP noise subspace fitting method presented in [19]. It begins by estimating the array manifold vectors of the sources of opportunity using PROS algorithm detailed in [20]. It then estimates sensor locations using Fistas and Manikas's global pilot calibration algorithm. Next it re-estimates the directions of arrival (DOAs) using Viberg and Swindlehurst's MAP noise subspace fitting method. This scheme then iterates to provide coarse sensor location calibration which allows Weiss and Friedlander's self-calibration method to then be used under a small perturbation approximation. Flanagan and Bell improve their algorithm in [21] by making it more robust when sources of opportunity are close together. Note however that this approach requires two iteration loops, making it extremely computationally intensive. In [22], Chung and Wan propose a method offering faster convergence to improve computational efficiency by modifying Weiss and Friedlander's maximum likelihood approach using the SAGE algorithm [23]. In addition to these approaches, in [24], Fuhrman developed a maximum likelihood approach for estimating complex gain uncertainties using point or diffused sources. Furthermore, Wijnholds and van der Veen develop a self-calibration approach in [25] which estimates the gain of direction-dependent sensors in an array using the weighted subspace fitting (WSF) algorithm. Recently, biology-inspired optimisation procedures have also been used for the purposes of array self-calibration. These techniques are based on optimisation procedures observed in nature and appear to offer a much improved performance over traditional approaches. Self-calibration using a maximum likelihood cost function has been successfully applied in [26] using the genetic algorithm (GA) which is based on the optimal coding of a population of strings (e.g. chromosomes) and also in [27] using particle swarm optimisation (PSO) which is based on the swarming of bees.

In this chapter, an array auto-calibration approach is described which is capable of estimating geometrical (array shape), gain and phase uncertainties associated with a large or small aperture array of sensors. By changing the array reference point to be at each of the array elements, it is shown how a number of metrics may be extracted from the received array data and utilised within a set of linear equations to estimate the array uncertainties. Note that these uncertainties are estimated analytically, unlike the self-calibration methods (i.e. no optimisation procedure is required) and do not require pilot sources to be present at known locations, unlike the pilot calibration methods. One further attractive feature of this algorithm is that it can even operate in the presence of large geometrical uncertainties (i.e. those that are several wavelengths in size).

The remainder of this chapter is structured as follows:

In Section 7.2 the received array signal model and its statistics will be presented in general terms. Furthermore, the *spherical* wave manifold vector and the concept of changing the array reference point will be introduced. Next, in Section 7.3, an array auto-calibration algorithm will be presented which allows geometrical, gain and phase uncertainties associated with the array of sensors to be estimated. Following this, in Section 7.4, computer simulation studies will be used to investigate the performance of the approach for small and large aperture array geometries. Finally, in Section 7.5 the chapter is concluded.

7.2 Signal model

Consider a sensor array (array of nodes) with Cartesian coordinates described by the matrix $\mathbf{r} \in \mathcal{R}^{3 \times N}$ defined as follows:

$$\mathbf{r} = [\underline{r}_1, \underline{r}_2, \ldots, \underline{r}_N] = \left[\underline{r}_x, \underline{r}_y, \underline{r}_z\right]^T, \tag{7.1}$$

where $\underline{r}_i$ for $i = 1, 2, \ldots, N$ denotes the location of the i-th element in the array with respect to a *known* array reference point and $\underline{r}_x$, $\underline{r}_y$, $\underline{r}_z \in \mathcal{R}^{N \times 1}$ denote vectors describing the x, y and z coordinates of the array elements respectively. In addition, the array elements have a gain and phase described by

$$\underline{g} = [g_1, g_2, \ldots, g_N]^T \in \mathcal{R}^{N \times 1}, \tag{7.2}$$

$$\underline{\varphi} = [\varphi_1, \varphi_2, \ldots, \varphi_N]^T \in \mathcal{R}^{N \times 1}, \tag{7.3}$$

where g_i and φ_i for $i = 1, 2, \ldots, N$ denote the gain and phase associated with the i-th array element respectively. Let us assume that the array operates in the presence of a single transmitter (Tx) at location $\underline{r}_m \in \mathcal{R}^3$ (Cartesian coordinates) where the transmitted signal arrives at the input of the array via L paths (multipaths). Consider that the ℓ-th path arrives at the array from direction (θ_ℓ, ϕ_ℓ) and with channel propagation parameters β_ℓ and τ_ℓ representing the complex path gain and path delay respectively. Note that θ_ℓ and ϕ_ℓ represent the azimuth and elevation angles associated with the ℓ-th path. Let us assume that the L paths are arranged such that

$$\tau_1 \leq \tau_2 \leq \cdots \leq \tau_\ell \leq \cdots \leq \tau_L \tag{7.4}$$

with τ_1 denoting the line of sight (LOS), i.e. the direct path. Furthermore, the path coefficient β_ℓ models the effects of path losses and shadowing on the ℓ-th path, in addition to random phase shifts due to reflection; it also encompasses the effects of the phase offset between the modulating carrier at the transmitter and the demodulating carrier at the receiver, as well as differences in the transmitter powers. For instance, for a free space propagation

$$\beta_\ell = \left(\frac{c}{4\pi F_c \rho_\ell} K \right)^a \sqrt{P_{\text{tx}}} \exp(j\psi) \tag{7.5}$$

where ρ_ℓ denotes the ℓ-th path length, F_c is the carrier frequency, c is the velocity of light, a represents the path loss exponent and P_{tx} is the power of the transmitter. The parameter K denotes other system parameters that, without any loss of generality, are ignored here by taking $K = 1$. Finally, ψ denotes the random phase (uniformly distributed over 0°-360°) introduced by the channel.

Under the narrowband assumption, the impulse response (vector) of the SIMO (single-input multiple-output) multipath channel is

$$\text{SIMO impulse response: } \underline{h}(t) = \sum_{\ell=1}^{L} \beta_\ell \underline{S}_\ell \delta(t - \tau_\ell) \tag{7.6}$$

where the vector $\underline{S}_\ell \triangleq \underline{S}\left(\|\underline{r}_m\|, \theta_\ell, \phi_\ell \right) \in \mathcal{C}^{N \times 1}$ is the array response vector (array manifold vector) and $\delta(t)$ denotes a "delta" function. Based on the above discussion, a frequency selective SIMO wireless channel of L paths can be represented as shown in Fig. 7.1. Thus, for a Tx-baseband message signal $m(t)$, with $\text{pdf}_m = \mathcal{NC}(\mu_m = 0, P_m = 1)$, it is clear that the received baseband signal-vector $\underline{x}(t) \in \mathcal{C}^{N \times 1}$ can be modelled as follows:

$$\underline{x}(t) = \underline{h}(t) \circledast m(t) + \underline{n}(t)$$

$$= \sum_{\ell=1}^{L} \beta_\ell \underline{S}_\ell m(t - \tau_\ell) + \underline{n}(t). \qquad (7.7)$$

In the case of a single path (LOS), this is simplified to:

$$\underline{x}(t) = \beta \cdot \underline{S} \cdot \underbrace{m(t - \tau)}_{\triangleq m(t)} + \underline{n}(t) \qquad (7.8)$$

where the subscript 1 has been dropped for notational convenience and $m(t - \tau)$ has been redefined[2] as $m(t)$ (i.e. the time t has been reset to 0 at the Rx-array). Based on Eq. (7.8), Fig. 7.1 is simplified to Fig. 7.2. Note that this also shows the carrier at the Tx (point A1) and Rx (point A8) as well as the bandpass noise vector $\underline{n}_{\mathrm{BP}}(t) \in \mathcal{C}^{N \times 1}$, i.e.

$$\text{at point A6 } \underline{n}_{\mathrm{BP}}(t) \in \mathcal{C}^{N \times 1} : \mathrm{pdf}_n = \mathcal{NC}(\underline{0}_N, \sigma_n^2 \mathbb{I}_N). \qquad (7.9)$$

By controlling both amplitude and phase at point A8, at point B the signal at the reference point may be set to be $m(t)$ while at the i-th sensor

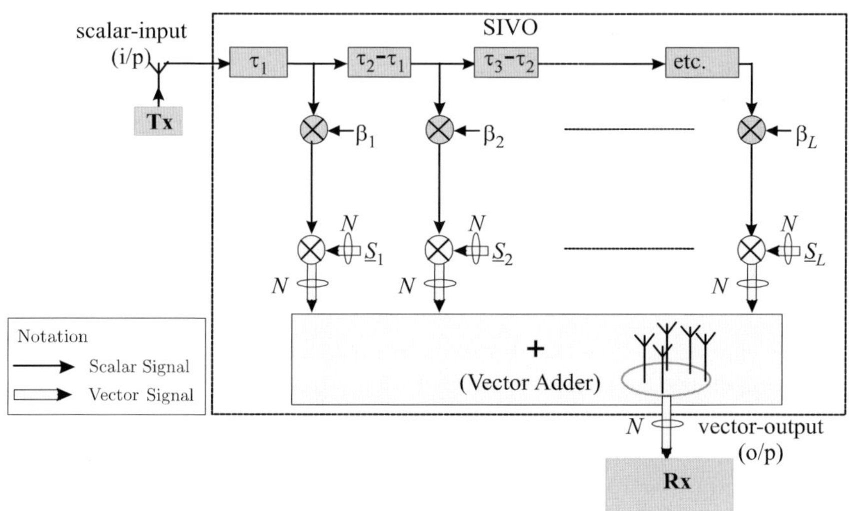

Fig. 7.1   Wireless SIMO multipath channel modelling.

---

[2]Note that the power of $m(t)$ is assumed equal to 1 as the Tx power has been incorporated into β.

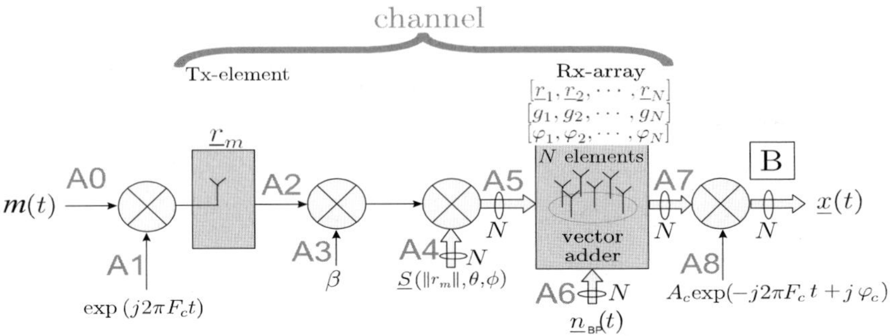

Fig. 7.2   Wireless SIMO single path model.

it would be $S_i m(t)$, where S_i is defined as the i-th element of the manifold vector $\underline{S}$. This implies that at point B, we have copies of the same signals but with amplitudes and phases measured with respect to the chosen reference point giving the well-known array processing equation,

$$\underline{x}(t) = \underline{S}m(t) + \underline{n}(t), \tag{7.10}$$

where $\underline{n}(t) \in \mathcal{C}^{N \times 1}$ is assumed to be zero mean additive white Gaussian noise with the $N \times N$ covariance matrix

$$\mathbb{R}_{nn} = \sigma_n^2 \mathbb{I}_N, \tag{7.11}$$

with σ_n^2 denoting the noise power. Note that the amplitude and phase control at downconversion does not change the signal-to-noise ratio (SNR). That is,

$$\mathrm{SNR}_{\text{at point B}} = \mathrm{SNR}_{\text{at point A7}} \tag{7.12}$$

This chapter is concerned with the estimation of $\mathbf{r}$, $\underline{g}$, and $\underline{\varphi}$ under the assumption that they are completely unknown. However, the nominal values $\widehat{r}_k \forall k$ (see Table 7.1) may be known.[3]

In addition, in contrast to Fig. 7.2, the array elements/nodes in the auto-calibration approach will operate as transceivers. Hence, the Tx sensor will be one of the N elements of the array while the remaining $N-1$ will form the Rx-array. This will be further discussed in Section 7.3.

---

[3]Notation: $\widehat{(.)}$ denotes "nominal values" (known) and $\widetilde{(.)}$ denotes "uncertainties" (unknown).

Table 7.1   Array element locations: nominal locations and uncertainties.

Sensor	Geometry	Gain	Phase
1st	$\underline{r}_1 = \widehat{\underline{r}}_1 + \widetilde{\underline{r}}_1$	$\underline{g}_1 = \widehat{\underline{g}}_1 + \widetilde{\underline{g}}_1$	$\underline{\varphi}_1 = \widehat{\underline{\varphi}}_1 + \widetilde{\underline{\varphi}}_1$
2nd	$\underline{r}_2 = \widehat{\underline{r}}_2 + \widetilde{\underline{r}}_2$	$\underline{g}_2 = \widehat{\underline{g}}_2 + \widetilde{\underline{g}}_2$	$\underline{\varphi}_2 = \widehat{\underline{\varphi}}_2 + \widetilde{\underline{\varphi}}_2$
$\vdots$	$\vdots$	$\vdots$	$\vdots$
N-th	$\underline{r}_N = \widehat{\underline{r}}_N + \widetilde{\underline{r}}_N$	$\underline{g}_N = \widehat{\underline{g}}_N + \widetilde{\underline{g}}_N$	$\underline{\varphi}_N = \widehat{\underline{\varphi}}_N + \widetilde{\underline{\varphi}}_N$

7.2.1   *Array manifold vector*

In the case that the source is in the *near-far field* of the array, $\underline{S}$ obeys the spherical wave propagation model. Otherwise, if it is in the *far field* of the array (i.e. $\|\underline{r}_m\| \to \infty$), the vector $\underline{S}$ obeys the plane wave propagation model. The array response vector (or manifold vector) for a signal arriving from $(\|\underline{r}_m\|, \theta, \phi)$ can be expressed (see [2]) as follows:

$$
\underline{S} \triangleq
\begin{cases}
\text{for plane wave propagation:} \\
\underline{S}(\theta, \phi) = \underline{g} \odot \exp\left(j\underline{\varphi}\right) \odot \exp\left(-j\mathbf{r}^T \underline{k}(\theta, \phi)\right) \\[2mm]
\text{for spherical wave propagation:} \\
\underline{S}(\|\underline{r}_m\|, \theta, \phi) = \underline{g} \odot \exp\left(j\underline{\varphi}\right) \odot \|\underline{r}_m\|^a \\
\qquad\qquad \odot \underline{\rho}^{-a} \odot \exp\left(-j\frac{2\pi F_c}{c}\left(\|\underline{r}_m\| \cdot \underline{1}_N - \underline{\rho}\right)\right)
\end{cases}
\tag{7.13}
$$

with

$$
\underline{\rho} = \sqrt{\|\underline{r}_m\|^2 \cdot \underline{1}_N + \underline{r}_x^2 + \underline{r}_y^2 + \underline{r}_z^2 - \frac{\|\underline{r}_m\| c}{\pi F_c}\mathbf{r}^T \underline{k}(\theta, \phi)}.
\tag{7.14}
$$

Here, a is a constant scalar which is assumed to be known and represents the path loss exponent measured with respect to the array reference point. In addition, the vector $\underline{k}(\theta, \phi) \in \mathcal{R}^{3 \times 1}$ is the wavenumber vector, i.e.

$$
\underline{k}(\theta, \phi) = \frac{2\pi F_c}{c} \underbrace{\begin{bmatrix} \cos(\theta)\cos(\phi) \\ \sin(\theta)\cos(\phi) \\ \sin(\phi) \end{bmatrix}}_{\triangleq \underline{u}(\theta, \phi)}
\tag{7.15}
$$

with $\underline{u}(\theta, \phi) \in \mathcal{R}^{3 \times 1}$ denoting the unity norm vector pointing towards the direction of the transmitting source. Equation 7.13 provides an important relationship between the array manifold vector, the geometry of the array and the source location with respect to the array reference point. The

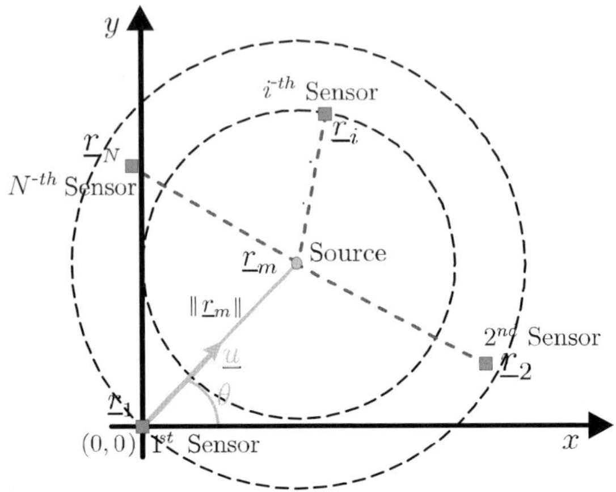

Fig. 7.3   A representative example in $\mathcal{R}^2$ space demonstrating the spherical wave propagation of a source to the N sensors/nodes. The azimuth angle θ and the range of the source, measured from the array reference point, are also shown.

parameter $\underline{\rho} = \underline{\rho}\left(\|\underline{r}_m\|, \theta, \phi\right) \in \mathcal{R}^{N \times 1}$ is the unknown vector of ranges from the source to each of the array elements, i.e.

$$\underline{\rho} = \left[\|\underline{r}_m - \underline{r}_1\|, \|\underline{r}_m - \underline{r}_2\|, \ldots, \|\underline{r}_m - \underline{r}_N\|\right]^T \in \mathcal{R}^{N \times 1}. \tag{7.16}$$

Note that Eq. (7.14) follows directly from Eq. (7.16) and can easily be proved via the cosine rule. With reference to Fig. 7.3, it should be emphasised that in the case of the spherical wave model in Eq. (7.13), the array response vector (array manifold vector) is a function of the array geometry (given by Eq. (7.1)), the range of the source from the reference point $\|\underline{r}_m\|$ and its direction (θ, ϕ) — where without any loss of generality it is assumed in this chapter that $\phi = 0$, i.e. all sensors lie in $\mathcal{R}^2$ space. The auto-calibration approach to be described later in this chapter is based upon a spherical wave propagation model, assuming that the aperture of the array (i.e. the maximum distance between any two array sensors) is sufficiently large relative to the source so that the source operates in the near-far field of the array. Since the transmitter is one of the array elements, this assumption is valid.

7.2.2   *Changing the array reference point*

The auto-calibration approach to be presented in this chapter utilises the concept of changing the array reference point to calibrate the array. Two methods of changing the array reference point will now be described:

(1) "geometric" and
(2) "approximate".

It is important to note that changing the reference point has no effect on the second order statistics of the received signal $\underline{x}(t)$ under the planewave propagation model. It only has an effect if the transmitter is in the *near-far* field of the array (i.e. spherical wave propagation model).

7.2.2.1 *Geometric case*

Consider now that the location of the array reference point is changed to be at the k-th sensor of the array. This implies that A_c and φ_c are set such that $m(t)$ is received at the k-th sensor and hence the new reference point is at $\underline{r}_k \in \mathcal{R}^{3\times 1}$ and the array geometry $\mathbf{r}$ is now set with respect to this. Therefore, the new array geometry is

$$\text{New array coordinates } = \mathbf{r}_k = \mathbf{r} - \underline{r}_k \underline{1}_N^T \tag{7.17}$$

while the new location (azimuth, elevation, range) of the source is set relative to this new reference point, i.e. $(\|\underline{r}_m - \underline{r}_k\|, \theta_k, \phi_k)$.

By adding to Eq. (7.8) a subscript indicating that the reference point is at the k-th array element, the Rx-array received baseband signal $\underline{x}_k(t) \in \mathcal{C}^N$ (i.e. measured with respect to $\underline{r}_k$) is

$$\underline{x}_k(t) = \underline{S}_k m(t) + \underline{n}_k(t). \tag{7.18}$$

Here, $\underline{S}_k$ denotes the manifold vector as described in Eq. (7.13) with $\|\underline{r}_m\|$ substituted by $\|\underline{r}_m - \underline{r}_k\|$.

7.2.2.2 *Approximate case*

In the geometric case, the reference point is changed by controlling A_c and φ_c associated with the local oscillator at the array receiver. In addition to this, an approximate method for changing the array reference point was devised in [28]. This method is based on dividing data received from the array. In particular, consider that the signal $\underline{x}(t)$ modelled by Eq. (7.10) is received by the array. The array reference point may be approximately changed to be at the k-th array element in order to create the signal $\underline{x}_k(t)$, by performing a division of $\underline{x}(t)$ by the signal received at this sensor. Hence,

$$\underline{x}_k(t) \approx \frac{1}{x_k(t)}\underline{x}(t) \tag{7.19}$$

$$= \frac{1}{S_k m(t) + n_k(t)}\underline{x}(t) \tag{7.20}$$

where $x_k(t)$, S_k and $n_k(t)$ denote the k-th element of the vectors $\underline{x}(t)$, $\underline{S}$ and $\underline{n}(t)$. Considering that L snapshots are collected from the array, this can be achieved using practical data by

$$\mathbb{X}_k = \mathbb{X}\oslash(\underline{1}_N \underline{\mathrm{row}}_k^T(\mathbb{X})) \tag{7.21}$$

where the operator $\underline{\mathrm{row}}_k(\cdot)$ yields the k-th row of the argument matrix in column vector form. Hence, the transpose $\underline{\mathrm{row}}_k^T(\cdot)$ of the operator yields the k-th row of the argument matrix. The main drawback to this approach is that it is clear from Eq. (7.20) that the presence of the noise component in the denominator (i.e. in the k-th sensor) will introduce a distortion. As a result, this method is only suitable at high SNR. However, consider that a "carrier only" tone is used as a calibration signal, which implies that there is no message signal. In this special case, consider that by taking the average of the signal received at the k-th sensor,

$$\mathcal{E}\left\{x_k(t)\right\} = S_k$$

where S_k is the k-th element of $\underline{S}$, the noise term can be "filtered". Hence, division of the array signal by the average of the signal at the k-th array element will provide an approach to change the reference point which is equivalent to the geometric approach. Considering that L snapshots are collected from the array, this can be achieved using practical data by

$$\mathbb{X}_k = \frac{L}{\underline{1}_L^T \underline{\mathrm{row}}_k(\mathbb{X})}\mathbb{X}. \tag{7.22}$$

It is important to note that this "noise filtering" approach is only suitable for "carrier only" transmissions but provides a mechanism to change the reference point which is unaffected by noise. In this case, the reference point can be changed with a performance comparable to the geometrical approach and does not require A_c and φ_c to be controlled.

7.3 Array auto-calibration

This section is concerned with the estimation of the geometrical and gain and phase uncertainties existing simultaneously in a sensor array. It will be assumed that there is no multipath interference or other uncertainties. As opposed to other data-based calibration techniques in the literature, no external sources will be required to calibrate the array (pilots or sources of opportunity). Instead, elements in the array will operate as transceivers which will be used to auto-calibrate it. In this case, the range of the transmitting element from the array reference point will always be small compared to the array aperture. Hence, the plane wave propagation assumption is no longer valid and spherical wave propagation should be considered. In this case, the notion of changing the array reference point via the methods proposed in Section 7.2.2 can be applied. By changing the array reference point, it is shown how a number of metrics may be extracted from the received array data and utilised within a set of linear equations to estimate the array uncertainties. It will be demonstrated that the proposed approach can operate even in the presence of large geometrical uncertainties (i.e. those that are several wavelengths in size).

The covariance matrix of $\underline{x}(t)$ given by Eq. (7.10), which is related to the original reference point, is denoted by $\mathbb{R} \in \mathcal{C}^{N \times N}$ and is defined by

$$\mathbb{R} = \mathcal{E}\left\{\underline{x}(t)\underline{x}(t)^H\right\}, \tag{7.23}$$

or

$$\mathbb{R} = \underline{S}\ \underline{S}^H + \mathbb{R}_{nn}. \tag{7.24}$$

Performing the eigenvector decomposition of $\mathbb{R}$, it is clear from Eq. (7.24) that

$$\lambda_0 = \max\left(\text{eig }\mathbb{R}\right) - \sigma_n^{(0)^2} \tag{7.25}$$

$$= \|\underline{r}_m\|^{2a}\ \underline{1}_N^T\left(\underline{g}^2 \odot \underline{\rho}^{-2a}\right) \tag{7.26}$$

where λ_0 is defined as the signal eigenvalue with respect to the original reference point, $\max\left(\text{eig }\mathbb{R}\right)$ denotes the maximum eigenvalue associated with the matrix $\mathbb{R}$ and $\sigma_n^{(0)^2}$ denotes the noise power.

By changing the array reference point to be at the i-th array element, $\mathbb{R}$ becomes $\mathbb{R}_i$ such that

$$\mathbb{R}_i = \mathcal{E}\left\{\underline{x}_i(t)\underline{x}_i(t)^H\right\} \tag{7.27}$$

where $\underline{x}_i(t)$ is given by Eq. (7.18). This implies that

$$\lambda_i = \max\left(\text{eig } \mathbb{R}_i\right) - \sigma_n^{2^{(i)}} \tag{7.28}$$

$$= \|\underline{r}_m - \underline{r}_i\|^{2a}\, \mathbf{1}_N^T \left(\underline{g}^2 \odot \underline{\rho}^{-2a}\right) \tag{7.29}$$

where λ_i is defined as the signal eigenvalue when the reference point is at the i-th array element and $\sigma_n^{2^{(i)}}$ denotes the corresponding noise power.

Note that Eqs. (7.25) and (7.28) require an estimate of the noise power $\sigma_n^{2^{(0)}}$ or $\sigma_n^{2^{(i)}}$. Assuming the narrowband case, this can be achieved by taking the average of the non-principal eigenvalues of the corresponding covariance matrix. The narrowband assumption implies $m(t)$ is slow varying across the array such that $m(t - \tau_1) \approx m(t - \tau_2) \approx \cdots \approx m(t - \tau_N)$ where τ_i is the propagation delay associated with the i-th array element and it always holds under a "carrier only" tone. In the more general wideband case, which implies that $m(t - \tau_1) \neq m(t - \tau_2) \neq \cdots \neq m(t - \tau_N)$, this assumption does not apply. However, in this case, signal eigenvalues may be estimated by using the trace of the eigenvalues of the covariance matrix as described in [2]. It is important to note that the signal eigenvalue provides a metric which is a function of the range from the source to the reference point. By changing the reference point to be at the different array elements, different metrics can be extracted. However, changing the array reference point only has an effect on the second-order statistics of the received signal under a spherical wave propagation model, having no effect under a plane wave propagation model. Further to this, it is well known that in the case of a single transmitter the eigenvector $\underline{E}_0 \in \mathcal{C}^{N \times 1}$ corresponding to the principal eigenvalue of the received covariance matrix $\mathbb{R}$, in theory spans the same linear subspace as the array manifold vector. Hence,

$$\underline{S} \in \{\underline{E}_0\} \tag{7.30}$$

$$= \|\underline{S}\| \cdot \underline{E}_0. \tag{7.31}$$

Under spherical wave propagation, assuming the eigenvector $\underline{E}_0$ is chosen to have a unit norm, it is easy to see from Eq. (7.13) that,

$$\|\underline{S}\| = \|\underline{r}_m\|^a \cdot \left\|\underline{g} \odot \underline{\rho}^{-a}\right\|. \tag{7.32}$$

It is important to note that this eigenvector provides a metric which is a function of the array manifold vector (i.e. the location of the source, the geometry of the array and the gain and phase of the array elements).

Consider now that the N-sensor array presented in Section 7.2 has an unknown geometry and gain and phase parameters. Assume also without any loss of generality that the first array element is located at a known (relative to the reference point) location $\underline{r}_1$ and has a unity gain and zero phase. This assumption helps with fixing the absolute orientation of the whole array in real physical space, because if only the range to the first sensor is known, the shape and geometry can be estimated but not the array orientation. Furthermore, without any loss of generality, consider for notational convenience that the N array elements are arranged such that their azimuth angles $\theta_1, \theta_2, \cdots, \theta_N$, measured anticlockwise with respect to the positive x-axis, increase with $\theta_1 < \theta_2 < \cdots < \theta_N$ measured over the range $[0, 2\pi]$. In this section, up to N array elements transmit in turn from unknown locations to the other $N - 1$ elements. These other elements operate as a single entity by forming an array receiver of $N - 1$ elements. Hence, in terms of Fig. 7.2, the dimensionality at points A4, A5, A6, A7, A8 and B is now $N - 1$ rather than N. In addition, the transmitting array element is now at point A2 and the point A5 provides the signal received at the other $N - 1$ array elements.

By changing the array reference point to be at each of the array elements, signal eigenvalues can be extracted from the received signal covariance matrix which are a function of the range between the transmitting array element and the other $N - 1$ elements. Having extracted these signal eigenvalues for all transmissions, they can be used to estimate the true geometry of the array. Then, by extracting the principal eigenvector of the received covariance matrix for each transmission, having estimated the array geometry (and hence knowing the location of the transmitting array element and the array geometry formed by the other $N - 1$ elements), the true gain and phase of the array elements can also be estimated following Eq. (7.31). The auto-calibration approach is split into three phases:

- array measurement phase,
- shape estimation phase and
- gain and phase calibration phase.

It is important to note that since the array geometry is estimated using eigenvalues alone, the approach can be employed even if the location uncertainties are large (i.e. several wavelengths in size). This is particularly useful for large aperture geometries where errors will likely be more significant.

7.3.1 *Measurement phase*

Consider that all array elements transmit in turn and the other $N - 1$ array elements form an array receiver. When the m-th array element transmits, data is initially collected from the array using the original (initial) reference point (0,0,0). The second-order statistics of the signals received by the array are constructed and the signal eigenvalue defined by λ_{0m} is extracted as well as the eigenvector $\underline{E}_{0m}$ corresponding to the principal eigenvalue. Here, λ_{0m} is defined as the *primary* signal eigenvalue. Note that, in contrast to the previous sections and for the remainder of this chapter, two subscripts are used in the notation of signal eigenvalues, eigenvectors and manifold vectors (amongst others) with the first subscript denoting the reference point and the second denoting the transmitting array element. In addition, in this section, for notational convenience, the extracted eigenvector $\underline{E}_{0m}$ is made to have length $N \times 1$ despite the array receiver having only $N-1$ elements by inserting a zero as the m-th element. This effectively denotes an array response of zero at the transmitting array element.

Now consider that when the m-th element is transmitting, the array reference point is changed to be at each of the $N-1$ elements of the array receiver. When the array reference point is at the j-th element, the second order statistics of the signals received from the array are again constructed and the signal eigenvalue defined by λ_{jm} is extracted. By changing the reference point to be at all $N-1$ elements of the array receiver and constructing the received covariance matrix in each case, a total of N signal eigenvalues $\lambda_{0m}, \lambda_{1m}, \lambda_{2m}, \ldots, \lambda_{(m-1)m}, \lambda_{(m+1)m}, \ldots, \lambda_{Nm}$ can be extracted. Considering all array elements transmit, a matrix of eigenvectors $\mathbb{E} \in \mathcal{C}^{N \times N}$ and a matrix of eigenvalues $\mathbf{\Lambda} \in \mathcal{R}^{N \times N}$ may be constructed as,

$$\mathbb{E} = [\underline{E}_{01}, \underline{E}_{02}, \cdots, \underline{E}_{0N}] \tag{7.33}$$

$$\mathbf{\Lambda} = \begin{bmatrix} \lambda_{01}, & \lambda_{12}, & \lambda_{13}, & \ldots & \lambda_{1N} \\ \lambda_{21}, & \lambda_{02}, & \lambda_{23}, & \ldots & \lambda_{2N} \\ \lambda_{31}, & \lambda_{32}, & \lambda_{03}, & \ldots & \lambda_{3N} \\ \lambda_{41}, & \lambda_{42}, & \lambda_{43}, & \ldots & \lambda_{4N} \\ \vdots & \vdots & \vdots & \ddots & \vdots \\ \lambda_{N1}, & \lambda_{N2}, & \lambda_{N3}, & \cdots, & \lambda_{0N} \end{bmatrix}^{\frac{1}{2a}}. \tag{7.34}$$

7.3.2 *Array shape estimation phase*

With reference to Eqs. (7.26) and (7.29), by taking the ratios of signal eigenvalues at different reference points when a given array element is transmitting, ratios of ranges between the transmitting element and the other $N - 1$ array elements can be constructed. This is intuitive because dividing signal eigenvalues cancels many of the common components in Eqs. (7.26) and (7.29). Hence, using the matrix $\mathbf{\Lambda}$, the matrix $\mathbb{K} \in \mathcal{R}^{N \times N}$ containing these ratios may be defined as

$$\mathbb{K} = \mathbf{\Lambda} \oslash \left(\underline{1}_N \cdot \mathrm{diag}\left(\mathbf{\Lambda}\right)^T \right) \tag{7.35}$$

and re-expressed as follows

$$\mathbb{K} = \begin{bmatrix}
1, & \dfrac{\|r_2 - r_1\|}{\|r_2\|}, & \cdots, & \dfrac{\|r_N - r_1\|}{\|r_N\|} \\[2ex]
\dfrac{\|r_1 - r_2\|}{\|r_1\|}, & 1, & \cdots, & \dfrac{\|r_N - r_2\|}{\|r_N\|} \\[2ex]
\dfrac{\|r_1 - r_3\|}{\|r_1\|}, & \dfrac{\|r_2 - r_3\|}{\|r_2\|}, & \cdots, & \dfrac{\|r_N - r_3\|}{\|r_N\|} \\[2ex]
\dfrac{\|r_1 - r_4\|}{\|r_1\|}, & \dfrac{\|r_2 - r_4\|}{\|r_2\|}, & \cdots, & \dfrac{\|r_N - r_4\|}{\|r_N\|} \\[2ex]
\vdots & \vdots & \ddots & \vdots \\[2ex]
\dfrac{\|r_1 - r_N\|}{\|r_1\|}, & \dfrac{\|r_2 - r_N\|}{\|r_2\|}, & \cdots, & 1
\end{bmatrix}. \tag{7.36}$$

where the columns are denoted $\underline{\mathcal{K}}_1 \triangleq$, $\underline{\mathcal{K}}_2 \triangleq$, $\ldots$, $\underline{\mathcal{K}}_N \triangleq$.

Here it is clear that the i-th row of the matrix $\mathbb{K}$ denotes the ratio of true ranges from the i-th array element to each of the other array elements and to the original reference point. The rank 1 matrix $\mathbb{D} \in \mathcal{R}^{N \times N}$ and the rank N matrix $\mathbb{Q} \in \mathcal{R}^{N \times N}$ are also constructed for notational convenience such that

$$\mathbb{D} = \mathbb{K} \oslash \mathbb{K}^T \tag{7.37}$$

$$\mathbb{Q} = \left[\underline{Q}_1, \underline{Q}_2, \cdots, \underline{Q}_N\right]$$

$$= \mathbb{K} - \underline{1}_N \underline{1}_N^T. \tag{7.38}$$

Here, the matrices $\mathbb{D}$ and $\mathbb{Q}$ are known and the matrix $\mathbb{D}$ can be re-expressed as follows

$$\mathbb{D} = \begin{bmatrix} 1, & \dfrac{\|\underline{r}_1\|}{\|\underline{r}_2\|}, & \cdots & \dfrac{\|\underline{r}_1\|}{\|\underline{r}_N\|} \\[2mm] \dfrac{\|\underline{r}_2\|}{\|\underline{r}_1\|}, & 1, & \cdots & \dfrac{\|\underline{r}_2\|}{\|\underline{r}_N\|} \\[2mm] \dfrac{\|\underline{r}_3\|}{\|\underline{r}_1\|}, & \dfrac{\|\underline{r}_3\|}{\|\underline{r}_2\|}, & \cdots & \dfrac{\|\underline{r}_3\|}{\|\underline{r}_N\|} \\[2mm] \dfrac{\|\underline{r}_4\|}{\|\underline{r}_1\|}, & \dfrac{\|\underline{r}_4\|}{\|\underline{r}_2\|}, & \cdots & \dfrac{\|\underline{r}_4\|}{\|\underline{r}_N\|} \\[2mm] \vdots & \vdots & \ddots & \vdots \\[2mm] \dfrac{\|\underline{r}_N\|}{\|\underline{r}_1\|}, & \dfrac{\|\underline{r}_N\|}{\|\underline{r}_2\|}, & \cdots & 1 \end{bmatrix} \tag{7.39}$$

where the columns are $\triangleq \mathcal{D}_1$, $\triangleq \mathcal{D}_2$, ..., $\triangleq \mathcal{D}_N$.

Hence, the matrix $\mathbb{D}$ denotes the ratio of the norms of the array element locations with respect to the original reference point. Furthermore, the matrix $\mathbb{Q}$ corresponds to the differences in ranges between array elements and the original reference point.

A matrix $\mathbb{B}_\vartheta \in \mathcal{R}^{N \times N}$ of angles between array elements can be defined as follows

$$\mathbb{B}_\vartheta \triangleq \begin{bmatrix} 1, & \underline{u}_1^T \underline{u}_2, & \underline{u}_1^T \underline{u}_3, & \cdots & \underline{u}_1^T \underline{u}_N \\[1mm] \underline{u}_2^T \underline{u}_1, & 1, & \underline{u}_2^T \underline{u}_3, & \cdots & \underline{u}_2^T \underline{u}_N \\[1mm] \underline{u}_3^T \underline{u}_1, & \underline{u}_3^T \underline{u}_2, & 1, & \cdots & \underline{u}_3^T \underline{u}_N \\[1mm] \underline{u}_4^T \underline{u}_1, & \underline{u}_4^T \underline{u}_2, & \underline{u}_4^T \underline{u}_3, & \cdots & \underline{u}_4^T \underline{u}_N \\[1mm] \vdots & \vdots & \vdots & \ddots & \vdots \\[1mm] \underline{u}_N^T \underline{u}_1, & \underline{u}_N^T \underline{u}_2, & \underline{u}_N^T \underline{u}_3, & \cdots & 1 \end{bmatrix}. \tag{7.40}$$

where the columns are $\triangleq \cos \underline{\vartheta}_1$, $\triangleq \cos \underline{\vartheta}_2$, $\triangleq \cos \underline{\vartheta}_3$, ..., $\triangleq \cos \underline{\vartheta}_N$.

Equivalently, as all array elements lie in the same $\mathcal{R}^2$ space,

$$\mathbb{B}_\vartheta = \begin{bmatrix} 1, & \cos\left(\theta_1 - \theta_2\right), & \cdots \cos\left(\theta_1 - \theta_N\right) \\ \cos\left(\theta_2 - \theta_1\right), & 1, & \cdots \cos\left(\theta_2 - \theta_N\right) \\ \cos\left(\theta_3 - \theta_1\right), & \cos\left(\theta_3 - \theta_2\right), & \cos\left(\theta_3 - \theta_N\right) \\ \cos\left(\theta_4 - \theta_1\right), & \cos\left(\theta_4 - \theta_2\right), & \cos\left(\theta_4 - \theta_N\right) \\ \vdots & & \ddots & \vdots \\ \cos\left(\theta_N - \theta_1\right), & \cos\left(\theta_N - \theta_2\right), & \cdots & 1 \end{bmatrix}. \tag{7.41}$$

Considering that the second array element is transmitting from an *unknown* location and the reference point is at the original (0,0,0) point, from Eq. (7.14),

$$\underbrace{\begin{bmatrix} \|\underline{r}_1 - \underline{r}_2\|^2 \\ 0 \\ \|\underline{r}_3 - \underline{r}_2\|^2 \\ \vdots \\ \|\underline{r}_N - \underline{r}_2\|^2 \end{bmatrix}}_{\triangleq \underline{\rho}_m} = \|\underline{r}_2\|^2 \cdot \underline{1}_N + \begin{bmatrix} \|\underline{r}_1\|^2 \\ \|\underline{r}_2\|^2 \\ \|\underline{r}_3\|^2 \\ \vdots \\ \|\underline{r}_N\|^2 \end{bmatrix}$$

$$-2\|\underline{r}_2\| \begin{bmatrix} \|\underline{r}_1\| \\ \|\underline{r}_2\| \\ \|\underline{r}_3\| \\ \vdots \\ \|\underline{r}_N\| \end{bmatrix} \odot \begin{bmatrix} \underline{u}_1^T \underline{u}_2 \\ \underline{u}_2^T \underline{u}_2 \\ \underline{u}_3^T \underline{u}_2 \\ \vdots \\ \underline{u}_N^T \underline{u}_2 \end{bmatrix}. \tag{7.42}$$

Dividing both sides of Eq. (7.42) by $\|\underline{r}_2\|^2$, it is easy to prove that,

$$\underline{Q}_2^2 + 2\underline{Q}_2 = \underline{D}_2^2 - 2\underline{D}_2 \odot \cos\underline{\vartheta}_2 + \underline{\delta}_2 \tag{7.43}$$

$$\cos\underline{\vartheta}_2 = \frac{1}{2}\left(\underline{D}_2^2 - \underline{Q}_2^2 - 2\underline{Q}_2 + \underline{\delta}_2\right) \oslash \underline{D}_2 \tag{7.44}$$

where $\underline{\delta}_m \in \mathcal{R}^{N\times1}$ denotes a vector of zeros with the m-th element equal to one. Assuming the direction to the first array element θ_1 represents the known[4] array orientation angle, it is easy to see that the angle of all the

---

[4]Note that if the orientation angle is unknown then it is still possible to estimate the array shape by arbitrarily setting $\theta_1 = 0$. This will give the array geometry independently of the array axis.

other array elements can be estimated using this set of equations alone. However, for all N elements transmitting it follows that

$$\mathbb{B}_\vartheta = \tfrac{1}{2}\left(\mathbb{D}\odot\mathbb{D} - \mathbb{Q}\odot\mathbb{Q} - 2\mathbb{Q} + \mathbb{I}_N\right)\oslash\mathbb{D}. \tag{7.45}$$

Any row of the matrix $\mathbb{B}_\vartheta$ may be used to obtain an estimate of the angle of the array elements. These estimates may be averaged over the different rows. Using this notion, the following set of linear equations may be constructed to obtain a least squares estimate of the directions of the array elements,

$$\mathbb{H}_\theta\underline{\theta} = \underline{b}_\theta \Rightarrow \underline{\theta} = \mathbb{H}_\theta^{\#}\underline{b}_\theta \tag{7.46}$$

where $\mathbb{H}_\theta \in \mathcal{R}^{N\times(N-1)}$, $\underline{b}_\theta \in \mathcal{R}^{N\times1}$ and $\underline{\theta}\in\mathcal{R}^{(N-1)\times1}$ are defined as,

$$\mathbb{H}_\theta = \left(\underline{1}_{N-1},\underline{1}_{N-1}\underline{1}_{N-1}^T - N\mathbb{I}_{N-1}\right)^T \tag{7.47}$$

$$\underline{b}_\theta = \arccos\left(\mathbb{B}_\vartheta\right)^T\underline{1}_N - \theta_1\underline{1}_N + N\theta_1\underline{\delta}_1 \tag{7.48}$$

$$\underline{\theta} = [\theta_2,\theta_3,\cdots,\theta_N]^T. \tag{7.49}$$

Assuming $\theta_1 < \theta_2 < \cdots < \theta_N$, the sign of the elements in the matrix $\arccos\left(\mathbb{B}_\vartheta\right)$ over the range $[0,2\pi]$ should be made negative in the upper triangle of the matrix $\arccos\left(\mathbb{B}_\vartheta\right)$ and positive elsewhere to resolve the cosine ambiguity. Solving Eq. (7.46) will provide a least squares solution of the direction of all array elements. Note that this approach may be used to estimate azimuth and elevation angles or the "cone" angles (see Chapter 5 of [3]) in the case of 3D array geometries.

To estimate the array shape, the range between the array elements with respect to the original reference point must also be estimated. Knowing the range to the first array element and using $\underline{\mathcal{D}}_1$ as defined in Eq. (7.36),

$$\underline{\rho}_0 = \begin{bmatrix} \|\underline{r}_1\| \\ \|\underline{r}_2\| \\ \|\underline{r}_3\| \\ \vdots \\ \|\underline{r}_N\| \end{bmatrix} = \|\underline{r}_1\|\,\underline{\mathcal{D}}_1. \tag{7.50}$$

Hence, using the full matrix $\mathbb{D}$, it follows that

$$\underline{\rho}_0 = \tfrac{1}{N}\,\|\underline{r}_1\|\cdot\left(\mathbb{D}\oslash\left(\underline{1}_N\cdot\underline{\mathrm{row}}_1^T\left(\mathbb{D}\right)\right)\right)\underline{1}_N. \tag{7.51}$$

Having estimated the angle and range of each of the array elements with respect to the original reference point, the array geometry can now be inferred. It is important to note that no nominal values have been used to estimate this geometry. Instead, the geometry is estimated directly. This would not be the case if a pilot/self-calibration approach were employed.

7.3.3 *Complex gain estimation phase*

With reference to Eq. (7.31), when the m-th array element is transmitting from an unknown location and the other elements form an array receiver, the extracted eigenvector $\underline{E}_{0m} \in \mathcal{C}^{N \times 1}$ (with a zero inserted as the m-th element and ignoring finite averaging effects) is related to the array manifold vector associated with the array receiver by

$$\underline{E}_{0m} = \|\underline{S}_{0m}\|^{-1} \cdot \underline{S}_{0m} \tag{7.52}$$

where in this section, $\underline{S}_{0m}$ is redefined to be an N element vector with the m-th array element (corresponding to the transmitting element) set to be 1. Hence,

$$
\begin{aligned}
\underline{S}_{0m} &= \|\underline{r}_m\|^a \, \underline{\rho}_m^{-a} \odot \left(\mathbb{I}_N - \underline{\delta}_m \underline{\delta}_m^T\right) \underline{g} \\
&\quad \odot \exp\left(j \left(\mathbb{I}_N - \underline{\delta}_m \underline{\delta}_m^T\right) \underline{\varphi}\right) \\
&\quad \odot \exp\left(-j \frac{2\pi F_c}{c} \left(\|\underline{r}_m\| \cdot \underline{1}_{N-1} - \underline{\rho}_m\right)\right) \tag{7.53} \\
&= \underline{K}_m^{-a} \odot \left(\mathbb{I}_N - \underline{\delta}_m \underline{\delta}_m^T\right) \underline{g} \\
&\quad \odot \exp\left(j \left(\mathbb{I}_N - \underline{\delta}_m \underline{\delta}_m^T\right) \underline{\varphi}\right) \\
&\quad \odot \exp\left(j \frac{2\pi F_c}{c} \|\underline{r}_m\| \underline{Q}_m\right) \tag{7.54}
\end{aligned}
$$

where $\underline{\rho}_m \in \mathcal{R}^{N \times 1}$ describes the range from the transmitting array element to all N array elements. Here note that the m-th element of $\underline{\rho}_m$ is intuitively equal to zero. With reference to Eq. (7.54), the vector $\underline{E}_{0m}$ provides a set of equations related to the difference and ratio of ranges of the transmitting element to the original reference point and the other array elements (i.e. $\underline{Q}_m$ and $\underline{K}_m$) as well as the gain and phase of the array elements.

Consider that the second array element is transmitting from an unknown location. Hence,

$$\underline{E}_{02} = \|\underline{S}_{02}\|^{-1} \cdot \underline{S}_{02}. \tag{7.55}$$

Defining $E_{02}(1)$ as the first element of the extracted eigenvector of the received covariance matrix,

$$\|E_{02}(1)\|^{-1} \underline{E}_{02} = \underline{S}_{12}. \tag{7.56}$$

Moving the reference point back to the original reference point,

$$C_2 \cdot \underline{E}_{02} = \underline{S}_{02} \tag{7.57}$$

where

$$C_2 = \frac{\|\underline{r}_2\|^a}{\|\underline{r}_2 - \underline{r}_1\|^a} \cdot \exp\left(-j\frac{2\pi F_c}{c}\left(\|\underline{r}_2\| - \|\underline{r}_2 - \underline{r}_1\|\right)\right) \cdot \|E_{02}(1)\|^{-1}.$$

Hence combining Eq. (7.57) with Eq. (7.54),

$$\left(\mathbb{I}_N - \underline{\delta}_2\underline{\delta}_2^T\right)\left(\underline{g}\odot\exp\left(j\underline{\varphi}\right)\right) = C_2 \cdot \underline{E}_{02}\odot\underline{K}_2^a$$
$$\odot\exp\left(-j\frac{2\pi F_c}{c}\|\underline{r}_2\|\,\underline{Q}_2\right). \tag{7.58}$$

Considering all N array elements transmit from unknown locations, it can easily be inferred that

$$\underbrace{\begin{bmatrix} \mathbb{O}_{N,N} \\ \mathbb{I}_N - \underline{\delta}_2\underline{\delta}_2^T \\ \mathbb{I}_N - \underline{\delta}_3\underline{\delta}_3^T \\ \vdots \\ \mathbb{I}_N - \underline{\delta}_N\underline{\delta}_N^T \end{bmatrix}}_{\triangleq\,\mathbb{H}_g\in\mathcal{C}^{NN\times N}} \left(\underline{g}\odot\exp\left(j\underline{\varphi}\right)\right) = \mathrm{vec}(\mathbb{B}_g) \tag{7.59}$$

where the matrix $\mathbb{B}_g \in \mathcal{C}^{N\times N}$ is formed by

$$\mathbb{B}_g = (\mathbb{E}\cdot\mathrm{diag}\,(\underline{C}))\odot\mathbb{K}^a$$
$$\odot\exp\left(-j\frac{2\pi F_c}{c}\mathbb{Q}\cdot\mathrm{diag}\left(\underline{\rho}_0\right)\right) \tag{7.60}$$

$\text{vec}(\mathbb{B}_g)$ is its column based vectorised form and the vector $\underline{C} \in \mathcal{C}^{N \times 1}$ can be constructed by

$$\underline{C} = [1, C_2, C_3, \cdots, C_N]^T \tag{7.61}$$

$$= \exp\left(j \frac{2\pi F_c}{c} \underline{\text{row}}_1(\mathbb{Q}) \odot \underline{\rho}_0\right) \oslash$$

$$\left(\underline{\text{row}}_1(\mathbb{E}) \odot \underline{\text{row}}_1(\mathbb{K})^a + \underline{\delta}_1\right). \tag{7.62}$$

Here, $\underline{\delta}_1$ is used to prevent an indeterminant in the first element of the vector $\underline{C}$. Hence, the gain and phase of the array elements can be estimated by

$$\boxed{\underline{g} \odot \exp(j\underline{\varphi}) = \mathbb{H}_g^{\#} \cdot \text{vec}(\mathbb{B}_g)} \tag{7.63}$$

where $\mathbb{H}_g^{\#}$ is the pseudo-inverse of the matrix $\mathbb{H}_g$.

This approach is also valid if the gain of the array elements is directional, provided that the unknown gain uncertainty is non-directional. In this case, the vector $\underline{g}$ should be split into the known directional component and unknown non-directional component. After the known directional component is incorporated into the matrix $\mathbb{H}_g$, the unknown non-directional component can be estimated as normal. It should be also noted that the gain and phase of the array elements may be estimated independently of one another by taking the magnitude and phase components of the matrix $\mathbb{B}_g$ respectively and solving the same set of linear equations.

For a given array of N sensors, the auto-calibration approach is now described in a step-by-step format:

STEP-1 All array elements transmit in turn. When the i-th element transmits, the other $N-1$ elements form an array receiver and the array reference point is moved from its original point to be at each element of this array. When the reference point is at its original position and each of the array elements, the signal eigenvalue of the received covariance matrix is extracted. Repeating this process, construct the matrix $\mathbf{\Lambda}$ in Eq. (7.34). When the reference point is at the original reference point, extract the eigenvector corresponding to the principal eigenvalue of the received covariance matrix and construct the matrix $\mathbb{E}$ in Eq. (7.33).

STEP-2 Using the matrix $\mathbf{\Lambda}$, construct the matrices $\mathbb{K}$, $\mathbb{Q}$ and $\mathbb{D}$ using Eqs. (7.35), (7.38) and (7.39) respectively.

Table 7.2   True location (in metres), gain (in volts) and phase (in degrees) of the small aperture circular array.

	x	y	gain	phase
Rx1	0.7500	-1.2990	1	$0°$
Rx2	1.5000	0.0000	1.4050	$5.8964°$
Rx3	0.7500	1.2990	1.0930	$2.1369°$
Rx4	-0.7500	1.2990	0.8900	$0.4518°$
Rx5	-1.5000	0.0000	1.3890	$7.6028°$
Rx6	-0.7500	-1.2990	1.2430	$-6.5409°$

STEP-3 Using the matrices $\mathbb{D}$ and $\mathbb{Q}$, solve Eq. (7.46) to estimate the angle of the array element locations.

STEP-4 Using the matrix $\mathbb{D}$, solve Eq. (7.51) to estimate the range of the array elements from the origin.

STEP-5 Using the matrices $\mathbb{K}$, $\mathbb{Q}$ and $\mathbb{E}$, solve Eq. (7.59) to estimate the gain and phase of the array elements.

7.4   Performance evaluation

7.4.1   *Small aperture array*

Consider a small aperture uniform circular array of $N = 6$ elements with half-wavelength intersensor spacing and a diameter (aperture) of 3 m. Assume for the purposes of calibration that the array elements are transceivers operating at $F_c = 100$ MHz and transmit "carrier only" tones. The unknown true location of the array elements and their unknown gain and phase are described in Table 7.2 (Rx1 is assumed known).

The above geometry is shown in Fig. 7.4.

It is important to note that even though the array aperture is small, the transmitters for the purposes of the auto-calibration approach are the array elements themselves which inherently implies that the array manifold vector $\underline{S}$ (defined by Eq. (7.13)) obeys spherical wave propagation.

To allow for a comparative study of the performance of the auto-calibration approach with other calibration approaches based on imprecisely known array parameters instead of completely unknown array parameters, consider that the known nominal locations of the array elements and their known nominal gain and phase are described in Table 7.3.

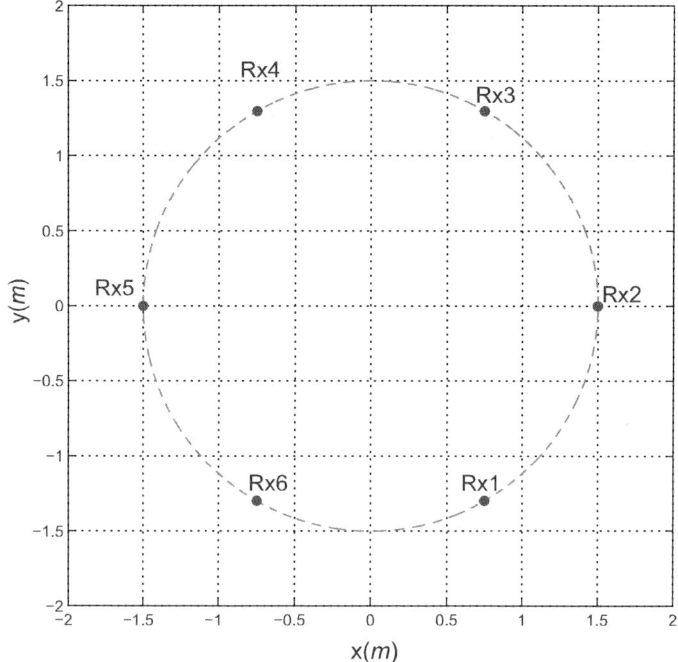

Fig. 7.4   The small aperture circle geometry consisting of $N = 6$ sensor nodes arranged in a circle with the Cartesian coordinates in Table 7.2.

Table 7.3   Nominal location (in metres), gain (in volts) and phase (in degrees) of the small aperture array.

	x	y	gain	phase
Rx1	3.7500	1.7010	1	0°
Rx2	4.5000	3.0000	1	0°
Rx3	2.2500	1.7010	1	0°
Rx4	3.7500	4.2990	1	0°
Rx5	2.2500	4.2990	1	0°
Rx6	1.5000	3.0000	1	0°

It is important to note that here the location uncertainties are small ($<10\%$ of a wavelength) and that these nominal values are not used by the auto-calibration approach. In this section it is assumed that array geometries lie in $\mathcal{R}^2$ space and that there is no multipath or fading effects

Table 7.4   Estimation error in location (in metres), gain (in volts) and phase (in degrees) of the small aperture array after the auto–calibration approach.

$L = 10000$; SNR= 20 dB			
Geometrical Approach (metres $(\times 10^{-7})$, volts $(\times 10^{-4})$ or degrees $(\times 10^{-6})$)			
Error	x	y	**gain**	**phase**
Rx1	0	0	0	$0°$
Rx2	0.0334	0.0193	0.0379	$634.92°$
Rx3	0	-0.0527	0.1666	$50.644°$
Rx4	0.1101	-0.0636	0.1255	$-1300°$
Rx5	0.1369	0.0790	-0.0704	$507.07°$
Rx6	0	0.0527	0.1063	$-716.21°$

in the channel. In addition, a free space path loss model is assumed (i.e. $a = 1$).

Initially, the performance of the auto-calibration approach for the small aperture circular array shown in Fig. 7.4 is analysed under an SNR of 20 dB using $L = 10,000$ snapshots. Calibration signals transmitted by the sensor nodes are *carrier-only* 100 MHz tones. The geometrical and gain and phase estimation errors after applying the auto-calibration approach when using the geometrical method of changing the array reference point are given in Table 7.4. It is clear that the auto-calibration approach can estimate the unknown array parameters to a very high degree of accuracy.

Monte-Carlo simulations are performed to investigate the performance of the auto-calibration approach under this setup. In particular, the root-mean-square error (RMSE) of the gain, phase and location estimates after the auto-calibration approach is investigated as a function of SNR under 200 noise realisations. This analysis is performed over SNR= $[0, 10, 20, 30, 40]$dB and $L = 10,000$ snapshots. Here, the SNR is the average SNR between each transmitter and receivers over the entire calibration process (i.e. based on $N(N-1)$ SNRs). The performance of the auto-calibration approach should also be compared to a pilot calibration approach similar to [6] but using two near-far field sources and extracting the signal eigenvectors of the received covariance matrix. Here the two pilot sources are located at ranges of 5 m and 10 m and angles of $\theta = 56°$ and $\theta = 134°$ and the nominal parameters in Table 7.3 are now utilised. The locations of the pilot sources are assumed to be known. Results for the sensor location (green) and gain (blue) and phase (red) estimation errors after the auto-calibration approach (solid lines) and

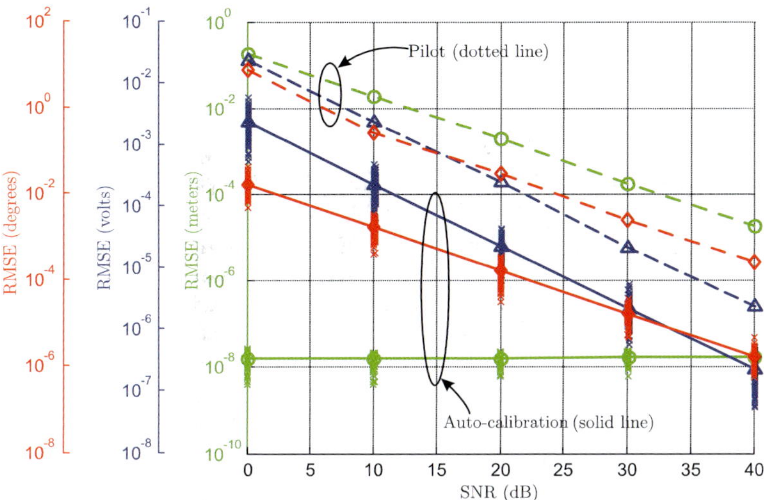

Fig. 7.5 RMSE vs SNR estimation performance of the auto-calibration and pilot calibration approaches for the small aperture array described in Table 7.2.

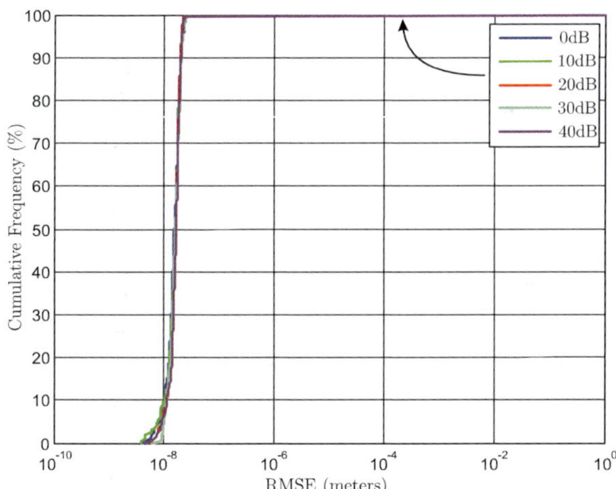

Fig. 7.6 Small aperture — *Location* estimation error: cumulative distribution of the array location estimation errors shown in Fig. 7.5 as a function of SNR over 200 realisations.

pilot calibration approach (dotted lines) are shown in Fig. 7.5. In addition, "crosses" showing the results of each of the 200 realisations for each SNR are also shown for the auto-calibration approach. It is clear that the auto-calibration method is able to estimate all types of array parameters reliably

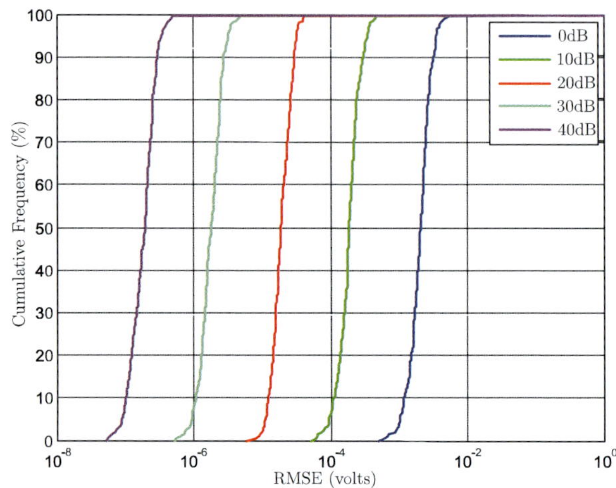

Fig. 7.7   Small aperture — *Gain* estimation error: cumulative distribution of the gain estimation errors shown in Fig. 7.5 as a function of SNR over 200 realisations.

Table 7.5   True location (in metres) of the large aperture circle array.

	$\underline{r}_x$	$\underline{r}_y$	$\underline{r}_z$
$\underline{r}_1^T$	216.5064	125	0
$\underline{r}_2^T$	0	250	0
$\underline{r}_3^T$	−216.5064	125	0
$\underline{r}_4^T$	−216.5064	−125	0
$\underline{r}_5^T$	0	−250	0
$\underline{r}_6^T$	216.5064	−125	0

and to a very high accuracy. The performance exceeds the pilot calibration approach, particularly at low SNR. This is attributed to the use of eigenvalues instead of eigenvectors as they can be estimated more rapidly and more accurately using a smaller number of snapshots.

The cumulative distribution of the 200 realisations for each SNR in Fig. 7.5 is given for the location and gain estimation error after the auto-calibration approach. These are shown in Figs. 7.6 and 7.7 respectively. It is clear that under 10,000 snapshots, the array elements' locations can be estimated to a very high accuracy apparently independent of SNR since only eigenvalues are used. However, since the gain estimation uses eigenvectors, the performance here decreases and has more of a dependence

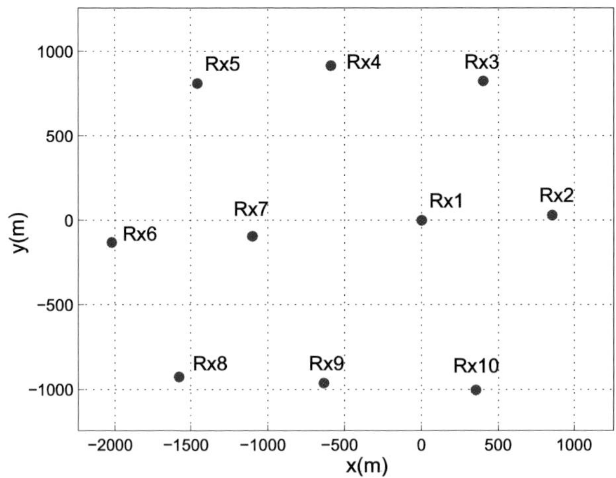

Fig. 7.8   The random geometry consisting of $N = 10$ sensor nodes arranged randomly with the Cartesian coordinates in Table 7.6.

on SNR. As expected, the CDF of the phase estimation at different SNRs is comparable to that of the gain.

7.4.2   *Large aperture array*

Consider a large aperture *circle* geometry of $N = 6$ nodes and diameter (aperture) 500 m. In particular, the array coordinates in metres are given in Table 7.5. This geometry provides a similar array shape to the geometry in Table 7.2 (as shown in Fig. 7.4) with the difference that the aperture is now much bigger.

Also consider a *random* geometry of $N = 10$ nodes as shown in Fig. 7.8 forming an array with an aperture of 2859.8 m, whose array coordinates are given in metres in Table 7.6.

Unlike in the small aperture case that was investigated earlier, in the large aperture simulation studies presented here it will be assumed that the gain and phase of the array elements are fully known such that

$$\underline{g} = \underline{1}_N, \tag{7.64}$$

$$\underline{\varphi} = \underline{0}_N. \tag{7.65}$$

Hence, only the measurement and the array shape estimation stages of the auto-calibration algorithm need to be employed. However, the algorithm would still work if these parameters also had to be estimated.

Table 7.6   True location (in metres) of the random array.

	$\underline{r}_x$	$\underline{r}_y$	$\underline{r}_z$
$\underline{r}_1^T$	0	0	0
$\underline{r}_2^T$	853.4	29.5	0
$\underline{r}_3^T$	399.7	824.4	0
$\underline{r}_4^T$	−591.1	915.4	0
$\underline{r}_5^T$	−1457.2	810.9	0
$\underline{r}_6^T$	−2022	−129.7	0
$\underline{r}_7^T$	−1101.5	−93.4	0
$\underline{r}_8^T$	−1578.2	−925.1	0
$\underline{r}_9^T$	−635.6	−961.3	0
$\underline{r}_{10}^T$	352.7	−1002.9	0

Array auto-calibration is applied under an SNR of 30 dB using $L = 1000$ snapshots. As for the small aperture case, calibration signals transmitted by the array elements are *carrier-only* 100 MHz tones.

For the large aperture circular geometry, the error in the array element locations after calibration when using the geometrical and approximate (with noise filtering) methods of changing the reference point is given in Table 7.7.

For the random geometry, the error in the array element locations after calibration when using the geometrical and approximate (with noise filtering) methods of changing the reference point is given in Table 7.8.

The RMSE of the sensor location errors after the array calibration approach is now analysed as a function of SNR. The number of snapshots used is fixed at $L = 10000$ and SNR $= [0, 20, 30, 40, 50]$ dB. For each SNR chosen (taken to be the average of the $N(N-1)$ SNRs at the receiver inputs over all transmissions), the auto-calibration algorithm is run over 250 realisations. For each realisation, the average norm error associated with the unknown array element locations is plotted (red dots). The root mean squared estimation error over 250 realisations is also plotted (blue solid line). Results for the geometrical method of changing the reference point as well as the approximate methods (both with and without noise filtering) are presented in Figs. 7.9, 7.10 and 7.11 respectively for each large aperture geometry. It is clear that the geometrical and noise filtered approximate methods provide a near-perfect performance independent of SNR for both geometries. However, the approximate method of changing the reference point without noise filtering causes significant errors to arise

Table 7.7 Estimation error in location (in metres) of the large aperture circle array after the auto-calibration approach.

$L = 1000$; SNR= 30 dB			
Geometrical Approach (m or degrees) $\times 10^{-12}$			
error	x	y	**distance**	**angle**
Rx1	0	0	0	$0°$
Rx2	0	0	0	$0°$
Rx3	0.0114	0.0644	0	$0°$
Rx4	−0.0796	0.1506	0.05684	$0°$
Rx5	−0.1023	0.0853	0	$0°$
Rx6	−0.0744	−0.0085	0	$0.02544°$
Approximate Approach: Noise Filtering (m or degrees) $\times 10^{-12}$			
error	x	y	**distance**	**angle**
Rx1	0	0	0	$0°$
Rx2	0	0	0	$0°$
Rx3	−0.341	0.430	0.3411	$0.0509°$
Rx4	−0.568	1.933	0.5116	$0.2036°$
Rx5	−1.251	0.966	0.1705	$0.2036°$
Rx6	−0.744	−0.313	0.3126	$0.1781°$

at low SNR for the reasons described in Section 7.2.2. In Fig. 7.10 a solid line shows the point at which the auto-calibration approach fails to work under this method of changing the reference point.

A cumulative distribution (frequency) of the sensor location error over 250 Monte-Carlo realisations for a given SNR is now given in Figs 7.12, 7.13 and 7.14 for the different methods of changing the array reference point. It is clear that the geometric and noise filtered approximate methods offer a near-perfect performance.

7.4.3 *A representative example of the effects of uncertainties on a large aperture array*

Consider that the two large aperture arrays whose geometries are given in Tables 7.5 and 7.6 operate in the presence of a single source transmitting at $F_c = 100$ MHz from polar coordinates ($\rho = 4$ km, $\theta = 30°$). Hence, the array elements and the source are assumed to lie in $\mathcal{R}^2$ space, i.e.

Table 7.8 Estimation error in location (in metres) of the random array after the auto-calibration approach.

$L = 1000$; SNR= 30 dB				
Geometrical Approach (m or degrees) $\times 10^{-12}$				
error	x	y	distance	angle
Rx1	0	0	0	$0°$
Rx2	0	0	0	$0°$
Rx3	-2.30	-0.91	0.9095	$0.0763°$
Rx4	-0.40	0.23	0.4547	$0.0127°$
Rx5	-1.59	-1.82	0.9095	$0.0509°$
Rx6	-2.73	-3.18	0.4547	$0.0763°$
Rx7	-0.23	-0.80	0.2274	$0.0509°$
Rx8	-1.82	-3.52	0.9095	$0.0763°$
Rx9	-0.23	-1.53	0.3411	$0.6616°$
Rx10	0.45	18.25	0.2274	$0.5343°$
Approximate Approach: Noise Filtering (m or degrees) $\times 10^{-12}$				
error	x	y	distance	angle
Rx1	0	0	0	$0°$
Rx2	0	0	0	$0°$
Rx3	-4.0	4.1	3.865	$0.127°$
Rx4	-0.5	0.5	0.682	$0.025°$
Rx5	-3.1	3.9	4.547	$0°$
Rx6	-5.7	-2.0	2.274	$0.102°$
Rx7	2.3	-4.3	4.093	$0.102°$
Rx8	-5.0	-1.0	4.547	$0.051°$
Rx9	1.7	56.7	0.568	$3.282°$
Rx10	-4.3	-114.0	0.455	$3.384°$

$\underline{r}_z = \underline{0}_N$, $\underline{\widehat{r}}_z = \underline{0}_N$, $\underline{\widetilde{r}}_z = \underline{0}_N$ and $\phi = 0$. Now consider that there are geometrical uncertainties associated with these locations, given in Table 7.9. The resulting MUSIC spectra for the two arrays are presented in Fig. 7.15 in the presence o these uncertainties. Here, the spherical MUSIC cost function is denoted by

$$\xi(\rho, \theta) \triangleq \frac{\widehat{\underline{S}}^{H}(\rho, \theta)\widehat{\underline{S}}(\rho, \theta)}{\widehat{\underline{S}}^{H}(\rho, \theta)\mathbb{E}_n^H \mathbb{E}_n \widehat{\underline{S}}(\rho, \theta)} \tag{7.66}$$

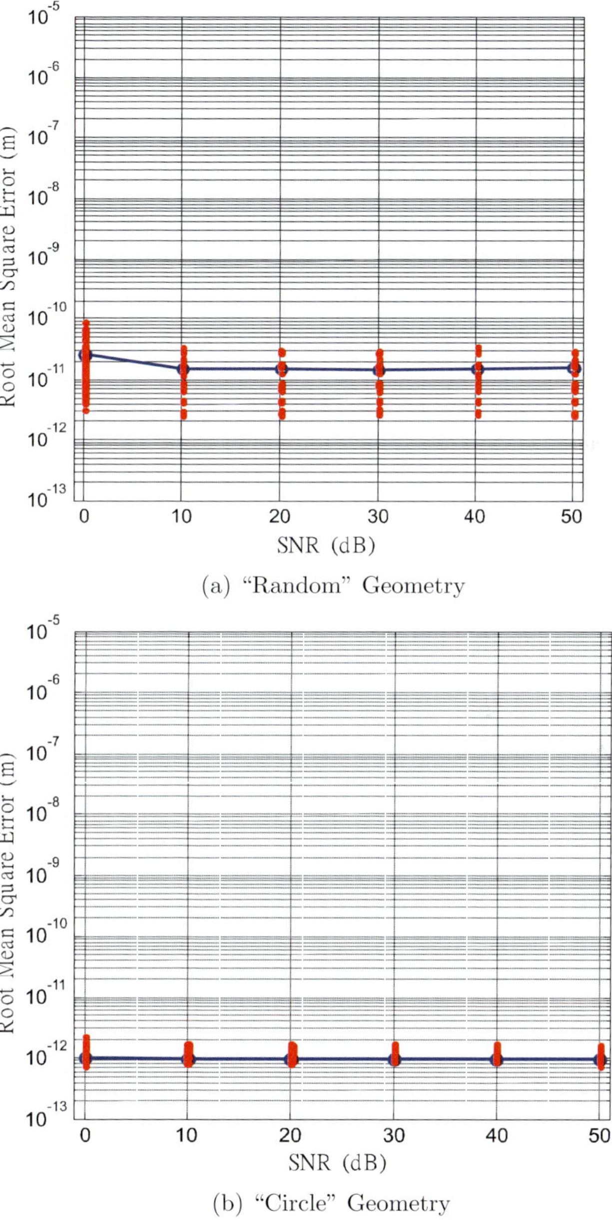

(a) "Random" Geometry

(b) "Circle" Geometry

Fig. 7.9   Geometric method: SNR vs RMSE performance under geometric method of changing the array reference point for the two large aperture geometries in Tables 7.5 and 7.6. Results are taken over 250 realisations with $L = 10000$ snapshots.

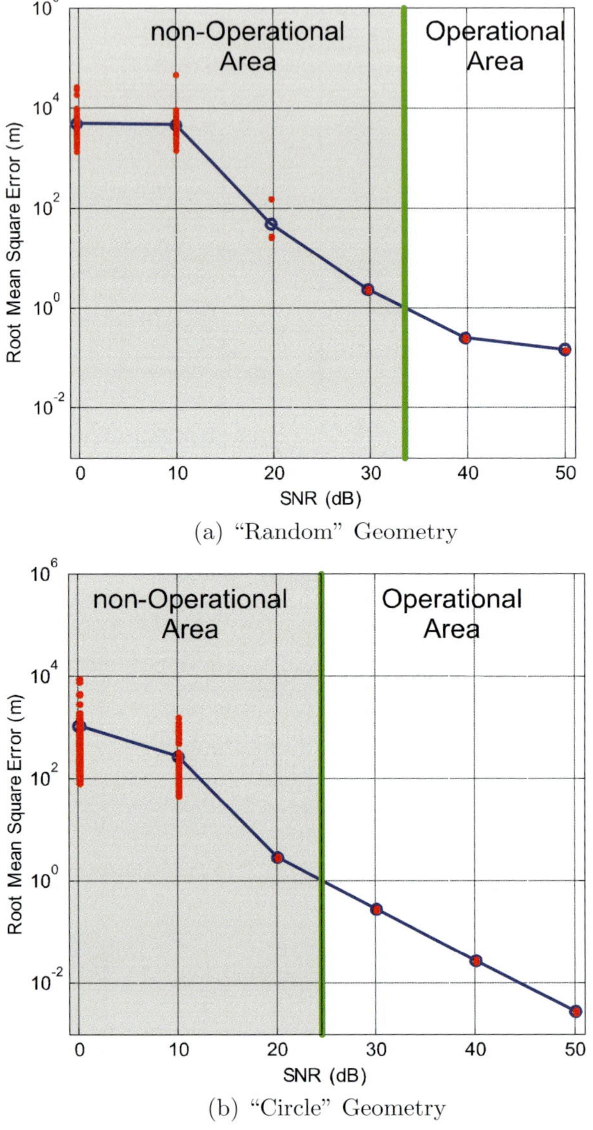

(a) "Random" Geometry

(b) "Circle" Geometry

Fig. 7.10 Approximate method without noise filtering: SNR vs RMSE performance under approximate method of changing the array reference point without noise filtering for the two large aperture geometries in Tables 7.5 and 7.6. Results are taken over 250 realisations with $L = 10000$ snapshots. Operational area: localisation RMSE <1 m.

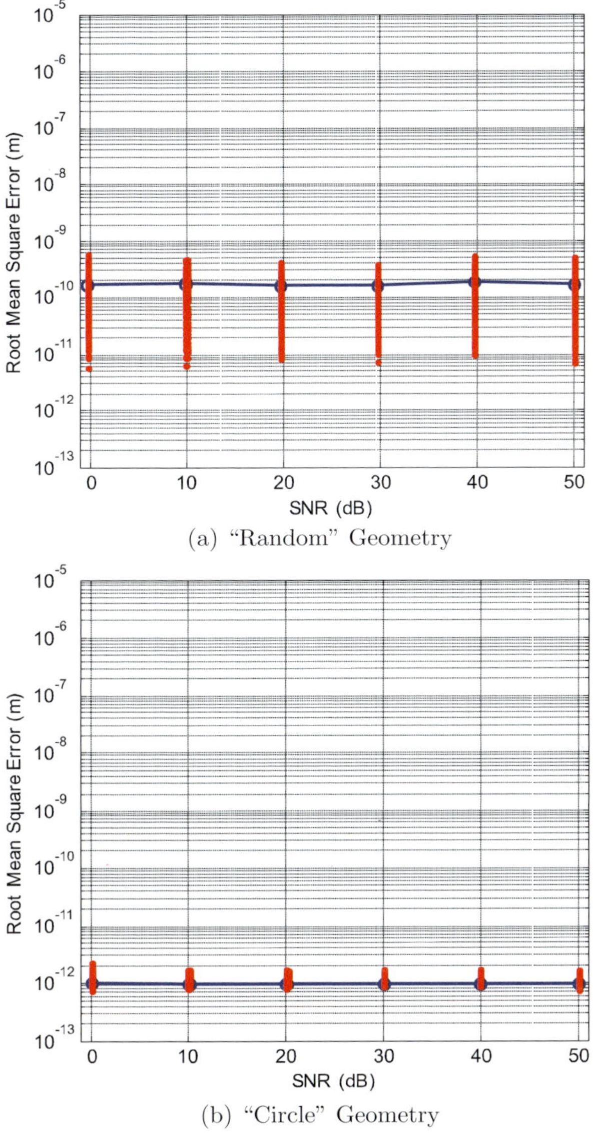

(a) "Random" Geometry

(b) "Circle" Geometry

Fig. 7.11 Approximate method with noise filtering: SNR vs RMSE performance under approximate method of changing the array reference point with noise filtering for the two large aperture geometries in Tables 7.5 and 7.6. Results are taken over 250 realisations with $L = 10000$ snapshots.

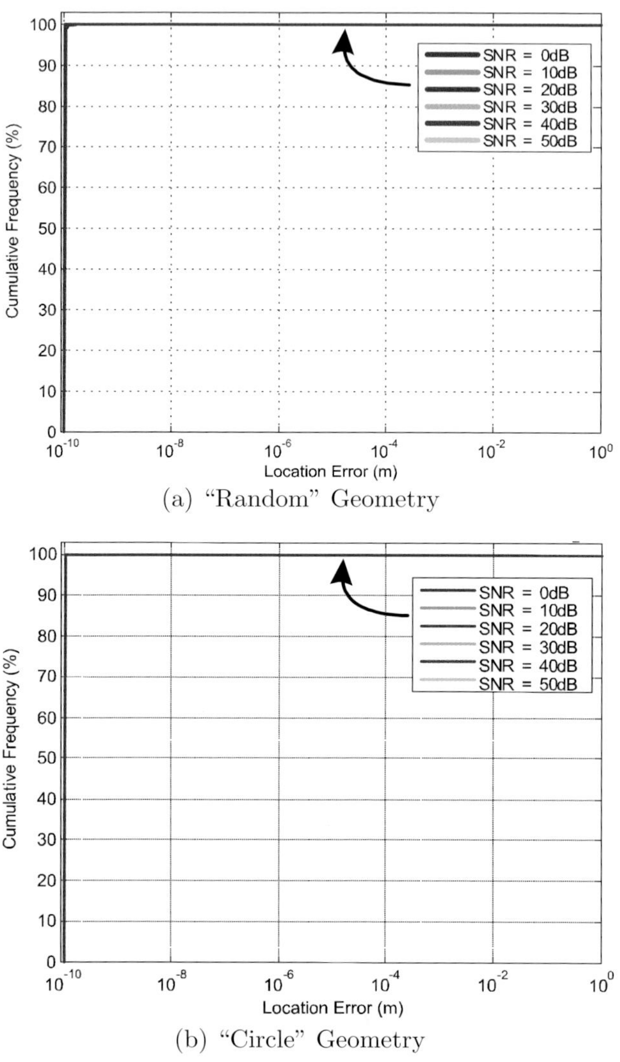

(a) "Random" Geometry

(b) "Circle" Geometry

Fig. 7.12 Cumulative distribution for geometric approach to changing the reference point for the two large aperture geometries in Tables 7.5 and 7.6. Results are taken over 250 realisations with $L = 10000$ snapshots.

where $\widehat{\underline{S}}(\rho, \theta)$ is the nominal spherical wave manifold vector, i.e. the manifold vector with uncertainties, using the array elements with geometrical uncertainties $[\widetilde{\underline{r}}_x, \widetilde{\underline{r}}_y, \widetilde{\underline{r}}_z]^T$ and $\mathbb{E}_n$ denotes the matrix whose columns are the noise eigenvectors of the array data covariance matrix $\mathbb{R}_{xx}$.

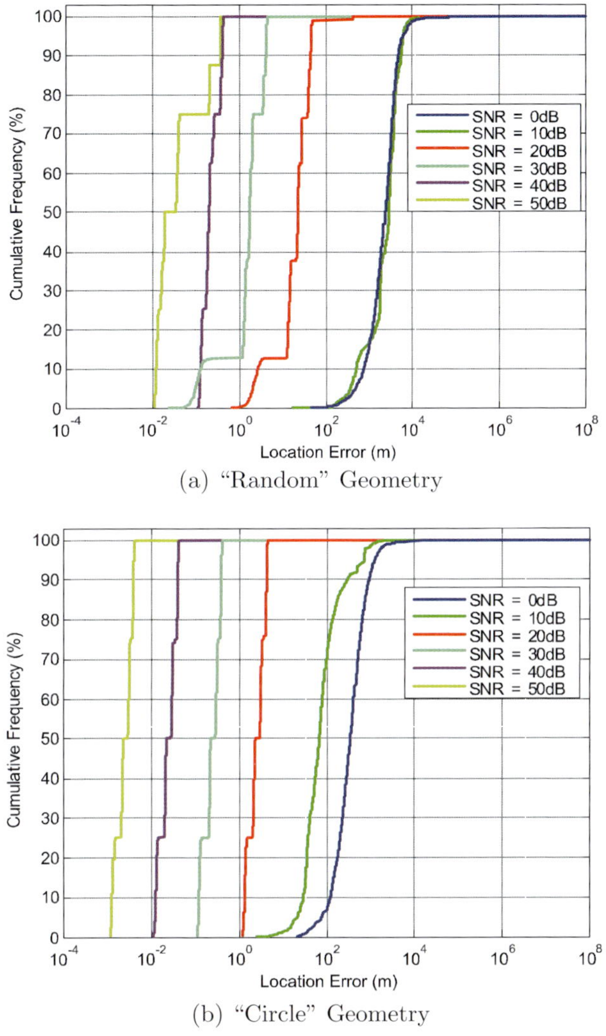

(a) "Random" Geometry

(b) "Circle" Geometry

Fig. 7.13   Cumulative distribution for approximate approach to changing the reference point without noise filtering for the two large aperture geometries in Tables 7.5 and 7.6. Results are taken over 250 realisations with $L = 10000$ snapshots.

It is clear that the presence of geometrical uncertainties severely degrades the performance of the spherical MUSIC algorithm. The array auto-calibration procedure is then performed on each array and the MUSIC algorithm is again applied giving the spectra in Fig. 7.16. This shows a significant improvement in the MUSIC localisation which now has great accuracy.

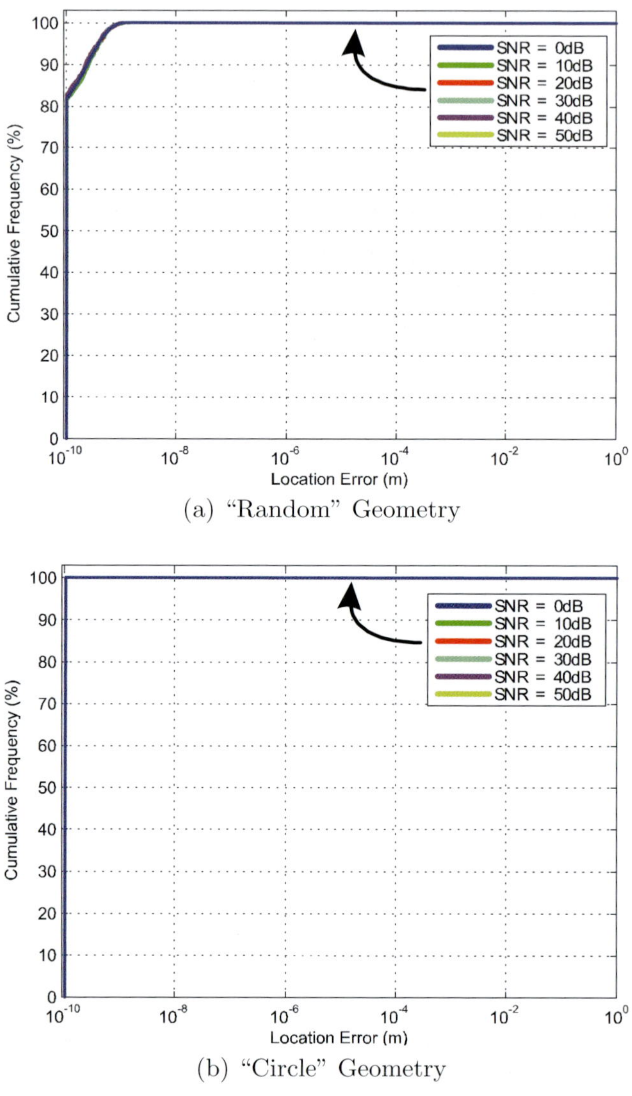

(a) "Random" Geometry

(b) "Circle" Geometry

Fig. 7.14   Cumulative distribution for approximate approach to changing the reference point with noise filtering for the two large aperture geometries in Tables 7.5 and 7.6. Results are taken over 250 realisations with $L = 10000$ snapshots.

Table 7.9   Array element geometrical uncertainties as
a result of the GPS location accuracy.

Random	LA Circle	$\widetilde{\underline{r}}_x$ (m)	$\widetilde{\underline{r}}_y$ (m)
$\widetilde{\underline{r}}_1$	$\widetilde{\underline{r}}_1$	0	0
$\widetilde{\underline{r}}_2$	$\widetilde{\underline{r}}_2$	0	0
$\widetilde{\underline{r}}_3$	$\widetilde{\underline{r}}_3$	-7.9870	9.2890
$\widetilde{\underline{r}}_4$	$\widetilde{\underline{r}}_4$	8.4430	-6.3780
$\widetilde{\underline{r}}_5$	$\widetilde{\underline{r}}_5$	0.2820	5.9890
$\widetilde{\underline{r}}_6$	$\widetilde{\underline{r}}_6$	4.3090	-4.8340
$\widetilde{\underline{r}}_7$		-2.2680	5.8640
$\widetilde{\underline{r}}_8$		9.8930	1.3700
$\widetilde{\underline{r}}_9$		8.1560	8.4570
$\widetilde{\underline{r}}_{10}$		-8.7640	-6.2890

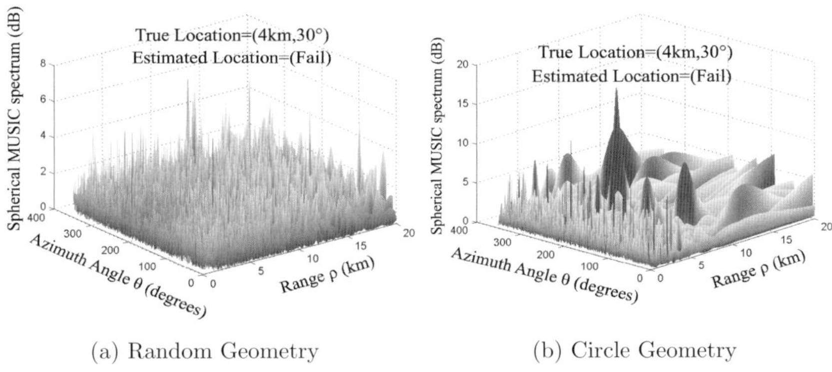

(a) Random Geometry (b) Circle Geometry

Fig. 7.15   Before auto-calibration: MUSIC-based localisation using a random and a circular array (see Tables 7.6 and 7.5) with the geometrical uncertainties given in Table 7.9. Parameters: SNR=20 dB and $L = 1000$ snapshots.

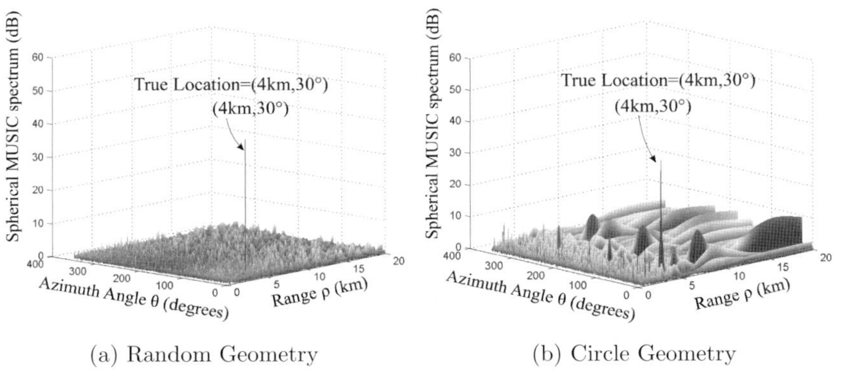

(a) Random Geometry (b) Circle Geometry

Fig. 7.16   After auto-calibration: MUSIC-based localisation for a random and a circular array (see Tables 7.6 and 7.5) after "auto-calibration" for SNR=20 dB and L=1000 snapshots.

7.5 Summary and conclusions

In this chapter, an array auto-calibration approach has been presented. By exploiting the sensor nodes as an array system, the array element locations were estimated by exploiting the spherical wave array manifold vector and changing the array reference point. In contrast to other techniques in the literature, this was based on allowing the nodes to operate as transceivers, removing the need for any external sources to be introduced for the calibration process.

Simulation studies showed that the array auto-calibration approach is more accurate than the pilot calibration approach at low SNR and that the geometrical approach to changing the array reference point results in an extremely low sensor location error after calibration for the large aperture array geometries, which is largely independent of SNR. However, if an approach based on division of data is used to change the array reference point, it is clear that noise filtering is imperative unless the SNR is high.

References

[1] G. Mao and B. Fidan, *Localization Algorithms and Strategies for Wireless Sensor Networks*. Information Science Reference, 2009.

[2] A. Manikas, Y. I. Kamil, and M. Willerton, "Source localization using sparse large aperture arrays," *IEEE Transactions on Signal Processing*, vol. 60, no. 12, pp. 6617–6629, Dec. 2012.

[3] A. Manikas, *Differential Geometry in Array Processing*. London, UK: Imperial College Press, 2004.

[4] A. Swindlehurst and T. Kailath, "A performance analysis of subspace-based methods in the presence of model errors. Part I. The MUSIC algorithm," *IEEE Transactions on Signal Processing*, vol. 40, no. 7, pp. 1758–1774, Jul. 1992.

[5] ——, "A performance analysis of subspace-based methods in the presence of model error. Part II—Multidimensional algorithms," *IEEE Transactions on Signal Processing*, vol. 41, no. 9, pp. 2882–2890, Sep. 1993.

[6] N. Fistas and A. Manikas, "A new general global array calibration method," in *IEEE International Conference on Acoustics, Speech and Signal Processing*, vol. 4, Apr. 1994, pp. IV/73–IV/76.

[7] A. Manikas and N. Fistas, "Modelling and estimation of mutual coupling between array elements," in *IEEE Conference on Acoustics, Speech and Signal Processing (ICASSP)*, Apr. 1994, pp. 553–556.

[8] K. Stavropoulos and A. Manikas, "Array calibration in the presence of unknown sensor characteristics and mutual coupling," in *Proc. EUSIPCO*, vol. 3, Sep. 2000, pp. 1417–1420.

[9] G. Efstathopoulos and A. Manikas, "A blind array calibration algorithm using a moving source," in *5th IEEE Sensor Array and Multichannel Signal Processing Workshop*, Jul. 2008, pp. 455–458.

[10] A. Leshem and M. Wax, "Array calibration in the presence of multipath," *IEEE Transactions on Signal Processing*, vol. 48, no. 1, pp. 53–59, Jan. 2000.

[11] A. Mengot and A. Manikas, "Global calibration of CDMA-based systems," in *European Signal Processing Conference*, Florence, Italy., Sep. 2005.

[12] S. E. Dosso, M. R. Fallat, B. J. Sotirin, and J. L. Newton, "Array element localization for horizontal arrays via Occams inversion," *The Journal of the Acoustical Society of America*, vol. 104, no. 2, p. 846, Aug. 1998.

[13] Y. Rockah and P. Schultheiss, "Array shape calibration using sources in unknown locations—Part I: Far-field sources," *IEEE Transactions on Acoustics, Speech and Signal Processing*, vol. 35, no. 3, pp. 286–299, Mar. 1987.

[14] ——, "Array shape calibration using sources in unknown locations—Part II: Near-field sources and estimator implementation," *IEEE Transactions on Acoustics, Speech, and Signal Processing*, vol. 35, no. 6, pp. 724–735, Jun. 1987.

[15] A. Paulraj and T. Kailath, "Direction of arrival estimation by eigenstructure methods with unknown sensor gain and phase," in *IEEE International Conference on Acoustics, Speech, and Signal Processing*, Apr. 1985, pp. 640–643.

[16] A. Weiss and B. Friedlander, "Array shape calibration using sources in unknown locations—a maximum likelihood approach," *IEEE Transactions on Acoustics, Speech, and Signal Processing*, vol. 37, no. 12, pp. 1958–1966, Dec. 1989.

[17] A. J. Weiss and B. Friedlander, "Array shape calibration using eigenstructure methods," in *IEEE Twenty-Third Asimolar Conference in Signals, Systems and Computers*, Oct./Nov. 1989, pp. 925–929.

[18] B. Flanagan and K. Bell, "Array self calibration with large sensor position errors," in *IEEE Thirty-Third Asilomar Conference on Signals, Systems, and Computers*, vol. 1, Oct. 1999, pp. 258–262.

[19] M. Viberg and A. Swindlehurst, "A Bayesian approach to auto-calibration for parametric array signal processing," *IEEE Transactions on Signal Processing*, vol. 42, no. 12, pp. 3495–3507, Dec. 1994.

[20] C.-Y. Tseng, D. Feldman, and L. Griffiths, "Estimation of signal steering vectors in uncalibrated arrays," in *IEEE Asilomar Conference on Signals, Systems and Computers*, vol. 2, Nov. 1993, pp. 1062–1066.

[21] B. Flanagan and K. Bell, "Improved array self calibration with large sensor position errors for closely spaced sources," in *IEEE Sensor Array and Multichannel (SAM) Signal Processing Workshop*, Mar. 2000, pp. 484–488.

[22] P. Chung and S. Wang, "Array self-calibration using SAGE algorithm," in *5th IEEE Sensor Array and Multichannel Signal Processing Workshop*, Jul. 2008, pp. 165–169.

[23] J. Fessler and A. Hero, "Space-alternating generalized expectation-maximization algorithm," *IEEE Transactions on Signal Processing*, vol. 42, no. 10, pp. 2664–2677, Oct. 1994.

[24] D. Fuhrmann, "Estimation of sensor gain and phase," *IEEE Transactions on Signal Processing*, vol. 42, no. 1, pp. 77–87, Jan. 1994.

[25] S. Wijnholds and A.-J. van der Veen, "Multisource self-calibration for sensor arrays," *IEEE Transactions on Signal Processing*, vol. 57, no. 9, pp. 3512–3522, Sep. 2009.

[26] J. Zhuo, C. Sun, and J. Feng, "Towed array shape self-calibration based on blind signal separation," in *Oceans 2005 - Europe*, vol. 1, Jun. 2005, pp. 599–604.

[27] S. Wang, P. Chung, and B. Mulgrew, "Array shape self-calibration using particle swarm optimization and decaying diagonal loading," in *Sensor Signal Processing for Defence (SSPD 2010)*, London, Oct. 2010, pp. 1–5.

[28] M. Willerton and A. Manikas, "Virtual linear array modelling of a planar array," in *Mathematics in Defence*, vol. 2, no. 5, 2011, pp. 1–5.

Chapter 8

Robust Beamforming to Pointing Errors

Jie Zhuang[1] and Athanassios Manikas

Communications and Array Processing,
Department of Electrical and Electronic Engineering,
Imperial College London

A ubiquitous task with array signal processing and wireless communications is interference cancellation. This can be achieved by using a beamformer which places nulls in the interference directions while maintaining relatively high gain in the direction of the desired signal. However, when pointing errors occur, the performance of the traditional Wiener–Hopf beamformer (or Capon beamformer) may deteriorate significantly. In this chapter, a robust array processor is presented which not only reduces the effect of pointing errors but also offers asymptotically complete interference cancellation. Prior to the beamforming, a linear subspace can easily be found using the knowledge of the nominal manifold vector. Then it is shown that the manifold vector corresponding to the desired signal can be obtained by using this linear subspace and the signal subspace together. To eliminate much of the pointing error effects, the vector space projections (VSP) method is utilised to find a new manifold vector. Furthermore, we can obtain the power of the desired signal in a single-step computation. By using the estimated manifold vector and power of the desired signal, we can excise the effects due to the desired signal to form the desired-signal-absent covariance matrix. Then a weight vector orthogonal to the interference subspace can be constructed. Simulation results demonstrate the

---

[1] Jie Zhuang was with Imperial College London and is now with the School of Communication and Information Engineering, University of Electronic Science and Technology of China (UESTC).

superior performance of this interference cancellation beamformer relative to other existing methods.

8.1   Introduction

Let us consider an array with N sensors operating in the presence of one desired and M interference signals. It is assumed that the $(M + 1)$ signals are narrowband, uncorrelated to each other and located in the far field of the array. The received signal vector can be expressed as

$$\underline{x}(t) = \underline{S}(\theta_d)m_d(t) + \sum_{i=1}^{M}\underline{S}(\theta_i)m_i(t) + \underline{n}(t) \tag{8.1}$$

where the subscript d stands for the desired signal and θ_i and $m_i(t)$ represent, respectively, the direction of arrival (DOA) and the complex envelope of the i-th interference signal. In Eq. 8.1 $\underline{n}(t) \in \mathcal{C}^N$ is the additive white complex Gaussian noise of zero mean and covariance $\sigma_n^2\mathbb{I}_N$. The vector $\underline{S}(\theta)$ denotes the array response at the DOA (azimuth) θ, which is also known as the array manifold vector or array steering vector. The true manifold information of the desired signal, denoted by $\underline{S}_d \triangleq \underline{S}(\theta_d)$ for notational simplicity, is not perfectly known in many practical applications [1], e.g. in local scattering and non-stationarity environment, for an uncalibrated array or in the presence of pointing errors (or look-direction errors). It has been pointed out in [2] that even a small mismatch between the actual manifold vector and its nominal version may lead to a substantial performance degradation in conventional adaptive beamformers. In this chapter, we focus on the case of pointing errors only. That is, the DOA of the desired signal θ_d is not perfectly known and therefore

$$\underline{S}_d \neq \underline{S}(\theta_0) \text{ when } \theta_d \neq \theta_0 \tag{8.2}$$

where θ_0 is the presumed DOA used by the array processor.

However, it can be found that $\underline{S}_d$ lies in a known p-dimensional (where $p > 1$) linear subspace $\mathcal{L}[\mathbb{H}]$ and thus can be expressed as a linear combination of the basis vectors of $\mathcal{L}[\mathbb{H}]$ [3, 4]. This means that the manifold vector $\underline{S}_d$ can be given by

$$\underline{S}_d = \mathbb{H}\underline{b} \tag{8.3}$$

where the vector $\underline{b}$ is the *linear combination vector*. Note that $\underline{b}$ and $\underline{S}_d$ are unknown in a practical situation, while $\mathbb{H}$ can be easily obtained. Next, two examples will be presented to demonstrate how to obtain the matrix $\mathbb{H}$.

In the first example, the true (unknown) manifold vector is expanded using a Taylor series around the nominal (assumed) manifold vector $\underline{S}(\theta_0)$ as [3]

$$\underline{S}_d = \underline{S}(\theta_0 + \Delta\theta_d) = \underline{S}(\theta_0) + \sum_{k=1}^{\infty} \frac{(\Delta\theta_d)^k}{k!} \frac{\partial^k \underline{S}(\theta)}{\partial \theta^k}\bigg|_{\theta=\theta_0} \tag{8.4}$$

with

$$\Delta\theta_d = \theta_d - \theta_0 \tag{8.5}$$

where the nominal DOA θ_0 is known. In practice, the first two derivatives are enough to satisfy the desired precision. Hence by defining the matrix $\mathbb{H}_1$ as

$$\mathbb{H}_1 \triangleq [\underline{S}_0, \dot{\underline{S}}(\theta_0), \ddot{\underline{S}}(\theta_0)] \tag{8.6}$$

where $\dot{\underline{S}}(\theta_0)$ and $\ddot{\underline{S}}(\theta_0)$ denote the first and second derivatives with respect to θ_0, the manifold vector given by Eq. 8.4 can be approximated as follows

$$\underline{S}_d \approx \mathbb{H}_1 \underbrace{\left[1, \Delta\theta_d, \frac{(\Delta\theta_d)^2}{2} \right]^T}_{\triangleq \underline{b}_1} \tag{8.7}$$

where the vector $\underline{b}_1$ is unknown. Note that $\mathbb{H}_1$ can be obtained prior to beamforming assuming that the nominal direction θ_0 and array geometry information are known.

Alternatively, applying the Taylor series expansion around the true manifold vector $\underline{S}_d$ and retaining the first two order terms, one obtains the following approximations:

$$\underline{S}(\theta_0) = \underline{S}(\theta_d - \Delta\theta_d) \tag{8.8a}$$

$$\approx \underline{S}_d - \Delta\theta_d \dot{\underline{S}}(\theta_d) + \frac{(\Delta\theta_d)^2}{2} \ddot{\underline{S}}(\theta_d)$$

$$\underline{S}(\theta_0 \pm \Delta\theta) = \underline{S}(\theta_d \pm \Delta\theta - \Delta\theta_d) \tag{8.8b}$$

$$\approx \underline{S}_d + (\pm\Delta\theta - \Delta\theta_d)\dot{\underline{S}}(\theta_d) + \frac{(\pm\Delta\theta - \Delta\theta_d)^2}{2} \ddot{\underline{S}}(\theta_d)$$

where $\Delta\theta$ denotes the expected DOA range and can be pre-selected even if the true DOA is unavailable/unknown. Generally the value of $\Delta\theta$ is set such that $|\Delta\theta \pm \Delta\theta_d|$ is insignificant so that the above approximations are valid. Using Eq. (8.8) it is easy to prove that $\underline{S}_d$ may be rewritten as

$$\underline{S}_d \approx \frac{(\Delta\theta)^2 - (\Delta\theta_d)^2}{(\Delta\theta)^2} \underline{S}(\theta_0) + \frac{(\Delta\theta_d)^2 + \Delta\theta\Delta\theta_d}{2(\Delta\theta)^2} \underline{S}(\theta_0 + \Delta\theta)$$

$$+ \frac{(\Delta\theta_d)^2 - \Delta\theta\Delta\theta_d}{2(\Delta\theta)^2} \underline{S}(\theta_0 - \Delta\theta)$$

$$= \underbrace{[\underline{S}(\theta_0), \underline{S}(\theta_0 + \Delta\theta), \underline{S}(\theta_0 - \Delta\theta)]}_{\triangleq \mathbb{H}_2} \underbrace{\begin{bmatrix} \frac{(\Delta\theta)^2 - (\Delta\theta_d)^2}{(\Delta\theta)^2} \\ \frac{(\Delta\theta_d)^2 + \Delta\theta\Delta\theta_d}{2(\Delta\theta)^2} \\ \frac{(\Delta\theta_d)^2 - \Delta\theta\Delta\theta_d}{2(\Delta\theta)^2} \end{bmatrix}}_{\triangleq \underline{b}_2}. \tag{8.9}$$

Hence $\underline{S}_d$ belongs to the known linear subspace $\mathcal{L}[\mathbb{H}_2]$ but the associated linear combination vector $\underline{b}_2$ is unknown.

In this chapter, the unknown vector $\underline{b}$ is assumed to stay unchanged during the observation interval of L successive snapshots, although it may vary in the next L snapshots. The effect of the desired signal in the data covariance matrix $\mathbb{R}_{xx}$ is given by

$$\begin{aligned} \mathbb{R}_{dd} &= \mathcal{E}\{(m_d(t)\underline{S}_d)\,(m_d(t)\underline{S}_d)^H\} \\ &= \mathcal{E}\{m_d(t)m_d^*(t)\}\underline{S}_d\underline{S}_d^H \\ &= \sigma_d^2 \mathbb{H}\underline{b}\,\underline{b}^H\mathbb{H}^H \end{aligned} \tag{8.10}$$

where σ_d^2 is the power of the desired signal. It is clear that $\mathbb{R}_{dd}$ is a rank-1 matrix. Under the assumption that the M interference signals are uncorrelated with the desired signal, the second-order model for $\underline{x}(t)$ can be written as

$$\mathbb{R}_{xx} = \mathcal{E}\{\underline{x}(t)\underline{x}^H(t)\} = \sigma_d^2 \mathbb{H}\underline{b}\,\underline{b}^H\mathbb{H}^H + \mathbb{R}_i + \sigma_n^2\mathbb{I}_N \tag{8.11}$$

where the rank-M matrix $\mathbb{R}_i$ represents the interference effects and may have the following form

$$\mathbb{R}_i = \sum_{i=1}^{M}\sigma_i^2\underline{S}(\theta_i)\underline{S}^H(\theta_i) \tag{8.12}$$

where σ_i^2 denotes the power of the i-th interference signal. In practical applications, however, the theoretical covariance matrix $\mathbb{R}_{xx}$ is often unavailable and instead we use its sample estimate

$$\hat{\mathbb{R}}_{xx} = \frac{1}{L}\sum_{l=1}^{L}\underline{x}(t_l)\underline{x}^H(t_l) \tag{8.13}$$

where $\underline{x}(t_l)$ denotes the snapshot vector received at time t_l and L is the number of the snapshots.

8.2 Estimation of the linear combination vector using signal subspace

Let $\mathbb{S}_I$ represent the matrix consisting of the interference manifold vectors, i.e. $\mathbb{S}_I = [\underline{S}(\theta_1), \ldots, \underline{S}(\theta_M)]$. In [3], it is assumed that $\mathbb{S}_I$ is exactly known and the authors derive a maximum likelihood estimator of the vector $\underline{b}$, given by

$$\underline{b} = \beta \underline{\mathcal{P}}\{(\mathbb{V}^H \mathbb{V})^{-1} \mathbb{V}^H \mathbb{R}_{xx} \mathbb{V}\} \tag{8.14}$$

where the coefficient β is a normalization constant. The operator $\mathcal{P}\{\cdot\}$ stands for the principal eigenvector of the matrix argument between braces. Furthermore, the matrix $\mathbb{V}$ is defined as

$$\mathbb{V} = \mathbb{P}_{\mathbb{S}_I}^{\perp} \mathbb{H} \tag{8.15}$$

where the notation $\mathbb{P}_{\mathbb{A}}^{\perp} = \mathbb{I}_N - \mathbb{A}(\mathbb{A}^H \mathbb{A})^{-1} \mathbb{A}^H$ represents the projection operator onto the complement subspace of $\mathcal{L}[\mathbb{A}]$. The main problem in [3] is that perfect knowledge of the interference subspace is required for its implementation. Unfortunately, this assumption is not always valid in many practical applications because the interference subspace cannot be easily found.

A multi-rank minimum variance distortionless response (MVDR) beamformer is presented in [4] where the vector $\underline{b}$ is estimated by using a generalized sidelobe canceller (GSC) [5]. The second-order statistics of the output at the GSC are represented by the matrix $\mathbb{R}_{ee}$ which has the form

$$\mathbb{R}_{ee} = \left(\mathbb{H}^H \mathbb{R}_{xx}^{-1} \mathbb{H}\right)^{-1}. \tag{8.16}$$

Let us consider the simple case where only one interference is present. Then the matrix $\mathbb{R}_{ee}$ may be expressed as (using Eq. (25) in [4])

$$\mathbb{R}_{ee} = \sigma_d^2 \underline{b}\,\underline{b}^H + \eta \mathbb{H}^H \underline{S}(\theta_1) \underline{S}^H(\theta_1) \mathbb{H} + \sigma_n^2 \mathbb{I}_N \tag{8.17}$$

where $\underline{S}(\theta_1)$ represents the interference manifold vector, and η denotes the interference suppression level and has the form [4]

$$\eta = \frac{\sigma_1^2 \sigma_n^2}{\sigma_n^2 + \sigma_1^2 \underline{S}^H(\theta_1) \mathbb{P}_{\mathbb{H}}^{\perp} \underline{S}(\theta_1)}. \tag{8.18}$$

It can be seen that $\underline{b}$ approximates the principal eigenvector of $\mathbb{R}_{ee}$ (up to a scaling factor) if the desired signal subspace $\mathcal{L}[\mathbb{H}]$ and the interference subspace $\mathcal{L}[\underline{S}(\theta_1)]$ are well separated. However, when the two subspaces are relatively close, the interference effects in $\mathbb{R}_{ee}$ cannot be ignored any more. This implies that the principal eigenvector of $\mathbb{R}_{ee}$ contains significant components of the interference. Furthermore, the desired signal does not only contribute to the principal eigenvector but also to the rest of the eignvectors. Therefore this GSC-type estimator cannot offer a satisfying estimation when the interference subspace is located relatively close to the linear subspace $\mathcal{L}[\mathbb{H}]$.

Next, a method to estimate the combination vector using the knowledge of the signal subspace will be presented. The eigendecomposition of $\mathbb{R}_{xx}$ produces

$$\mathbb{R}_{xx} = \sum_{i=1}^{N} \lambda_i \underline{E}_i \underline{E}_i^H = \sum_{i=1}^{M+1} \lambda_i \underline{E}_i \underline{E}_i^H + \sum_{i=M+2}^{N} \lambda_i \underline{E}_i \underline{E}_i^H \tag{8.19}$$

where the eigenvalues $\{\lambda_i, i = 1, \ldots, N\}$ are arranged in non-increasingly order (i.e., $\lambda_1 \geq \ldots \geq \lambda_N$) and $\underline{E}_i$ denotes the eigenvector corresponding to λ_i. The columns of $\mathbb{E}_s$ and $\mathbb{E}_n$ are, respectively, the eigenvectors associated with the largest $M+1$ eigenvalues (or the first $M+1$ dominant eigenvectors) and the remaining eigenvectors, i.e.,

$$\mathbb{E}_s \triangleq [\underline{E}_1, \underline{E}_2, \ldots, \underline{E}_{M+1}]$$

$$\mathbb{E}_n \triangleq [\underline{E}_{M+2}, \underline{E}_{M+3}, \ldots, \underline{E}_N] \tag{8.20}$$

The matrices $\mathbb{E}_s$ and $\mathbb{E}_n$ are also referred to as signal and noise subspace eigenmatrices respectively. Then a matrix $\mathbb{A}_\ell$ is used to collect the signal eigenvectors except $\underline{E}_\ell$, i.e.,

$$\mathbb{A}_\ell = [\underline{E}_1, \ldots, \underline{E}_{\ell-1}, \underline{E}_{\ell+1}, \ldots, \underline{E}_{M+1}] \tag{8.21}$$

where $\underline{E}_\ell$ can be any $M+1$ signal eigenvector. Now a matrix $\mathbb{U}$ is defined as

$$\mathbb{U} = \mathbb{P}_{\mathbb{A}_\ell}^{\perp} \mathbb{H} \tag{8.22}$$

and the eigenvector $\underline{E}_\ell$ can be expressed as:

$$\begin{aligned}
\underline{E}_\ell &= \underline{E}_\ell (\underline{E}_\ell^H \underline{E}_\ell)^{-1} \frac{\underline{E}_\ell^H \underline{S}_d}{\underline{E}_\ell^H \underline{S}_d} \\
&= \frac{\underline{E}_\ell (\underline{E}_\ell^H \underline{E}_\ell)^{-1} \underline{E}_\ell^H \underline{S}_d}{\underline{E}_\ell^H \underline{S}_d} \\
&= \frac{\mathbb{P}_{\underline{E}_\ell} \underline{S}_d}{\underline{E}_\ell^H \underline{S}_d}
\end{aligned} \tag{8.23}$$

where the fact $(\underline{E}_\ell^H \underline{E}_\ell)^{-1} = 1$ has been used in Eq. (8.23). Note that the matrix $\mathbb{P}_{\mathbb{A}} = \mathbb{A}(\mathbb{A}^H \mathbb{A})^{-1} \mathbb{A}^H$ represents the projection operator onto the subspace $\mathcal{L}[\mathbb{A}]$. The vector $\underline{E}_\ell$ lies in the signal subspace, implying that $\underline{E}_\ell$ is a linear combination of the manifolds of the desired signal and all interferences. Hence $\underline{E}_\ell^H \underline{S}_d \neq 0$ holds. Let $\mathbb{P}_n$ denote the projection onto the noise subspace, i.e. $\mathbb{P}_n = \mathbb{E}_n \mathbb{E}_n^H$. Then the following expressions are true:

$$\mathbb{P}_n \underline{S}_d = \underline{0}_N$$
$$\mathbb{P}_{\underline{E}_\ell} + \mathbb{P}_{\mathbb{A}_\ell} + \mathbb{P}_n = \mathbb{I}_N. \tag{8.24}$$

By pre-multiplying Eq. (8.23) with $\mathbb{U}^H$ and then using Eq. (8.24) we have

$$\mathbb{U}^H \underline{E}_\ell = \mathbb{U}^H \frac{\mathbb{P}_{\underline{E}_\ell} \underline{S}_d}{\underline{E}_\ell^H \underline{S}_d}$$

$$= \frac{1}{\underline{E}_\ell^H \underline{S}_d} \mathbb{U}^H \left(\mathbb{I} - \mathbb{P}_{\mathbb{A}_\ell} - \mathbb{P}_n \right) \underline{S}_d$$

$$= \frac{1}{\underline{E}_\ell^H \underline{S}_d} \left(\mathbb{U}^H \underline{S}_d - \mathbb{U}^H \mathbb{P}_{\mathbb{A}_\ell} \underline{S}_d \right)$$

$$= \frac{1}{\underline{E}_\ell^H \underline{S}_d} \left(\mathbb{U}^H \underline{S}_d - \mathbb{H}^H \mathbb{P}_{\mathbb{A}_\ell}^\perp \mathbb{P}_{\mathbb{A}_\ell} \underline{S}_d \right)$$

$$= \frac{1}{\underline{E}_\ell^H \underline{S}_d} \mathbb{U}^H \underline{S}_d \tag{8.25}$$

where the equations $\mathbb{P}_{\mathbb{A}_\ell}^\perp = (\mathbb{P}_{\mathbb{A}_\ell}^\perp)^H$ and $\mathbb{P}_{\mathbb{A}_\ell}^\perp \mathbb{P}_{\mathbb{A}_\ell} = \mathbb{O}_{N \times N}$ have been utilised. Finally, pre-multiplying (8.25) by $(\mathbb{U}^H \mathbb{U})^{-1}$ and then inserting Eqs. (8.3) and (8.22), one obtains

$$(\mathbb{U}^H \mathbb{U})^{-1} \mathbb{U}^H \underline{E}_\ell = \frac{1}{\underline{E}_\ell^H \underline{S}_d} (\mathbb{U}^H \mathbb{U})^{-1} \mathbb{U}^H \underline{S}_d$$

$$= \frac{1}{\underline{E}_\ell^H \underline{S}_d} (\mathbb{U}^H \mathbb{U})^{-1} \mathbb{H}^H \mathbb{P}_{\mathbb{A}_\ell}^\perp \mathbb{H} \underline{b}$$

$$= \frac{1}{\underline{E}_\ell^H \underline{S}_d} (\mathbb{U}^H \mathbb{U})^{-1} \mathbb{H}^H \mathbb{P}_{\mathbb{A}_\ell}^\perp \mathbb{P}_{\mathbb{A}_\ell}^\perp \mathbb{H} \underline{b}$$

$$= \frac{1}{\underline{E}_\ell^H \underline{S}_d} (\mathbb{U}^H \mathbb{U})^{-1} \mathbb{U}^H \mathbb{U} \underline{b}$$

$$= \frac{1}{\underline{E}_\ell^H \underline{S}_d} \underline{b} \tag{8.26}$$

where the idempotence property of projection operator $\mathbb{P}^{\perp}_{\mathbb{A}_\ell} = \mathbb{P}^{\perp}_{\mathbb{A}_\ell}\mathbb{P}^{\perp}_{\mathbb{A}_\ell}$ has been used. Now the vector $\underline{b}$ can be expressed as

$$\underline{b} = \beta(\mathbb{U}^H\mathbb{U})^{-1}\mathbb{U}^H\underline{E}_\ell = \beta\mathbb{U}^\dagger\underline{E}_\ell \tag{8.27}$$

where $\beta \triangleq \underline{E}_\ell^H\underline{S}_d$ and $(\cdot)^\dagger$ denotes the pseudo-inverse operation.

In the derivation of (8.27), it is required that the signal eigenvectors $[\underline{E}_1,\ldots,\underline{E}_{M+1}]$ are available, which can be easily estimated and thus it can be seen as quite a weak assumption compared to that in [3] where perfect knowledge of interference subspace $\mathcal{L}[\mathbb{S}_I]$ is assumed.

Let us consider the situation when we overestimate the signal-subspace dimension. Then the dimensionality of the noise subspace is reduced. The desired manifold vector $\underline{S}_d$ is still orthogonal to the projector $\mathbb{P}_n$ with reduced dimensionality, implying that the above derivation is valid in this case as well. Therefore the method used in (8.27) is insensitive to signal-subspace overestimation. However, when we underestimate the signal-subspace dimension, the estimated noise subspace would contain a signal component which renders $\mathbb{P}_n$ no longer orthogonal with $\underline{S}_d$. Therefore, signal-subspace underestimation may result in the failure of the estimator in (8.27).

It can be seen that the estimator in (8.27) is independent of the interference manifold vectors and therefore, theoretically, offers accurate estimation even if the interferences are very close to the desired signal. In comparison, the estimator presented in [4] is sensitive to the interferences close to the desired signal.

There are two possible factors that may restrict the accuracy of the estimator in (8.27). The first is the finite snapshot effect. If the number of snapshots acquired to estimate $\widehat{\mathbb{R}}$ in (8.13) is quite small, the orthogonality between $\widehat{\mathbb{P}}_n$ and $\underline{S}_d$ may be impaired, which leads to $\widehat{\mathbb{P}}_n\underline{S}_d \neq \underline{0}$. Another factor is the accuracy of the subspace $\mathcal{L}[\mathbb{H}]$. Since the matrix $\mathbb{H}$ is formed by discarding the higher orders, the desired manifold vector $\underline{S}_d$ may be not completely located in $\mathcal{L}[\mathbb{H}]$. In such case, the actual manifold may be written as

$$\underline{S}_d = \mathbb{H}\underline{b} + \underline{\Delta} \tag{8.28}$$

where $\underline{b} = \mathbb{H}^\dagger\underline{S}_d$ and $\underline{\Delta}$ lies in the complement subspace of $\mathcal{L}[\mathbb{H}]$. Taking the effect of finite snapshots into account and inserting (8.28) into (8.25) and then (8.26), we have

$$\widehat{\underline{b}} = \beta\left(\underline{b} + \widehat{\mathbb{U}}^\dagger\underline{\Delta} - \widehat{\mathbb{U}}^\dagger\widehat{\mathbb{P}}_n\underline{S}_d\right). \tag{8.29}$$

8.3 Estimation of the desired signal manifold via vector space projections (VSP)

It has been established that the signal manifold vectors (both the desired signal and the interference signals) belong to the so-called signal subspace $\mathcal{L}[\mathbb{E}_s]$ where the columns of $\mathbb{E}_s$ are the signal eigenvectors of the covariance matrix $\mathbb{R}_{xx}$. Also, it is found that the desired signal manifold vector $\underline{S}_d$ lies in the linear subspace $\mathcal{L}[\mathbb{H}]$. Now two different subspaces associated with the desired signal are available and thus it is intended to make the most of this information to remove the effects of pointing errors as much as possible.

The approach presented here to eliminate pointing errors employs the theorem of *sequential* VSP to find the intersection of the two constraint subspaces. The simplest form of VSP, suggested by von Neumann in [6], is an alternating projection in which the iterative projections are performed to find the intersection of two Hilbert subspaces. This work has been generalized to more than two closed subspaces and also for convex sets by Stark and Yang [7] so that the VSP method has found applications in a wide range of practical engineering problems. For instance, the applications of the projection methods in array signal processing have been considered in [8–10]. A fundamental theorem of VSP states the following:

Theorem 8.1. *Let* $\mathcal{C}_1, \mathcal{C}_2, \ldots, \mathcal{C}_m$ *represent* m *closed convex sets in a Hilbert space* $\mathcal{H}$*, and let* $\mathbb{P}_{\mathcal{C}_i}$ *denote the projection onto* $\mathcal{C}_i$*,* $i = 1, \ldots, m$*. If the intersection* $\mathcal{C}_0 \overset{\triangle}{=} \bigcap_{i=1}^{m} \mathcal{C}_i$ *is non-empty and the dimension of* $\mathcal{H}$ *is finite, then the sequence* $(\mathbb{P}_{\mathcal{C}_1} \mathbb{P}_{\mathcal{C}_2} \ldots \mathbb{P}_{\mathcal{C}_m})^k \underline{a}_0$ *converges strongly to a point in the solution set* $\mathcal{C}_0$*,* $\mathbb{P}_{\mathcal{C}_0} \underline{a}_0$*, for every non-zero point* $\underline{a}_0 \in \mathcal{H}$*.*

Proof. See [7]. $\qquad\qquad\square$

To place this theorem in the context of the pointing error elimination, the key is how to define the appropriate constraint sets $\{\mathcal{C}_i\}$ to describe the available information. Under the above assumptions, two constraints can now be imposed on $\underline{S}_d$

$$\mathcal{C}_1 = \{\underline{S} : \underline{S} \in \mathcal{L}[\mathbb{E}_s]\} \tag{8.30a}$$

$$\mathcal{C}_2 = \{\underline{S} : \underline{S} \in \mathcal{L}[\mathbb{H}]\}. \tag{8.30b}$$

The unknown manifold $\underline{S}_d$ can be viewed as the intersection of $\mathcal{C}_0 \overset{\triangle}{=} \mathcal{C}_1 \bigcap \mathcal{C}_2$. Following Theorem 8.1, with the start point $\underline{S}(\theta_0)$ (the nominal manifold vector) the sequence $\{\underline{a}_k\}$ generated by

$$\underline{a}_{k+1} = \mathbb{P}_{\mathcal{C}_2} \mathbb{P}_{\mathcal{C}_1} \underline{a}_k \tag{8.31}$$

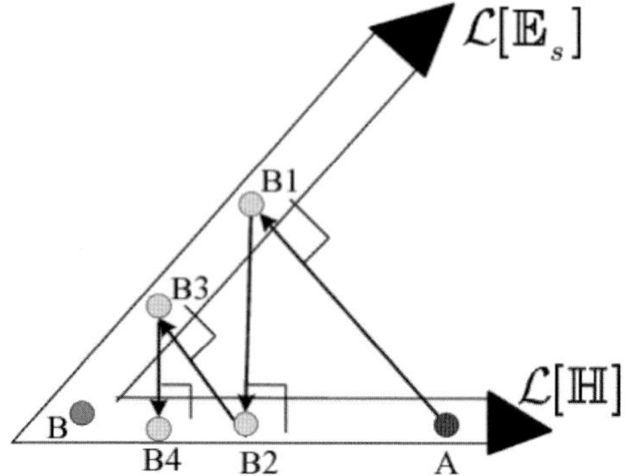

Fig. 8.1   Illustration of the convergence of the VSP method. The point A represents the nominal manifold. The point B represents the true manifold.

converges to $\underline{S}_d$ as $k \to \infty$. The projection operations are defined as

$$\mathbb{P}_{\mathcal{C}_1} = \mathbb{E}_s \mathbb{E}_s^H$$
$$\mathbb{P}_{\mathcal{C}_2} = \mathbb{H}(\mathbb{H}^H \mathbb{H})^{-1} \mathbb{H}^H. \tag{8.32}$$

This method is schematically depicted in Fig. 8.1, where it is shown that the nominal manifold vector (i.e. point A in Fig. 8.1) is expected to converge to the actual manifold (point B) after some iterations.

Here a solution using only one step is derived to avoid the iterative alternating projection. Note that if $k \to \infty$. then Eq. (8.31) becomes

$$\underline{a}_\infty = \mathbb{P}_{\mathcal{C}_2} \mathbb{P}_{\mathcal{C}_1} \underline{a}_\infty. \tag{8.33}$$

The above equation implies that the final converged estimate $\underline{a}_\infty$ is given by the eigenvector (up to a scaling factor) of $\mathbb{P}_{\mathcal{C}_2} \mathbb{P}_{\mathcal{C}_1}$ with the corresponding eigenvalue equal to one. In addition, the maximum eigenvalue of $\mathbb{P}_{\mathcal{C}_2} \mathbb{P}_{\mathcal{C}_1}$ should be one. By using *Corollary 11* of [11] and the fact that the maximum eigenvalue of $\mathbb{P}_{\mathcal{C}_1}$ or $\mathbb{P}_{\mathcal{C}_2}$ is one, we have

$$eig_{\max}(\mathbb{P}_{\mathcal{C}_2} \mathbb{P}_{\mathcal{C}_1}) \le eig_{\max}(\mathbb{P}_{\mathcal{C}_2}) \max_{\substack{\underline{u} \\ \underline{u}^H \underline{u}=1}} (\underline{u}^H \mathbb{P}_{\mathcal{C}_1} \underline{u})$$

$$= \max_{\substack{\underline{u} \\ \underline{u}^H \underline{u}=1}} \left(\frac{\underline{u}^H \mathbb{P}_{\mathcal{C}_1} \underline{u}}{\underline{u}^H \underline{u}} \right)$$

$$= eig_{\max}(\mathbb{P}_{\mathcal{C}_1})$$

$$= 1 \tag{8.34}$$

where $eig_{\max}(\mathbb{A})$ denotes the maximum eigenvalue of $\mathbb{A}$. Once the nominal θ_0 and $\mathbb{E}_s$ are given, the desired signal manifold vector can be estimated by

$$\boxed{\widehat{\underline{U}}_s = \beta \underline{\mathcal{P}}\{\mathbb{P}_{\mathcal{C}_2}\mathbb{P}_{\mathcal{C}_1}\}}$$

(8.35)

where β satisfies $\|\widehat{\underline{U}}_s\|^2 = N$. In what follows, an equivalent but computationally effective method will be presented to replace the estimator in (8.35). First, we can perform the eigendecomposition on the rank-p matrix $\mathbb{P}_{\mathcal{C}_2}$ to obtain

$$\mathbb{P}_{\mathcal{C}_2} = \mathbb{E}_h \mathbb{E}_h^H$$

(8.36)

where the $N \times p$ matrix $\mathbb{E}_h$ collects the p dominant eigenvectors of $\mathbb{P}_{\mathcal{C}_2}$. By inspection of (8.33), the final estimate $\underline{a}_\infty$ must lie in the subspace spanned by the columns of $\mathbb{P}_{\mathcal{C}_2}$. This implies that we can represent the final estimate by a linear combination of $\mathbb{E}_h$ and therefore the vector $\underline{a}_\infty$ can be expressed in terms of $\mathbb{E}_h$ as

$$\underline{a}_\infty = \mathbb{E}_h \underline{d}$$

(8.37)

where $\underline{d}$ is a $p \times 1$ vector. Then inserting (8.37) and (8.36) into (8.33) produces

$$\underline{a}_\infty = \mathbb{P}_{\mathcal{C}_2}\mathbb{P}_{\mathcal{C}_1}\underline{a}_\infty$$

$$\Leftrightarrow \mathbb{E}_h \underline{d} = \mathbb{E}_h \mathbb{E}_h^H \mathbb{P}_{\mathcal{C}_1} \mathbb{E}_h \underline{d}$$

$$\Leftrightarrow \underline{d} = \mathbb{E}_h^H \mathbb{P}_{\mathcal{C}_1} \mathbb{E}_h \underline{d}.$$

(8.38)

Now, the vector $\underline{d}$ is the eigenvector of $\mathbb{E}_h^H \mathbb{P}_{\mathcal{C}_1} \mathbb{E}_h$, up to a scaling factor, associated with the eigenvalue equal to one. According to Theorem 2.8 in [12] that the matrix products $\mathbb{A}_1 \mathbb{A}_2$ and $\mathbb{A}_2 \mathbb{A}_1$ have the same non-zero eigenvalues, we obtain

$$eig_{nz}\left(\mathbb{E}_h^H \mathbb{P}_{\mathcal{C}_1} \mathbb{E}_h\right) = eig_{nz}\left(\mathbb{E}_h \mathbb{E}_h^H \mathbb{P}_{\mathcal{C}_1}\right) = eig_{nz}(\mathbb{P}_{\mathcal{C}_2}\mathbb{P}_{\mathcal{C}_1})$$

(8.39)

where $eig_{nz}(\mathbb{A})$ denotes for the non-zero eigenvalues of the matrix $\mathbb{A}$. The above equation shows that the two matrix products $\mathbb{E}_h^H \mathbb{P}_{\mathcal{C}_1} \mathbb{E}_h$ and $\mathbb{P}_{\mathcal{C}_2}\mathbb{P}_{\mathcal{C}_1}$ have the same non-zero eigenvalues and hence the largest eigenvalue of $\mathbb{E}_h^H \mathbb{P}_{\mathcal{C}_1} \mathbb{E}_h$ is also equal to one. Likewise, the vector $\underline{d}$ can be resolved in a single step as

$$\underline{d} = \mathcal{P}\{\mathbb{E}_h^H \mathbb{P}_{\mathcal{C}_1} \mathbb{E}_h\}$$

(8.40)

and consequently the desired signal manifold vector can be estimated alternatively by

$$\boxed{\widehat{\underline{U}}_s = \beta \mathbb{E}_h \mathcal{P}\{\mathbb{E}_h^H \mathbb{P}_{\mathcal{C}_1} \mathbb{E}_h\}.}$$

(8.41)

It is worthwhile pointing out that in comparison with (8.35), the size of the matrix whose principal eigenvector we need to calculate is reduced from $N \times N$ to $p \times p$ by using (8.41), thereby reducing the computational complexity. This improvement is of interest when the array sensor number is quite large.

8.4   Desired signal power estimation

In [13], the power of the desired signal is found by searching the minimum of a cost function. The idea is that a temporary matrix $\mathbb{R}(\alpha)$ is formed by removing the noise effect and subtracting the effect of the desired signal with a real variable factor α from $\mathbb{R}_{xx}$, i.e.

$$\mathbb{R}(\alpha) = \mathbb{R}_{xx} - \sigma_n^2 \mathbb{I}_N - \alpha \underline{S}_d \underline{S}_d^H . \tag{8.42}$$

Then the estimated power is the largest possible value of α, which maintains $\mathbb{R}(\alpha)$ semipositive. However, searching the minimum of the cost function may involve expensive computations because a number of eigendecomposition operations are required.

In [14], the power estimation is solved by using the covariance fitting theory, which can be expressed as follows

$$\widehat{\sigma}_d^2 = \frac{1}{\underline{S}_d^H \mathbb{R}_{xx}^{-1} \underline{S}_d} \tag{8.43}$$

Let us consider the interference-absent case where the covariance matrix becomes $\mathbb{R}_{xx} = \sigma_d^2 \underline{S}_d \underline{S}_d^H + \sigma_n^2 \mathbb{I}_N$ and its inverse is readily obtained as

$$\mathbb{R}_{xx}^{-1} = \frac{1}{\sigma_n^2} \mathbb{I}_N - \frac{\sigma_d^2}{\sigma_n^2(\sigma_n^2 + N\sigma_d^2)} \underline{S}_d \underline{S}_d^H \tag{8.44}$$

Then the power estimation using (8.43) is given by

$$\widehat{\sigma}_d^2 = \sigma_d^2 + \frac{\sigma_n^2}{N} \neq \sigma_d^2 . \tag{8.45}$$

The above shows that the noise effect is neglected in (8.43).

In this section a single step power estimation is presented in which the noise effect is considered. The desired signal power σ_d^2 is equivalent to the largest σ^2 which satisfies

$$eig_i \left(\mathbb{R}_{xx} - \sigma^2 \underline{S}_d \underline{S}_d^H \right) \geq \sigma_n^2, \quad \forall i$$
$$\Leftrightarrow eig_i \left(\mathbb{R}_{xx} - \sigma_n^2 \mathbb{I}_N - \sigma^2 \underline{S}_d \underline{S}_d^H \right) \geq 0, \quad \forall i$$

$$\Leftrightarrow eig_i \left(\mathbb{E}_s^H (\mathbb{R}_{xx} - \sigma_n^2 \mathbb{I}_N) \mathbb{E}_s - \sigma^2 \mathbb{E}_s^H \underline{S}_d \underline{S}_d^H \mathbb{E}_s \right) \geq 0, \ \ \forall i$$

$$\Leftrightarrow eig_i \left(\mathbb{D}_s - \sigma^2 \mathbb{E}_s^H \underline{S}_d \underline{S}_d^H \mathbb{E}_s \right) \geq 0, \ \ \forall i$$

$$\Leftrightarrow eig_i \left(\mathbb{I}_{M+1} - \sigma^2 \mathbb{D}_s^{-\frac{1}{2}} \mathbb{E}_s^H \underline{S}_d \underline{S}_d^H \mathbb{E}_s \mathbb{D}_s^{-\frac{1}{2}} \right) \geq 0, \ \ \forall i$$

$$\Leftrightarrow 1 - \sigma^2 \underline{S}_d^H \mathbb{E}_s \mathbb{D}_s^{-1} \mathbb{E}_s^H \underline{S}_d \geq 0$$

$$\Leftrightarrow \sigma^2 \leq \frac{1}{\underline{S}_d^H \mathbb{E}_s \mathbb{D}_s^{-1} \mathbb{E}_s^H \underline{S}_d} \tag{8.46}$$

with

$$\mathbb{D}_s = \begin{bmatrix} \lambda_1 - \sigma_n^2, & 0, & \ldots, & 0 \\ 0, & \lambda_2 - \sigma_n^2, & \ldots, & 0 \\ \vdots & \vdots & \ddots & \vdots \\ 0, & 0, & \ldots, & \lambda_{M+1} - \sigma_n^2, \end{bmatrix}$$

where $\mathbb{D}_s^{-1/2}$ stands for the Hermitian square root of $\mathbb{D}_s^{-1}$. The operator $eig_i(\mathbb{A})$ represents the i-th largest eigenvalue of $\mathbb{A}$. In the above, the fact that the largest eigenvalue of the rank-one matrix $\mathbb{D}_s^{-\frac{1}{2}} \mathbb{E}_s^H \underline{S}_d \underline{S}_d^H \mathbb{E}_s \mathbb{D}_s^{-\frac{1}{2}}$ is $\underline{S}_d^H \mathbb{E}_s \mathbb{D}_s^{-1} \mathbb{E}_s^H \underline{S}_d$ is used. Thus the the desired signal power can be given by

$$\boxed{\sigma_d^2 = \frac{1}{\underline{S}_d^H \mathbb{E}_s \mathbb{D}_s^{-1} \mathbb{E}_s^H \underline{S}_d}} \tag{8.47}$$

Now let us consider the interference-absent case again. $\mathbb{D}_s$ becomes a scalar equal to $N\sigma_d^2$, and $\underline{S}_d^H \mathbb{E}_s \mathbb{E}_s^H \underline{S}_d = N$. Thus (8.47) offers the precise power estimation. Interestingly, [15] (see Lemma III.1 of [15]) proposed an equivalent power estimation method from the viewpoint of oblique projection.

8.5 Interference cancellation beamformer

In the well-known Wiener–Hopf beamformers, the array weight vector is chosen so that the array output SNIR is maximised. However, in a variety of practical applications, for instance, in the applications of radar and sonar, it is desirable to eliminate the interferences as much as possible [16]. These applications require complete interference cancellation rather than the maximisation of SNIR. Next, an interference canceller beamformer (ICB) will be presented.

First the ICB processor forms a desired-signal-absent covariance matrix as follows by estimating and subtracting the desired signal effects

$$\widehat{\mathbb{R}}_{i+n} = \mathbb{R}_{xx} - \widehat{\sigma}_d^2 \widehat{\underline{U}}_s (\widehat{\underline{U}}_s)^H \tag{8.48}$$

where $\widehat{\underline{U}}_s$ is obtained by Eq. (8.41) using the VSP method and the power $\widehat{\sigma}_d^2$ is computed by Eq. (8.47) where $\underline{S}_d$ is replaced by $\widehat{\underline{U}}_s$. Then we perform the eigendecomposition on $\widehat{\mathbb{R}}_{i+n}$ to obtain

$$\widehat{\mathbb{R}}_{i+n} = \sum_{i=1}^{N} \widehat{\lambda}_i \widehat{\underline{E}}_i \widehat{\underline{E}}_i^H$$

$$= \widehat{\mathbb{E}}_I (\widehat{\mathbb{D}}_I + \widehat{\sigma}_n^2 \mathbb{I}_M) \widehat{\mathbb{E}}_I^H + \widehat{\sigma}_n^2 \widehat{\mathbb{E}}_O \widehat{\mathbb{E}}_O^H \tag{8.49}$$

where $\widehat{\mathbb{D}}_I$ is a diagonal matrix with elements the eigenvalues $\widehat{\lambda}_1, \widehat{\lambda}_2, ..., \widehat{\lambda}_M$. Now the subspace spanned by $\widehat{\mathbb{E}}_I = [\underline{E}_1, \ldots, \underline{E}_M]$ is equivalent to the interference subspace $\mathcal{H}_I$ spanned by $[\underline{S}_1, \ldots, \underline{S}_M]$. The orthogonal projection operator of the interference subspace, denoted by $\mathbb{P}_I^\perp$, can be estimated by

$$\widehat{\mathbb{P}}_I^\perp = \mathbb{I}_N - \widehat{\mathbb{E}}_I \widehat{\mathbb{E}}_I^H. \tag{8.50}$$

The ICB weight vector can be constructed as

$$\underline{w}_{ic} = \frac{\widehat{\mathbb{P}}_I^\perp \widehat{\underline{U}}_s}{\sqrt{\widehat{\underline{U}}_s^H \widehat{\mathbb{P}}_I^\perp \widehat{\underline{U}}_s}}. \tag{8.51}$$

Using the actual DOA of the desired signal, the jammer output power will be zero because $\underline{w}_{ic}$ is orthogonal to the interference subspace. The total output power becomes

$$P_{out} = \sigma_d^2 \underline{S}_d^H \widehat{\mathbb{P}}_I^\perp \underline{S}_d + \sigma_n^2 \tag{8.52}$$

where the first term on the right-hand side of Eq. (8.52) is proportional to the input power of the desired signal. The array output SNR is given by

$$\text{SNR}_{out} = \frac{\sigma_d^2}{\sigma_n^2} \underline{S}_d^H \widehat{\mathbb{P}}_I^\perp \underline{S}_d. \tag{8.53}$$

By using the angle ψ between the subspace $\mathcal{H}_d$ (spanned by the manifold vector of the desired signal) and $\mathcal{H}_I^\perp$, which can be expressed as

$$\psi = \arccos \left(\frac{\sqrt{\underline{S}_d^H \mathbb{P}_I^\perp \underline{S}_d}}{\sqrt{\underline{S}_d^H \underline{S}_d}} \right), \tag{8.54}$$

the output SNR in Eq. (8.53) can be rewritten as

$$\boxed{\text{SNR}_{\text{out}} = \frac{\sigma_d^2}{\sigma_n^2}(\underline{S}_d^H\,\underline{S}_d)\cos^2\psi = \frac{\sigma_d^2}{\sigma_n^2}N\cos^2\psi.}$$

(8.55)

Equation (8.55) shows that the ability of the ICB to completely cancel the interferences is obtained at the cost of SNIR degradation or partial cancellation of the desired signal. If there is an interference very close to the desired signal direction then $\psi \to 90°$ and the output SNR deteriorates significantly, which may result in an unacceptable trade-off. However, this undesirable property is not restricted to the ICB. The Wiener–Hopf beamformer (and all other known beamformers) also suffers from this restriction when interferences are located close to the desired signal direction.

To summarise, the algorithm of the interference cancellation beamformer consists of the following steps.

(1) Calculate the matrix $\mathbb{H}$ and the projection operator $\mathbb{P}_{\mathcal{C}_2}$ using the nominal DOA θ_0. Perform the eigendecomposition on $\mathbb{P}_{\mathcal{C}_2}$ to obtain $\mathbb{E}_h$ (*off-line process*).
(2) Collect snapshots to estimate the data covariance matrix $\widehat{\mathbb{R}}_{xx}$. Perform the eigendecomposition on $\widehat{\mathbb{R}}_{xx}$ to obtain the signal-subspace eigenvectors $\mathbb{E}_s$ and the corresponding projection operator $\mathbb{P}_{\mathcal{C}_1}$.
(3) Estimate the steering vector $\widehat{\underline{U}}_s$ using Eq. (8.35) or Eq. (8.41) and the power $\widehat{\sigma}_d^2$ using Eq. (8.47) with $\underline{S}_d$ replaced by $\widehat{\underline{U}}_s$.
(4) Estimate the desired-signal-absent matrix $\widehat{\mathbb{R}}_{i+n}$ by Eq. (8.48).
(5) Perform the eigendecomposition on $\widehat{\mathbb{R}}_{i+n}$ to find the projection operator $\widehat{\mathbb{P}}_I^{\perp}$ using Eq. (8.50).
(6) Calculate the weight vector $\underline{w}_{ic}$ using Eq. (8.51).
(7) Weight the inputs of the array with $\underline{w}_{ic}$.

8.6 Performance analysis in the presence of pointing errors

Now consider the case $\widehat{\underline{U}}_s \neq \underline{S}_d$, i.e. when there is a mismatch between the steering vector and the real manifold vector of the desired signal. Then the desired-signal-absent covariance matrix contains some perturbations as depicted in Fig. 8.2, i.e.,

$$\begin{aligned}
\widehat{\mathbb{R}}_{i+n} &= \mathbb{R}_{xx} - \widehat{\sigma}_d^2\widehat{\underline{U}}_s(\widehat{\underline{U}}_s)^H \\
&= \mathbb{R}_{i+n} + \underbrace{(\sigma_d^2\underline{S}_d\underline{S}_d^H - \widehat{\sigma}_d^2\widehat{\underline{U}}_s(\widehat{\underline{U}}_s)^H)}_{\triangleq\widetilde{\mathbb{R}}_{i+n}}
\end{aligned}$$

(8.56)

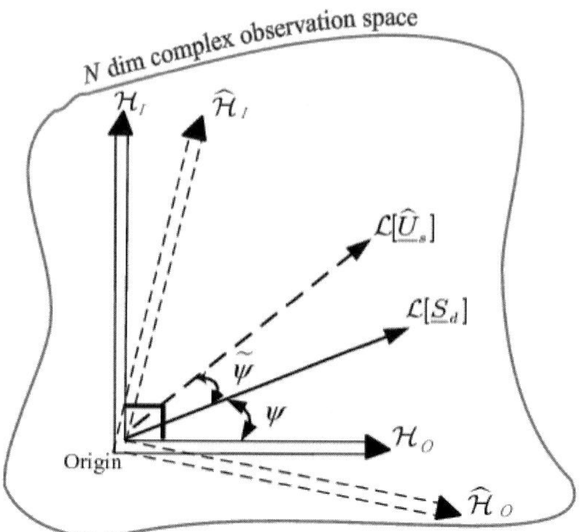

Fig. 8.2   Decomposition of the space by the columns of $\widehat{\mathbb{R}}_{i+n}$.

where the perturbation term $\widetilde{\mathbb{R}}_{i+n}$ is a rank-2 matrix if $\widehat{U}_s \neq S_d$.

Using the first-order subspace perturbation theory [17], the matrix of the perturbed signal eigenvectors are given by

$$\widehat{\mathbb{E}}_I = \mathbb{E}_I + \underbrace{\mathbb{E}_O \mathbb{E}_O^H \widetilde{\mathbb{R}}_{i+n}^H \mathbb{E}_I \mathbb{D}_I^{-1}}_{\triangleq \widetilde{\mathbb{E}}_I} \tag{8.57}$$

where $\mathbb{E}_O$ and $\mathbb{E}_I$ are the noise-subspace eigenvectors and signal-subspace eigenvectors of the actual desired-signal-absent covariance matrix $\mathbb{R}_{i+n}$. Then the perturbed orthogonal projection operator can be expressed by

$$\widehat{\mathbb{P}}_I^{\perp} = \mathbb{I}_N - \widehat{\mathbb{E}}_I \widehat{\mathbb{E}}_I^H$$
$$= \mathbb{P}_I^{\perp} + \underbrace{\mathbb{E}_I \widetilde{\mathbb{E}}_I^H + \widetilde{\mathbb{E}}_I \mathbb{E}_I^H + \widetilde{\mathbb{E}}_I \widetilde{\mathbb{E}}_I^H}_{\triangleq \widetilde{\mathbb{P}}_I^{\perp}}. \tag{8.58}$$

Substituting the above into Eq. (8.51) yields

$$\widehat{w}_{ic} = \frac{\widehat{\mathbb{P}}_I^{\perp} \widehat{U}_s}{\left| \widehat{\mathbb{P}}_I^{\perp} \widehat{U}_s \right|^2}. \tag{8.59}$$

Now the output interference power $P_{I_{out}}$ no longer equals zero due to pointing errors, because the weight vector is not orthogonal to the interference manifold, i.e. $\widehat{w}_{ic}^H S_i \neq 0$.

Consequently SNIR$_{\text{out}}$ in the presence of pointing errors can be given as

$$\text{SNIR}_{\text{out}} = \frac{\sigma_d^2 \left|\widehat{\underline{w}}_{ic}^H \underline{S}_d\right|^2}{P_{I_{out}} + \sigma_n^2 \left|\widehat{\underline{w}}_{ic}^H \widehat{\underline{w}}_{ic}\right|}$$

$$= \frac{\sigma_d^2 \left|\widehat{\underline{U}}_s^H \widehat{\mathbb{P}}_I^\perp \underline{S}_d\right|^2}{\displaystyle\sum_{i=1}^{M} \sigma_i^2 \left|\widehat{\underline{U}}_s^H \widetilde{\mathbb{P}}_I^\perp \underline{S}_i\right|^2 + \sigma_n^2 \left|\widehat{\underline{U}}_s^H \widehat{\mathbb{P}}_I^\perp \widehat{\underline{U}}_s\right|}$$

$$= \frac{\sigma_d^2 N \cos^2 \widehat{\psi}}{\dfrac{\displaystyle\sum_{i=1}^{M} \sigma_i^2 \left|\widehat{\underline{U}}_s^H \widetilde{\mathbb{P}}_I^\perp \underline{S}_i\right|^2}{\left|\widehat{\underline{U}}_s^H \widehat{\mathbb{P}}_I^\perp \widehat{\underline{U}}_s\right|} + \sigma_n^2} \tag{8.60}$$

where $\widehat{\psi}$ denotes the angle between $\underline{S}_d$ and $\widehat{\mathbb{P}}_I^\perp \widehat{\underline{U}}_s$. When the pointing errors are absent, the perturbation term $\widetilde{\mathbb{P}}_I^\perp$ is a matrix of zeros and $\widehat{\psi}$ equals ψ, which means the above equation reduces to Eq. (8.55).

8.7 Simulation results

Assume a uniform linear array (ULA) with $N = 10$ isotropic sensors and half-wavelength inter-sensor spacing, which operates in the presence of one desired signal and three interferences. Note that although here a ULA array is used the ICB is applicable to any arbitrary array geometry. Without any loss of generality the desired signal's DOA is kept at $90°$ in all the simulations presented in this section and, furthermore, all signals have unit power while the noise power is $\sigma_n^2 = 0.1$ (i.e. input SNR $= 10\,\text{dB}$). The matrix $\mathbb{H}$ is composed of three columns. That is the nominal manifold vector and its first two derivatives with respect to the DOA (see Eq. (8.6)).

Case 1: In the first scenario, we consider the case without pointing error (i.e. the pointing angle is $90°$). The DOAs of the three interferences are $60°$, $62°$ and $70°$. Three different methods are tested in this case, which are: Wiener–Hopf, Lee–Lee [18] and the ICB presented in Section 8.5. The weight vector of Wiener–Hopf is given by

$$\underline{w}_{\text{WH}} = \mathbb{R}_{xx}^{-1} \underline{U}_s \tag{8.61}$$

where $\underline{U}_s = \underline{S}(\theta_0)$ is the nominal manifold vector of the desired signal. Lee–Lee method is one of the so-called blocking matrix methods in which

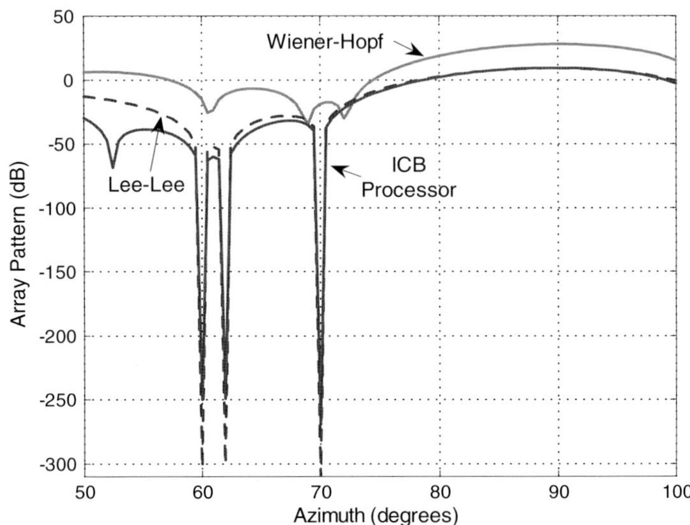

Fig. 8.3　Array pattern with two interferences close together ($60°$ and $62°$). The desired signal is at $90°$ and there is a third interference at $70°$.

a spatial filter is designed to block the desired signal and let all the interference signals pass through. Note that the design of this spatial filter is achieved strictly under the ULA assumption. Then the interference complement subspace, which can be easily found at the output of the blocking filter, is utilised to construct an interference cancellation beamformer, similar to that in Section 8.5.

The array patterns of these three methods are shown in Fig. 8.3, where it can be seen that the Wiener–Hopf processor cannot correctly resolve (distinguish) the two closely spaced interfering sources ($60°$ and $62°$) whereas it is possible by using the ICB weight vector $\underline{w}_{ic}$ or Lee–Lee weight. Also, the ICB method and Lee–Lee's method both place very deep nulls in the directions of the unknown interferences while allowing the desired signal to pass through. An added advantage is that these deep nulls can be used to estimate the locations of unknown interferences, because they are easily distinguishable from the rest of the array pattern. As displayed in Fig. 8.3, the Wiener–Hopf beamformer allows the interferences to pass through partially in order to maximise the output SNIR and thus the associated array pattern does not place deep nulls at the interference locations.

Case 2: In the second scenario, we compare the Wiener–Hopf ("modified" and "full") versions), the ICB and Lee–Lee beamformers when pointing

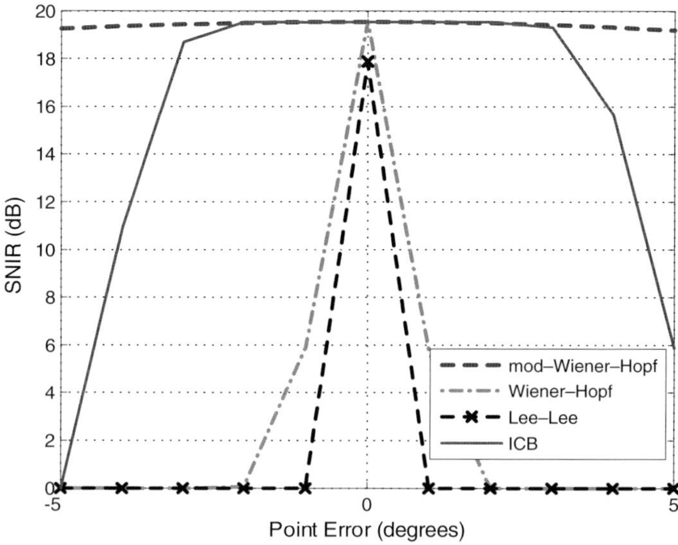

Fig. 8.4   Effects of pointing errors with noise level at $-10\,$dB. The actual direction of the desired signal maintained at $90°$.

Table 8.1 SNIR_{out} (unit: dB) with noise level at $-10\,$dB.

pointing angle	SNIR_{out} by simulation	SNIR_{out} by Eq.(8.60)
85^0	0	8.12
86^0	9.64	10.10
87^0	17.44	16.86
88^0	19.46	19.41
89^0	19.55	19.55
90^0	19.55	19.55
91^0	19.55	19.55
92^0	19.51	19.49
93^0	18.89	18.59
94^0	14.68	13.90
95^0	5.86	7.74

errors occur. The weight vector of the "full" Wiener–Hopf is given by Eq. (8.61), and the "modified" Wiener–Hopf by the following expression

$$\underline{w}_{\text{m-WH}} = \mathbb{R}_{i+n}^{-1}\underline{U}_s. \tag{8.62}$$

The DOAs of the three interferences are $70°$, $80°$ and $100°$. Though Fig. 8.4 shows that the ICB beamformer is more susceptible to pointing errors than the "modified" Wiener–Hopf solution, it outperforms the Lee–Lee and the "full" Wiener–Hopf processors. The SNIR of Lee–Lee without pointing error is less than the optimum SNIR. This can be explained by the fact that the dimensionality of the observation space is reduced in the Lee–Lee processor to construct the blocking matrix. In addition, Table 8.1 shows that the results computed by the analytical equations in the previous section are consistent with the simulation results when the pointing errors are small. If the pointing error is over $4°$, the approximation in Eq. (8.57) in the sense of the first-order subspace perturbation is invalid and therefore there are mismatches between the simulation results and the theoretical analysis.

Case 3: In the third scenario, three other robust beamformers are examined in the same environment as Case 2. The first is the traditional diagonal loading (DL) method with the weight vector [19]

$$\underline{w}_{DL} = \frac{(\mathbb{R}_{xx} + \mu\mathbb{I})^{-1}\underline{U}_s}{\underline{U}_s^H(\mathbb{R}_{xx} + \mu\mathbb{I})^{-1}\underline{U}_s} \tag{8.63}$$

where the diagonal loading factor $\mu = 10\sigma_n^2$. The second method is the SSP method [20] in which the weight vector is given as follows

$$\underline{w}_{SSP} = \mathbb{R}_{xx}^{-1}\mathbb{P}_{\mathcal{C}_1}\underline{S}(\theta_0). \tag{8.64}$$

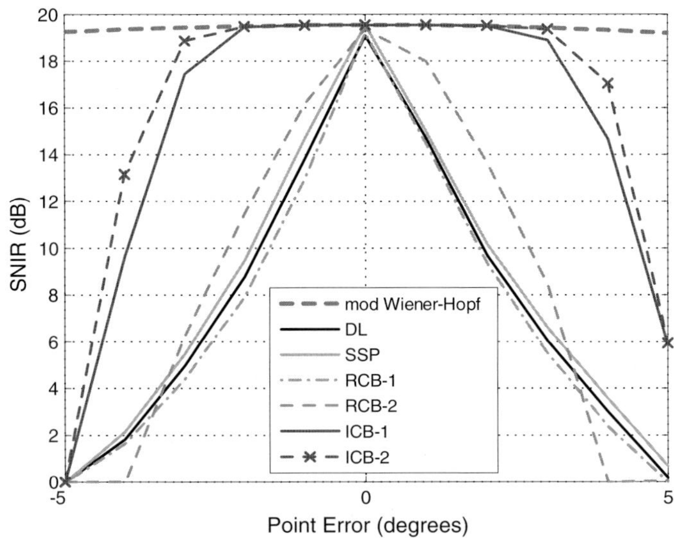

Fig. 8.5   Effects of pointing errors. The actual direction of the desired signal maintained at $90°$.

Two different robust Capon beamformers (RCBs) (with spherical constraint and with flat ellipsoidal constraint) presented in [21] are also examined. The weight vector of RCBs has the same form as that of traditional DL in Eq. (8.63), whereas the diagonal loading factor μ is chosen such that $\|(\mathbb{I} + \mu\mathbb{R})^{-1}\underline{S}(\theta_0)\|^2 = \varepsilon$ with the uncertainty level ε satisfying $\|\underline{S}_d - \underline{U}_s\|^2 \leq \epsilon$. The legend RCB-1 in Fig. 8.5 stands for the RCB with spherical constraint. Here the spherical boundary ϵ is equal to the larger of $\|\underline{S}(\theta_0) - \underline{S}(\theta_0 - \Delta\theta)\|^2$ and $\|\underline{S}(\theta_0) - \underline{S}(\theta_0 + \Delta\theta)\|^2$, where $\Delta\theta = |\theta_0 - \theta_d| + 0.5°$. The legend RCB-2 in Fig. 8.5 means the RCB with flat ellipsoidal constraint which means

$$\underline{S}(\theta_d) = \mathbb{B}\underline{v} + \underline{S}(\theta_0), \quad \|\underline{v}\| \leq 1. \tag{8.65}$$

Setting $\mathbb{B} = [\dot{\underline{S}}(\theta_0), \ddot{\underline{S}}(\theta_0)]$, the condition that $\underline{S}^H(\theta_0)(\mathbb{B}^\dagger)^H \mathbb{B}^\dagger \underline{S}(\theta_0) > 1$ (where $\dagger$ denotes the Moore–Penrose pseudo-inverse) does not hold. This means that RCB with $\mathbb{B} = [\dot{\underline{S}}(\theta_0), \ddot{\underline{S}}(\theta_0)]$ fails (see Eq. (45) of [21]). In order to satisfy this condition and simulate the second RCB method, $\mathbb{B}$ is set to be $\mathbb{B} = [\underline{S}(\theta_0) - \underline{S}(\theta_0 - \Delta\theta), \ \underline{S}(\theta_0) - \underline{S}(\theta_0 + \Delta\theta)]$. The legends ICB-1 and ICB-2 in Fig. 8.5 represent the ICB method using $\mathbb{H}_1 = [\underline{S}(\theta_0), \dot{\underline{S}}(\theta_0), \ddot{\underline{S}}(\theta_0)]$ and $\mathbb{H}_2 = [\underline{S}(\theta_0), \ \underline{S}(\theta_0) - \underline{S}(\theta_0 - \Delta\theta), \ \underline{S}(\theta_0) - \underline{S}(\theta_0 + \Delta\theta)]$ respectively. Figure 8.5 shows that the ICB method outperforms the other three methods when

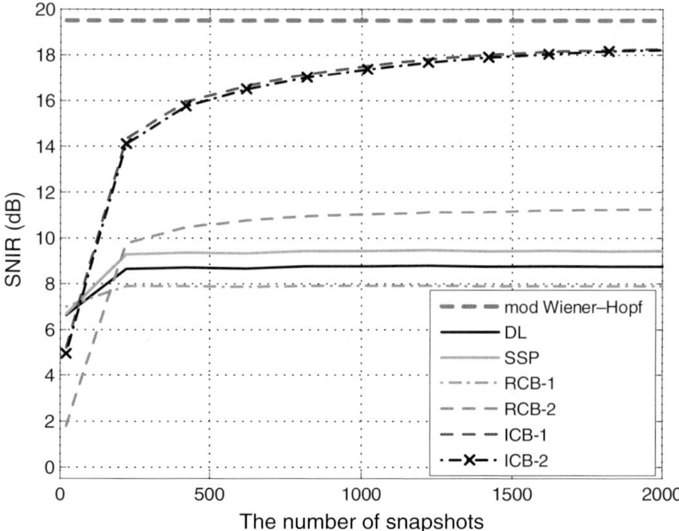

Fig. 8.6 The effect of finite snapshots. The actual direction of the desired signal maintained at 90° with the pointing angle 88°.

pointing errors happen. In addition, the ICB method is applicable to more practical applications than the RCB with flat ellipsoidal constraint because the condition $\underline{S}^H(\theta_0)(\mathbb{B}^\dagger)^H\mathbb{B}^\dagger\underline{S}(\theta_0) > 1$ is relaxed.

Case 4: Finally, the effect of finite snapshots is tested with the same simulation environment as Case 2. The pointing angle maintains at $88°$, i.e. $2°$ pointing error. The array output SNIRs of four methods versus the snapshot number are depicted in Fig. 8.6. This example indicates that the ICB method performs better than others when the number of snapshots is relatively large.

8.8 Summary and conclusions

When pointing errors occur, the true manifold vector of the desired signal differs from the nominal one. In this chapter it has been shown that the true manifold vector lies in a linear subspace spanned by the nominal manifold vector and its derivatives around the nominal DOA. Moreover, the true manifold vector also belongs to the signal subspace which is spanned by the dominant eigenvectors of the covariance matrix. Thus the desired signal manifold can be obtained using VSP, by means of which most of the pointing errors are eliminated. Furthermore, the power of the desired signal can be calculated in a single-step operation using the theory of covariance fitting. Finally, the estimate-and-subtract interference canceller beamformer can be constructed by using the estimates of manifold and power of the desired signal. This beamformer can completely cancel the interferences by maxmising the signal-to-interference power ratio (SIR) rather than SINR at the output of the beamformer and is robust to "pointing" errors.

References

[1] J. Li and P. Stoica, Eds., *Robust Adaptive Beamforming.* John Wiley & Sons, 2005.

[2] H. Cox, "Resolving power and sensitivity to mismatch of optimum array processors," *The Journal of the Acoustical Society of America,* vol. 54, no. 3, pp. 771–785, 1973.

[3] O. Besson, L. Scharf, and F. Vincent, "Matched direction detectors and estimators for array processing with subspace steering vector uncertainties," *IEEE Transactions on Signal Processing,* vol. 53, no. 12, pp. 4453–4463, Dec. 2005.

[4] A. Pezeshki, B. D. van Veen, L. L. Scharf, H. Cox, and M. L. Nordenvaad, "Eigenvalue beamforming using a multirank MVDR beamformer and subspace selection," *IEEE Transactions on Signal Processing*, vol. 56, no. 5, pp. 1954–1967, May 2008.

[5] L. Griffiths and C. Jim, "An alternative approach to linearly constrained adaptive beamforming," *IEEE Transactions on Antennas and Propagation*, vol. 30, no. 1, pp. 27–34, Jan. 1982.

[6] J. von Neumann, *Functional Operators II: The Geometry of Orthogonal Spaces.* Princeton University Press, Princeton, 1950.

[7] H. Stark and Y. Yang, *Vector Space Projections: A Numerical Approach to Signal and Image Processing, Neural Nets, and Optics.* New York, NY, USA: John Wiley & Sons, Inc., 1998.

[8] D. Sadler and A. Manikas, "Blind reception of multicarrier DS-CDMA using antenna arrays," *IEEE Transactions on Wireless Communications*, vol. 2, no. 6, pp. 1231–1239, Nov. 2003.

[9] Y. Yang and H. Stark, "Design of self-healing arrays using vector-space projections," *IEEE Transactions on Antennas and Propagation*, vol. 49, no. 4, pp. 526–534, Apr. 2001.

[10] J. Gu, H. Stark, and Y. Yang, "Wide-band smart antenna design using vector space projection methods," *IEEE Transactions on Antennas and Propagation*, vol. 52, no. 12, pp. 3228–3236, Dec. 2004.

[11] F. Zhang and Q. Zhang, "Eigenvalue inequalities for matrix product," *IEEE Transactions on Automatic Control*, vol. 51, no. 9, pp. 1506–1509, Sep. 2006.

[12] F. Zhang, *Matrix Theory: Basic Results and Technique.* Springer, 2011.

[13] P. D. Karaminas and A. Manikas, "Super-resolution broad null beamforming for cochannel interference cancellation in mobile radio networks," *IEEE Transactions on Vehicular Technology*, vol. 49, no. 3, pp. 689–697, May 2000.

[14] P. Stoica, Z. Wang, and J. Li, "Robust Capon beamforming," *IEEE Signal Processing Letters*, vol. 10, no. 6, pp. 172–175, Jun. 2003.

[15] M. McCloud and L. Scharf, "A new subspace identification algorithm for high-resolution DOA estimation," *IEEE Transactions on Antennas and Propagation*, vol. 50, no. 10, pp. 1382–1390, Oct. 2002.

[16] A. Haimovich and Y. Bar-Ness, "An eigenanalysis interference canceler," *IEEE Transactions on Signal Processing*, vol. 39, no. 1, pp. 76–84, Jan. 1991.

[17] F. Li and R. J. Vaccaro, "Unified analysis for DOA estimation algorithms in array signal processing," *Signal Processing*, vol. 25, no. 2, pp. 147–169, Nov. 1991.

[18] J.-H. Lee and C.-C. Lee, "Analysis of the performance and sensitivity of an eigenspace-based interference canceler," *IEEE Transactions on Antennas and Propagation*, vol. 48, no. 5, pp. 826–835, May 2000.

[19] H. Cox, R. Zeskind, and M. Owen, "Robust adaptive beamforming," *IEEE Transactions on Acoustics, Speech and Signal Processing*, vol. 35, no. 10, pp. 1365–1376, Oct. 1987.

[20] D. D. Feldman and L. J. Griffiths, "A projection approach for robust adaptive beamforming," *IEEE Transactions on Signal Processing*, vol. 42, no. 4, pp. 867–876, Apr. 1994.

[21] J. Li, P. Stoica, and Z. Wang, "On robust Capon beamforming and diagonal loading," *IEEE Transactions on Signal Processing*, vol. 51, no. 7, pp. 1702–1715, Jul. 2003.

Index